AMBER

The Golden Gem
of the Ages

Fourth Edition

To Glen & Spike
Best Wishes. and
continued interest in
Amber.

Patty C Rice
11-5-06

AMBER

The Golden Gem
of the Ages

FOURTH EDITION

PATTY C. RICE, PH.D.

2006

Printed in the United States of America

This book is printed on acid-free paper.

© 2006 Patty C. Rice, Ph.D..

ISBN: 1-4259-3849-3 (sc)

7/31/2006

PREFACE

Since I first became interested in the alluring golden gem known as amber more than thirty years ago, I have embarked on many exciting adventures as I continued my quest for knowledge about this fascinating substance. I began simply as a collector who loved the warm antique faceted beads my grandmother wore and called "amber." They were very long graduated, dark red, faceted beads, but were clear transparent red when held to the light. It wasn't until much later that I learned that amber was more often yellow! I once purchased some amber beads from a gem dealer for a friend and you can imagine my embarrassment a few weeks later when she hesitantly told me the beads were not real amber! When having them appraised, the gemologist found the faceted golden beads were only a form of plastic—not amber at all!

Since I had purchased the beads as "genuine amber," I reported the gemologist's findings to the dealer, who insisted the beads were definitely amber and could not be returned. With the zeal of an investigator, I began to search for literature that would help me identify genuine amber for myself so that I would not be duped again. Thus began my research resulting in the first edition of Amber: The Golden Gem of the Ages.

In 1975 the research itself was no easy task, since I found only short articles on amber in a few reference books. Much to my dismay, most books on amber were not published in English or were out of print in a library's rare book collection with much outdated information. Many authors in the past mixed folk beliefs and fantasy with scientific information and one author contradicting another—a problem that continues to this day.

Since 1980 there has been a renaissance in scientific techniques available for the study of amber and its inclusions. Therefore, this book will include new scientific discoveries related to amber, reference salient points made by today's scientists as they experiment with modern scientific equipment, and explain contradictions found in the historical literature.

Figure p.1. Dr. Norbert Vávra, from Institute of Paleontology in Austria, with the author at the Baltic Amber Symposium, *Warsaw, Poland, 1988.*

Figure p.2. Russian artist, Alexander Krylov, restorer for the Amber Room, with the author in 1996 at the Museum of Natural History Amber Exhibition.

My early research led me to venture to locations where amber was found in a quest to see for myself this fascinating substance in its original form. I wanted to see actual mining. I wanted to visit shops where workers polished the rough amber into pieces of shining jewelry. At that time, I seldom saw rough amber, because mainly finished jewelry was imported and offered for sale in the United States.

Because of all of my adventures to learn more about amber, I find the allure of amber to be not only "a treasure chest for molecular paleontologists," but a substance containing many scientific clues to the ancient past. Also, amber has greatly influenced people who lived in amber-producing areas. Amber brings with it the fascinating history of its use in healing and folk medicines. We find this information in Polish folk remedies even today.

In June of 2003 with great excitement, I visited the newly restored Amber Room in Tsarskoye Selo or Chatherine's Palace. It had taken many years for modern amber artists to replicate the room by studying old photographs because the original had been lost in 1945 after the World War II. Many theories remain today related to what actually happened at the end of the war.

Considering all the excitement relating to fossilized resins found in various locations around the world, I have attempted to update and reference comprehensive information and collect it together in this revised edition of Amber: The Golden Gem of the Ages.

I want to thank all of my acquaintances for their assistance in furthering my knowledge about amber and for sending me samples of amber from various locations. Many thanks to Leslie Burgess and Dr. Sarah Jastak-Burgess for

Figure p.3. Dr. Sarah Jastak-Burgess and Leslie Burgess with the author at the Artistry in Amber Exhibition *in Macomb Center for the Arts Gallery, Michigan, in 1985.*

their support in keeping *Amber: The Golden Gem of the Ages* in publication as it went through its earlier editions. I also want to thank my daughter, Diana, and her husband, David Odenwalder, for help in translating texts from German, Russian, and Polish. Also, thanks to Dr. Barbara Kosmowska-Ceranowicz, Dr. Stanley Woollams, and Dr. Susan W. Aber for reading and advising me during this revision. I especially want to thank Dr. George Poinar for providing me with actual photographs and articles about some of his significant research in amber. Most graciously, I thank all who have contributed photos and articles for the update of the fourth edition of *Amber: the Golden Gem of the Ages.*

—Patty C. Rice, Ph.D.

TEARS OF THE HELIADES

SISTERS OF THE SUN, BORN TO FATHER PHOEBUS,
GOD OF THE SUN AND MOTHER CLYMENE, A NYMPH. . .LIVE FOREVER NOW AS POPLAR
TREES, WEEPING   AMBER TEARS FOR THE LOSS OF THEIR BROTHER,
PHAETON. . .AS THEY MOURNED THEY FOUND THEMSELVES TURNED TO POPLAR TREES
WEEPING TEARS OF GOLDEN AMBER FOR EVERMORE . . .

Ovid

Their tears flow forth,
and from the new-formed boughs
amber distils and slowly hardens in the sun;
and far from the tree upon the waves is borne
to deck the Latin women.
---Ovid

CONTENTS

Dr. Brouwer, Dominican geologist, and author visit the Sierra de Agua mines in the Dominican Republic.

PART I

AMBER,
THE GOLDEN GEM
OF THE AGES

Biskupin, Poland

INTRODUCTION

Eurymachus
Received a golden necklace, richly wrought,
And set with amber beads, that glowed as if
With sunshine.
—Homer

Amber, a gemstone sought after by ancient Stone Age sun worshippers because its beautiful radiance resembled the sun's rays, well deserves the title, "golden gem of the ages." In early Greek and Roman civilizations, amber was so revered it was available only to nobility. Ladies of the Roman court desired it for its brilliant hue and for protection from evil spells that it was believed to bestow upon the wearer. In reverence to its talismanic powers, gladiators wore amber amulets when venturing into the coliseum. Throughout Europe, amber was worn as protection against various and sundry illnesses.

Although ancient man and the peoples of many later civilizations treasured amber as highly as gold, little was known of its origin until the age of science brought proof that it originated from the sticky resin that flowed from prehistoric trees. Few gems match amber in respect to its mode of creation, the depth of its history, and its transmission of aesthetic pleasure to man. None can match it in the range of human knowledge and scientific information its study reveals.

Over the centuries, amber primarily came into human hands from the seashores and outcroppings of amber-bearing strata near the Baltic Sea; however, smaller deposits have been found in other places throughout the world. By the late 1800s, the Baltic amber industry had become highly organized, with extensive mining taking place in East Prussia, a region that is now part of Poland, Lithuania, and the Kaliningrad Oblast—a roughly square landform, approximately 40 kilometers by 32 kilometers jutting out into the Baltic Sea called Kaliningrad District of the Russian republic. Over 16,168 tons of amber were produced in this region alone between 1876 and 1935, and from 1951 to 1986 over 17,700 tons were recovered. Geologic surveys by Eastern European scientists estimate the Baltic amber-bearing earth contains anywhere from 45 to 2677 grams of amber per cubic meter. Therefore, much amber is still being pro-

Figure 1.1. Baltic amber "seastone" has a semipolish as a result of churning salt water of the Baltic Sea.
below:
Figure 1.2. Amber "brack" is found in small chunks with a surface crust or shiny conchoidal fractured areas resulting from its long burial underground.

duced in the Baltic area and elsewhere.

During the 1800s, scientists began studying the geology, insects, and other evidences of fauna and flora of the past entrapped in amber resin as it flowed from trees in primeval forests in the "Amberland" of the Baltic. These researchers studied inclusions in amber, morphology of amber as it was found in nature, and geology of the land where amber was located to present a hypothesis of its origin and age. Early scientists classified amber as a *mineral* because it was found in the ground. However, modern scientists emphasize amber's organic nature as a tree resin, reclassifying it as one of the organic gemstones. Current researchers continue to focus on the study of fossil inclusions, the geology of amber-bearing beds, and especially the genetic relationships between ancient fossilized resins and those being produced by living trees today. New sophisticated techniques, such as infrared spectroscopy, X-ray diffraction, mass spectroscopy, and gas-liquid chromatography, are being used by modern scientists. These instruments provide paleo-botanists with a better understanding of the evolution of some present-day vegetation. Recent work on extracting ancient DNA (deoxyribonucleic acid—the building block for life or the molecular basis of inheritance) and bacteria from the stomachs of ancient bees by microbiologists, Raul J. Cano and Hendrik Poinar; and parasitologist, George O. Poinar, Jr. (Cano, Poinar, and Poinar 1992, 249–251) stimulated great fantasies for science fiction as illustrated by the movie, *Jurassic Park*. Actual scenes were shot inside amber mines in the Dominican Republic. In reality, amber from mining areas of the Dominican Republic is not old enough to contain dinosaur blood or dinosaur DNA.

Amber today is valued more than ever—not only by connoisseurs and collectors, but also by paleobotanists, paleo-zoologists, entomologists, geologists, archaeologists, and microbiologists, who continue to look for ancient bacteria that may be taken from a dormant state to a living state millions of years later. More realistically, paleobiologists find in amber an excellent means of tracing the development of our land and its inhabitants in the distant past.

AMBER AS A GEMSTONE

Rough or block amber is commonly found in irregular lumps or rounded nodules, also in grains, drops, and stalactitic masses. The pieces reflect the shapes initially assumed by outpourings of tree resins depending on whether or not the amber was formed on the exterior of the tree or formed as internal molds of cracks beneath the bark of the tree that once secreted resin. The variety of shapes and their significance are discussed in Chapter 6 in the section on amber as a fossil resin. Pieces are generally small, weighing up to about 200 grams, although considerably larger masses—in excess of several kilograms—have been found. The largest pieces on record weighed from 10 to as much as 20 kilograms. The largest single piece of Baltic amber ever discovered weighs 9.75 kilograms and is now in the Natural History Museum of Humboldt University in Berlin (Grimaldi 1996a, 50). Of all amber produced, however, only 5 to 6 percent of the pieces measure 30 millimeters, or a little over 1 inch, in diameter. All of the pieces, regardless of size, are used in some way in the amber jewelry industry.

Amber is often collected directly from the waters of the Baltic Sea since it is just buoyant enough to float in salt water. Typically, such floating amber receives a semipolish because of the action of waves and beach sand, which pummel these so-called "seastones" until they are divested of their natural crust and smoothed on all surfaces (see Fig. 1.1).

In contrast, mined amber, or pit amber, is covered with a dark brown crust that must be removed before the quality of the material within can be discovered (see Fig. 1.2). Major production of amber in the Baltic region is not from the sea, but from mining in the earth. The largest open-cut amber mining operation once operated in the Sambian peninsula, now the Kaliningrad Oblast (Almazjuvelirexport 1978, 26).

Because amber is fossilized tree resin, it is resinous in luster and lacks any crystalline structure as is commonly found in mineral gemstones. It otherwise resembles what one would expect from a hardened tree resin. Some recent X-ray work on amber does show crystalline organic compounds present. These structures are thought to be crystals of succinic acid or possibly other compounds (Encyclopedia Britannica 1971, s.v. "Amber").

Amber is commonly yellow to honey-colored, but within the range of these hues are many subtle variations from light yellow to dark brown—including lemon yellow, orange, reddish-brown, and hues that are almost white. Some are so pale that the amber seems colorless. Even greenish and bluish tints are found. These are among the

Figure 1.3. Varieties of Baltic amber in jewelry from Poland, Lithuania, Sambia, and Russia.

factors that make amber such a highly desirable and valued raw material in the jewelry trade and in folk art. Polish artists, from the Narew River region, are reported to have 80 names for varieties of amber based on its color, its level of translucence, and the character of its surfaces, the degree of weathering and even its suitability for working in folk art and rituals (Gierłowska 2004). (see Fig. 1.3).

Adam Chętnik, a Polish ethnographer, lists the categories of these names in his *Pocket Dictionary of Polish Amber Varieties* (1981, 31–8) as follows: Transparent amber *gems*, which is entirely transparent, pale yellow amber; *flames*, which is reddish amber whose color is reminiscent of fire; *honey*, which is honey-colored amber; and *clouded*, which is amber with concentrations of opaque areas that resemble clouds. Translucent amber includes *nebular*, in which opaque sections appear to resemble fluffy clouds and *woolly*, which are varieties with opaque sections appearing in strips or concentrations like strands of wool. Opaque amber varieties are divided into two groups—yellow and white. The opaque yellow includes: *beige* in reference to its color; *cabbage leaf*, where pale yellow or white streaks appear over a dark yellow background similar to veins of a cabbage leaf; *patchy*, which have different shades of white or yellow forming distinct patches of color; *marbled*, which have multicolored areas arranged in marble style; *mosaic*, which have multicolored portions in patterns resembling mosaics; *mixed*, which consist of several varieties all in one piece; *grainy* and *striped*, in which sections of various colors and degrees of translucence resemble the grain of wood. The opaque white varieties include *chalky* amber, which is chalky white in color; *bone* amber, which appears white with a hint of yellow similar to ivory; and unique specimens of *blue* amber.

In addition to variations in color, amber can be absolutely transparent or completely cloudy, and variations often can be found in all ranges within one specimen. The various colors are generally caused by light interference on minute gas bubbles or, less often, by stains of minerals or compounds resulting from the decomposition of organic debris enclosed in the amber. This type of amber is grouped into the *earthy* category, or that which is contaminated by plant matter, black plant detritus, and plant debris creating a greenish hue. It is believed that reddish hues result from oxidation. If amber is exposed to a radioactive substance for a month or more, its color darkens. Polish scientists have also studied a variety of "black" resins including processed ones sold on the Polish jewelry market. The investigators (Kosmowska-Ceranowicz and Migaszewski. 1988, 9–10) indicate that the color was produced by intense heat, which in natural resins, may be the effect of a lightning bolt! Figure 1.4 (pg. 8), adapted from Leciejewicz (1996, 15), shows the relationship of the varieties of Baltic amber to its primary structure and secondary structure or resulting from surface changes.

The red color caused by oxidation seldom penetrates completely. For example, amber found in Stone Age tombs, usually encrusted with a reddish-brown layer, is shown to be still yellow inside when it is cut through. In modern times, this aging can be produced by heating at a temperature of about 180-200°C under high pressure (Kosmowska-Ceranowicz and Migaszewski. 1988, 9–10). Amber can also be darkened by heating at a low temperature in an oven, which will induce a rich antique reddish-brown hue.

Amber is brittle and breaks with a conchoidal fracture surface that has a curved appearance, often with numerous concentric circular ridges. Large, clear pieces of amber often contain characteristic irregular fractures or cracks within them, and such flaws or "feathers" are not considered to detract from the piece's value. Some clear masses were formed from concentric layers of resin, one within another, like the layers of a hailstone, and are called *shelly* amber (Bauer 1904, 536). Fractures in this type of amber are likely to follow the concentric layering. Other lumps are compacted in a single mass and, therefore, are called *massive* amber. Specimens have been found representing all variations and gradations, from shelly to massive, and such structures are important in determining the manner in which any piece of rough amber can best be worked (see Fig. 1.5).

Although amber is soft enough to be cut with a pocket knife, it is scarcely scratched by a fingernail. When scraped with a knife, it tends to crumble into a powder. Parings or thin peels cannot be obtained. When chips of amber are burned, a smoke with a pleasant resinous odor of pine is emitted, and for this reason amber was burned as incense in temples in the Far East and in homes of peasants in the Baltic area. If a lump of amber is rubbed vigorously, not only does it give off a piney scent, but the friction causes it to produce enough negative static electrical charge to pick up small particles of tissue paper. The word *electricity* was actually derived from *elektron*, the Greek name for amber. Unlike mineral gemstones, amber is warm to the touch since it is a poor conductor of heat.

In the past, Baltic amber was considered by most gemologists and mineralogists to be the only "true amber." Baltic amber is the

Figure 1.5. Concentric layers "shelly" in amber. Size approx. 1.5 in., or 3 cm.

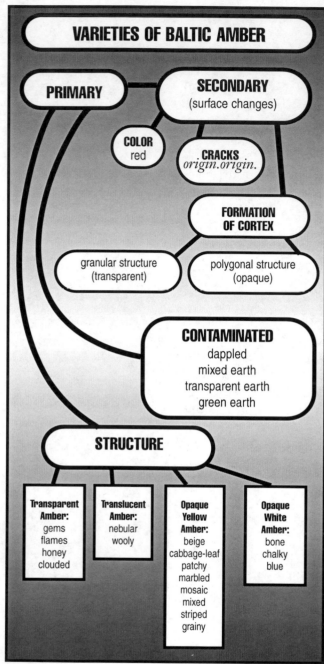

Figure 1.4. Varieties of Baltic amber. Reprinted with permission from Leciejewicz 1996.

most abundant of all ambers and is more often used for ornamental purposes since it has been a long-established industry, and it is said to take a better polish than other fossil resins. Currently, however, large quantities of amber are being mined in the Dominican Republic. Thus, in the Americas, amber from this location is readily available commercially and is crafted into beautiful ornaments by local craftsmen.

When considering amber as a gemstone, it must be remembered that until 1930, the largest production of amber was from deposits in the Baltic area, which today comprises the coastal regions of Denmark, Sweden, Northern Germany, Poland, Kaliningrad Peninsula, Lithuania, Latvia, and Estonia. The Danish amber belt lies on the western shore of the Baltic Sea and includes the Frisian Islands along the coastline of the North Sea (see Fig. 1.6).

In Poland, amber occurs along the entire Baltic coast about 10 to 20 feet below European sea level, being most abundant in the region of Gdańsk. In Lithuania, Latvia, and Estonia amber is also found occurring along the Baltic coast with smaller amounts found as one progresses north. The Kaliningrad Oblast, now belonging to Russia, is where the largest concentration of amber occurs. This area is reported to produce approximately two-thirds of the world's supply today.

The countries that import the largest quantities of Baltic amber are Germany and Japan. Before the break up of the Soviet Union, most Baltic amber imported into the United States came from either West Germany or Scandinavia. However, currently exporters from Poland, Russia, and Lithuania are establishing amber outlets in the United States, which will improve the accessibility of Baltic amber. After World War II, Russia restricted mining and commerce in the Kaliningrad region. All amber was sold through the official government commercial exporter,

Almazjuvelirexport of Moscow. They exported natural amber products of all shades within the yellow range and with a variety of textures, from opaque to transparent, as well as "fired" natural amber and pressed amber.

Elsewhere, small quantities of amber, or fossil resins, have been found in Sicily, Rumania, China, Burma, Thailand, Japan, Siberia, Canada, Mexico, and the United States. These sources produce fossil resins that differ only slightly in chemical composition from Baltic amber. More than 20 fossil resins, similar to Baltic succinite, but varying in geologic age, some physical and chemical properties, and perhaps botanical source, have been found in numerous localities throughout the world. Most have little commercial value as gemstones, but provide a rich source for collectors of fossil inclusions in amber of various ages.

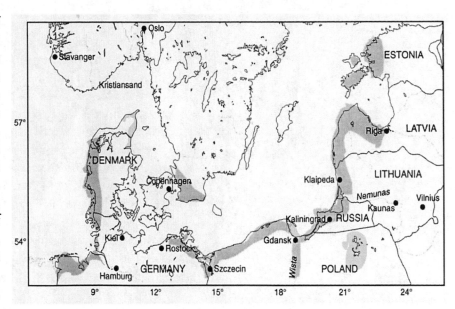

Figure 1.6. Occurrence of Baltic amber in various locations along the Baltic coast.

Antique shops today often provide evidence of the amber craftsmanship of the past. Frequently, one finds carefully faceted beads of clear, transparent golden-colored amber in estate sales and antique shops. These older styles of amber beads are darkened to a golden-orange color because of exposure to the atmosphere over the years. Beads fashioned during the early part of this century or before are often slightly crazed over the surface. Crazing is thought to result from long-sustained variations in temperature. Amber also tends to craze when it is subjected to extreme heat because of its low melting point and the escape of volatile substances. Crazing is an important feature to look for when examining old beads to determine whether they are true amber or imitations (see Fig. 1.7).

Another feature that is useful in distinguishing genuine faceted amber beads from impostors is that facets in the gem material tend to show signs of wear because of the gem's relative softness. The formerly sharp junctions between facets may be rounded. The holes through the beads also may be slightly chipped or irregular at their ends, resulting from constant rubbing of the stringing material and self-abrasion of the beads. Interestingly, the aging process adds a mellow beauty and a

Figure 1.7. Crazing on surface of amber faceted bead.

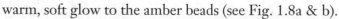

a

warm, soft glow to the amber beads (see Fig. 1.8a & b).

Many of the old faceted beads were produced by the German amber industry, which developed specific names for the jewelry trade based on variations in color shades as associated with different degrees of transparency. The two main categories were *transparent* (clear) amber and *cloudy* amber. German names were often picturesque using such terms as *flohmig* or *fatty, bastard, semibastard, osseous* or *bone*, and *foamy* or *frothy* amber (Bauer 1904, 538). All these varieties are used in making art objects and jewelry, with each type appealing to connoisseurs in various areas of the world at different times. Furthermore, each type has different capabilities for acquiring a polish, and it is this quality that dictates the use to which the pieces will be put in jewelry or artistic works.

Figure 1.8. a & b Antique faceted amber beads of German origin.

Transparent Amber

Pliny in *The Natural History* notes that:

> There are several kinds of amber. Of these the pale kind has the finest scent . . . The tawny is more valuable; and still more so it if is transparent, but the color must not be too fiery: not a fiery glare, but a mere suggestion of it, is what we admire in amber. The most highly approved specimens are the "Falernian," so called because they recall the color of the wine: they are transparent and glow gently, so as to have, moreover, the agreeable mellow tint of honey that has been reduced by boiling.

b

Transparent or clear amber takes a high polish and is in great demand for making faceted beads. Since the era of the amber guilds of the Middle Ages, amber beads—as meticulously faceted as any other transparent gemstone—were produced in great quantities by skilled craftsmen. During the 1930s, when amber was second only to diamonds in U.S. gem imports (Gemological Institute of America 1971, 2), much of the clear amber imported was in the form of faceted beads, and these can still be found in antique shops. U.S. importers tended to favor clear, golden varieties, and this is the hue most often thought of in the United States when speaking of amber (see Fig. 1.9).

The clear or partially clear shelly amber, with its loosely adhered layers, is rarely cloudy throughout and seldom contains cloudy layers as does massive amber. Both the structure and the resulting diaphaneity of amber are caused by the formation process at the time amber-producing resin was exuded from prehistoric trees. Transparent specimens range in hue from almost colorless to dark reddish-yellow. The "water-clear" amber is very rare. Most frequent are "yellow-clear" colors. The Polish names for these various hues translate to "gems, flames, honey, and clouded" (Leciejewicz 1996, 14).

Translucent Amber

Translucent or cloudy amber is highly prized in Europe for its muted beauty. The natural shape of the rough gem and the extent of its cloudiness are considered when pendants and beads are constructed. Woolly or fatty is slightly turbid and appears to have a fine dust suspended within the pieces (see Fig. 1.10). The German word, *flohmig*, was an older East Prussian term referring to the translucent yellowish fat of the goose. Thus, the term fatty amber was descriptive of the actual appearance of the amber. Translucent amber takes a high polish and provides beautiful, glistening gems with the appearance of whipped honey. When viewed under a microscope, a thin slice of translucent amber reveals an abundance of tiny bubbles, which interfere with light refraction and dilute the basic color, thus, producing the clouded appearance. Fatty amber is semitransparent, but not as turbid as the cloudy *bastard* amber (as it was called by the East Prussian industry), which is today referred to as opaque yellow amber.

Figure 1.9. Amber beads with insect inclusions. Courtesy Burgess-Jastak Collection.

Opaque Yellow Amber

Opaque cloudy amber is more turbid than fatty amber and comes in a variety of patterns that have been given various names. Currently in use in Poland are terms such as *beige, cabbage-leaf, patchy, marbled, mosaic, mixed, striped,* and *grainy*. The opaque cloudy, or bastard, amber takes a good polish. Again, the cloudy appearance is caused by a multitude of small bubbles within the mass. This amber is common in Baltic specimens and there are several degrees of cloudiness as expressed by the names given it. Variations in color also appear. Pieces may be whitish, yellowish, brownish-yellow or even reddish-brown. The term *egg yolk* amber was used for the yellow or brownish-yellow bastard amber within a clear ground mass. The East Prussian word for this type of amber was *Kumst*, which in German means *cabbage* or *sauerkraut*. The Polish amber industry today uses the term *cabbage-leaf* (Leciejewicz 1996, 15; see Fig. 1.11).

Figure 1.10. Baltic carving constructed of "fatty turbid" amber; hue is yellow to yellow turbid and is rather translucent. The carving has finely detailed feathers and features. In Poland, the symbol of the owl, or 'sowa,' indicates wisdom, alert watchfulness, or "Be on guard during evening hours." Size: Wt. 123 g, head to foot 10 cm, wing to wing 5.5 cm Courtesy Burgess-Jastak Collection.

Figure 1.11. Cloudy mottled yellow, "cabbage" patterned Baltic amber carved into a sculpture of a mother reading a book to her child, who is sitting at her knee. The full figure carving has detailed leaves, grapes, and a log on which the mother is sitting. The artist was Wiktor Matejczuk of Gdańsk, Poland. Size: Wt. 389 g, head to foot 11.7 cm, width flower to book 10.5 cm Courtesy Burgess-Jastak Collection.

The *patchy, marbled, mosaic, mixed, striped,* and *grainy* names apply to the "semibastard" amber that is between bastard and opaque white varieties in terms of diaphaneity. This type of amber displays patterns in hues that give the pieces their variety of names. It has the capability of receiving a polish similar to that of the opaque yellow or cloudy amber. The antique trade sometimes labeled this type of amber tomato amber because the process of aging caused it to deepen in color (see Fig. 1.12).

Opaque White Amber

Opaque white amber is divided into categories of bone, chalky, and blue. The bone, also called "osseous," is an opaque variety that is slightly softer than the others and consequently does not polish as well. The color is whitish-yellow to brown, and specimens give the general appearance of bone or ivory. The pieces can be variegated with osseous-clear or osseous-cloudy (see Fig. 1.13). This variety appears in forms with various classifications called *chalky, foamy,* or *frothy* amber. These contain a large amount of air bubbles that appear like solid foam when seen under an electron microscope.

In the past, it was thought that differences in turbidity in amber were caused by the presence of varying amounts of minute droplets of water. However, modern microscopic examination shows turbidity results from a vast number of microscopic gas bubbles enclosed within the gemstone. Between the gas bubbles, the resin is clear. Bubble inclusions interfere with light passage through the amber mass, resulting in dilutions of color, variations in color, and turbidity. Even the hue in rare blue, green, and grayish-green varieties of amber partially results from such enclosed bubbles.

Foamy amber is not as useful as other varieties to the jewelry industry, being opaque and very soft and almost incapable of receiving a polish. Occasionally, large, interesting lumps are used to form primitive and unusual sculptures or pendants. Pieces are also used in amber mosaics. Today, with the new amber dealers from Poland, Lithuania, and Russian entering the U.S. market, one can find unusual shaped beads and unique white pendants from foamy amber among their wares.

Blue varieties of opaque white amber are rare. The hues range from azure to sky blue. When blue amber is heated over a period of time in an oil with the same refractive index as amber, the piece clarifies, but the color changes to an ordinary yellow hue (Bauer 1904, 539). The blue color seems to appear near the surface of the amber piece and is often used in mosaics. Because of the scarcity of blue and green amber in Europe, these are generally favored over clear amber and fetch a higher price in most European markets. In contrast, transparent amber and clear amber with petal-like inclusions are currently most popular in the United States. This is mostly a result of a lack of exposure to the other varieties. This problem is being remedied with more liberal export laws in previous Eastern Block countries.

Earthy or Contaminated Amber

Earthy or contaminated amber varieties include dappled (black and white), mixed earth, transparent earth, green earth, and grayish-green in the Polish classification (see Fig. 1.14). Green amber is thought to have formed in marshy areas where decaying organic material may have influenced the color. All of these varieties contain organic debris. Inclusions in amber, both inorganic and organic, also influence the color. In foamy amber, pyrite is commonly enclosed in the form of thin lamellae that fill cracks, particularly in layered amber. If too much pyrite is present, it interferes with the lapidary treatment of the piece (Bauer 1904, 538).

Black and white amber, or a kind of bone amber dotted with black inclusions, is valued for its rarity. The black inclusions are organic material, generally decayed botanical debris (see Fig. 1.15). Finely divided particles of carbonized wood also appear black in color and sometimes are present abundantly in amber formed in boggy or marshy areas. Amber of this kind retains its typical coloring, within the yellow range, but carbonized plant debris often appears as black specks suspended in the amber. Paleobotanists believe particles of decayed wood found enclosed in amber probably are remains of ancient trees that were growing in the environs or perhaps were the same trees that exuded the original resin millions of years ago (Zalewska 1974, 57).

During the 1920s, amber necklaces with an insect embedded in each bead were especially in vogue. A fine example of a necklace from this period may be seen in the Hall of Gems of the Field Museum of Natural History in Chicago. Another more

Figure 1.12. "Bastard tomato," or sometimes called "egg yolk," amber pendant. The Baltic amber is opaque, yellow to orange, with swirls and patches of color. Size: Wt. 17.3 g, ht. 7 cm, w. 5.5 cm Patty Rice Collection.

Figure 1.13. White opaque bone amber carving depicting a lion. The Baltic amber rough weighing approximately 135 grams was purchased in Poland in 1986 and carved in Daniel McAuley's workshop in the Dominican Republic. The sculpture is highly polished with detailed carving and piercing to form the mouth, teeth, legs, and base. Carved in one piece. Size: Wt. 64 g, ht. from head to base 5.4 cm, w. from tail to nose 10.7 cm Courtesy Burgess-Jastak Collection.

Figure 1.14. Earthy green Baltic amber necklaces from Lithuania. Free-form, tumble-polished lumps, graduated in size, 40.6 mm to 25 mm, spaced with lemon-colored amber rhondels, 86.36 cm (34 inches) in length. Weight 143.4 g. Large green amber pendant is 70 mm by 100 mm Patty Rice Collection.

recent necklace of this type is in the University of Delaware Gallery from the Burgess-Jastak Amber Collection. Insect inclusions in Baltic amber are less common compared with those in Dominican amber. Typical Baltic amber insects tend to have a froth of bubbles obscuring one side of the specimen. (Further discussion of the flora and fauna found in amber is included in Chapter 6.)

AMBER AS A FOSSIL RESIN

Amber is one of the few gemstones of organic rather than mineral origin. Essentially, amber is a polymerized resin from prehistoric evergreens or other now-extinct species of resin-producing trees that flourished in large forests as far back as 120 million years ago in some areas.

Another name given by mineralogists to Baltic amber is *succinite*, a term based on the chemical analysis showing Baltic amber to contain at least 3 to 8 percent succinic acid, more than the amount found in any other comparable fossil resin. For this reason, presence of succinic acid was used to provide an important clue in identifying Baltic amber, and its presence or absence was used to specify probable places of origin of amber artifacts found in Stone Age tombs (Helm 1896, 51–7). However, modern technology has proven that this may not always be accurate in determining the origin of amber since more resins have been found that contain succinic acid.

Nevertheless, early studies in the 1970s investigating the botanical origin of resins used succinic acid content for identifying varieties of amber. Thus, based on its presence, amber resins were placed in two classes: *succinites*, those that contain succinic acid, and *retinites*, those that do not contain succinic acid. Today geologists and fossil fuel scientists refer to *resinite* in a much more general way, meaning any hardened resin, whether amber or copal. In fact, scientists from varying fields differ in when to label a hardened resin as actually becoming amber! In spite of this disagreement, recent studies have shown that the natural formation of amber—that is, its production by resinous prehistoric trees—was similar in all regions, and although the *formation* may have taken place in different epochs, the *processes* were comparable. Therefore, the presence or absence of succinic acid represents only a variation in chemical composition and not the existence of a distinctly different formation process.

Another problem developed as scientists from all over the world began studying amber and fossil resins. The naming and classification of fossil resins from various localities did—and continue to—

cause some difficulties between European scientists and American scientists and between different branches of science. Chemists and petrologists who study resins when investigating fossil fuels have developed classifications such as: Class 1 for those derived from polymers of *labdanoid diterpenes*, with Class 1*a* relating to those containing succinic acid and Class 1*b* relating to those without succinic acid, previously termed *retinites* (Anderson and Crelling 1995, xiii). (Further discussion of these scientific classifications can be found in Appendix A.)

Based on the geologic calendar, from the Eocene epoch to as late as the Miocene epoch of the Paleogene period, the northern parts of Europe (including regions now encompassing Finland, Scandinavia, the Baltic Sea, and northern Russia) are thought to have been one continuous land mass, labeled by geologists as the pre-Fennoscandian continent (see Table 1.1). It is believed to have extended from as far as Spitsbergen in the north to Iceland in the west, and to have covered parts of Greenland and North America, the British Isles, and northern France. To the south, it included land areas along the southern Baltic Sea. Between the Cambrian and Tertiary periods, the limits and boundaries of pre-Fennoscandia underwent many changes (Zalewska 1974, 58).

During the upper Eocene, gigantic forests yielding amber-producing resins grew on pre-Fennoscandia in a region now submerged beneath the Baltic Sea. The amber deposits indicated to earlier researchers (Conwentz 1890*b*, 81–151) that large amber-producing forests were approximately centered at latitude 55° N and longitude 19° to 20° E. However, it is currently accepted that the range of the forest also extended over a greater area, possibly as far south as the Black Sea (Komarow 1974, 59).

During the Eocene epoch, the climate of pre-Fennoscandia was warm and balmy, its weather ranging from temperate to subtropical. Adequate rainfall encouraged the growth of large tropical trees and luxuriant vegetation of ferns and mossy ground cover similar to that now found in some areas of South Asia. Numerous now-extinct species of pine, cedar, palm, oak, and cypress flourished in this so-called "amber-producing forest."

Figure 1.15a. Black and white opaque bone amber carved to depict a quail or eagle. Fully carved sculpture from one lump of amber. The black markings are from lignite and/ or pyrite inclusions. Artist was Wiktor Matejczuk, Gdańsk, Poland. Size: Wt. 400 g, ht. from head to foot 11.7 cm, w. beak to tail 15.2 cm Courtesy Burgess-Jastak Collection.

Figure 1.15b. (below) Samples of black and white varieties of Baltic amber in Muzeum Ziemi Collection, Warsaw, Poland.

Although paleobotanists are not in agreement as to the exact species of trees from which the resin flowed that formed amber, recent scientific studies indicate the probable existence of several resin-producing species, among which may be 20 to 40 varieties of coniferous trees, all of which produce resins having the capability of polymerizing or producing amber as the fossilized product. The capability of polymerizing is based on the *terpines* within the resin, in particular, the presence of *labdanoid diterpines*. The large number of tree varieties may also account for the many variations in color and diaphaneity of amber nodules. More recent research by botanists suggests perhaps ancient *Agathis*-related trees also grew in the ancient forests (Kucharska and Kwiatkowski 1978, 149–56).

Because more than one tree produced resins, and assuming Baltic amber's source to be several species of conifer trees, Conwentz (1890*b*) coined the name *Pinus succinifera* to include all such trees. *Pinus* is the Latin for the genus of pine tree, and the Latin word *succus* means "sap" or "gum."

In the past, it was thought that climatic changes or a disease must have caused these giant prehistoric trees—some as big as the California redwood—to exude an abundance of resin flow. Some authors were of the opinion that the trees virtually "bled to death" from disease (Rath 1971, 36). This view has been discredited by evidence of abundant and apparently unharmful resin flows from trees found growing in New Zealand (Zahl 1977, 423).

As the resin produced by the prehistoric trees flowed downward, it occasionally entrapped insects and plant pieces and eventually accumulated in masses of various sizes and shapes that later became buried in the soil below the trees. Thus, proof of the existence of other trees growing at the same time amber-producing resin was being exuded is also found in the remains of leaves, blossoms, twigs, bark, filaments, and other similar botanical debris trapped in amber. Such inclusions have been identified and suggest that relatives of the cypress, cedar, and sequoia, as well as other tropical plants now found in subtropical areas of Mexico, the Southwest and southern portions of the United States, and tropical parts of South Asia, once grew in the amber-producing forest. It is also believed that the original home of the American giant redwood (*Sequoia gigantea*) was in this immense forest. Other trees of this ancient forest include olives, chestnuts, camellias, magnolias, and the cinnamon tree, whose modern representative is now found only in Formosa, Japan, and China (Conwentz 1890*b*; Berry 1927, 268–78, 1930*b*).

The fossilization process of amber is different from that which converts wood into the stony substance known as petrified wood. In petrified wood, the woody cells are replaced wholly or in part by mineral substances. In contrast, amber retains basically the same organic substances present in the original resin exuded from the prehistoric trees. The polymerization process began as the resin was exposed to the air and the microscopic organisms and bacteria carried by the air. The volatile compounds that imparted *stickiness* to the resin escaped so slowly, the resin was prevented from cracking into numerous minute fragments as a result of shrinkage. Extreme conditions, such as pressure by the glacial covering, severe climatic changes, and submersion of the resin under salt water, took place over millions of years, causing the process of oxidation. During the lengthy time underground, molecules were forced to polymerize; that is, to rearrange themselves into longer chains of molecules. This caused a metamorphosis from a tacky resin to a solid, forming a compound with greater stability and hardness than the original substance and similar in appearance (and in some physical properties) to plastic. In no way is amber identical, chemically or otherwise, to plastic, though plastic often masquerades as amber to the unsuspecting public.

Modern technology provides the means not only for studying the past through inclusions in amber, but also for isolating and identifying individual resin components of amber. These are then compared to recent tree resin components to establish genetic relationships between fossil resins and recent resins. For example, the amber found in Chiapas, Mexico, has been studied with infrared spectroscopy and is believed to be related to the leguminous tree, *Hymenaea*, which produces the recent African resin known as *copal*, a material sometimes misrepresented as amber (Langenheim 1969, 1157–64). The ambers from the Dominican Republic and Chiapas, Mexico, are from the tree *Hymenaea pinta* whose nearest relative, *Trachylobium*, is found in Africa. The latter is the source of the Pleistocene/Recent copals as well as recent resins found in Tanzania and Madagascar (Langenheim, 1966, 201–10, 1969, 3–21, 1973, 5–38).

Pinus succinifera, the collective term used to represent a combination of ancient amber-producing trees of the Baltic region, was thought to have preserved its own cones, needles, and bark in the resin that flowed down its trunk. These physical features provided early scientists with a means for identifying and naming the tree and for estimating its size. Enormous pieces of amber, weighing as much as 15 kilograms, led paleobotanists to estimate that this stately conifer grew to a height of 100 feet or more.

Today's researchers find potential problems in dealing with the older theories and the infrared spectra of amber. Using modern technology to analyze the fossilized resin, scientists find a closer resemblance of amber's infrared spectra to the recent resin from an

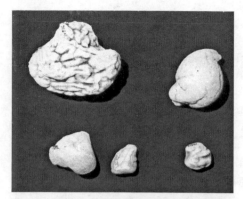

Figure 1.16. Natural shapes of amber as displayed in the Muzeum Ziemi Collection, Warsaw, Poland: a, Amber dripping forms or drops; b, Massive natural shapes of amber; c, natural shapes--external fissure fillings forms.

Agathis-like araucarian tree. This indicates that a tree similar to the *Agathis australis*, was the Baltic amber-resin producer in spite of the earlier morphological studies that indicated a *Pinus*-like conifer. However, no araucarian remains have been reported as inclusions in Baltic amber. Some theorists suggest the proposed tree, *Pinus succinifera*, was actually a now-extinct ancestor that shared characteristics with present-day pines and araucarians. It could have been chemically similar to araucarians but morphologically similar to pines (Larsson 1978, 192). Some researchers have suggested a cedar (*Cedrus atlantica*) or a Canadian larch (*Pseudolarix wheri*) as source trees. The problem of identifying the Baltic amber source tree is presented in Chapter 6.

The variations in places where the sticky exuded resin accumulated resulted in variations in the shape of the consolidated masses or lumps of today's amber. The most common morphological forms of natural amber are the drops and stalactites exuded from prehistoric trees during periods of ordinary resin production. During more abundant exudations, large incrustations in the form of streamlets and lumps resulted. When deposited in crevices or shivered parts of the trunks, amber formed flat, platelike incrustations, mostly without inclusions and of the purest material (see Fig. 1.16 and Fig. 1.17).

Over a thousand species of insects and arthropods have been found in amber. Most of the insects are now extinct; they are the ancestors of similar forms existing elsewhere in the world, and they illustrate the high degree of development in the evolution of insects even at this early period. Some species completely disappeared from the Baltic region, but their descendants can be found in warmer parts of the globe. Recently paleobiologists such as Poinar (1992, 350pp), while at the University of California, Berkeley, and Grimaldi (1996*a*, 126), from the Natural History Museum in New York, have studied insects in Dominican amber, identifying ancient representatives of most insect groups known today. Using technically advanced equipment available in the 1980s and 1990s, both scientists almost simultaneously were able to sequence the amino acids in proteins and the nucleotides in DNA from tissues within ancient insects preserved in Dominican amber and in one case Cretaceous age amber. Since DNA is the molecular basis of inheritance, these studies shed new light on the evolution of insects. In 1982 Poinar and Hess (1241–2) announced that, using high-magnification electron microscopic examination of tissue in an ancient insect in Baltic amber, they could see banded striations in the muscle. This announcement encouraged the early attempts to extract the DNA.

It wasn't until 1992 that Cano et al. (619–23) published that they isolated DNA from a bee in 25-million-year-old amber from the Dominican Republic. A month later, the American Museum of Natural History reported isolating fragments of DNA from a large primitive termite, *Mastotermes electrodominicus*, in Dominican amber. Since 1992 the search for ancient DNA has become a fervent topic. European scientists have not as yet isolated DNA in Baltic amber. It appears Baltic amber does not preserve the inner tissues as well as Dominican amber, which consistently shows fine preservation qualities and has more available fossil inclusions.

The DNA currently being isolated produces only fragments, or a mere fraction, of the entire gene. Yet the book, *Jurassic Park*, and the subsequent blockbuster film was based on recreating, or *cloning*, dinosaurs from DNA recovered from dinosaur blood inside an ancient insect. The insect would have had to have sucked blood from a dinosaur before being entrapped in the sticky amber-producing resin millions of years ago! So far this is still impossible, even though in June of 1993, DNA sequences were reported being extracted from a small nemonlychid weevil in 120- to 130-million-year-old amber from Lebanon. This is the most ancient DNA yet to be sequenced (Kharin 1995, 49–50). These discoveries, plus recoveries of DNA from five different laboratories, all of which used Dominican amber inclusions, illustrate the very special properties of amber-producing resin for desiccating or dehydration of insect tissues that is needed to embalm the tissue and preserve it for eons. The new possibilities present scientists with an ethical dilemma: "Should the ancient insect specimens be destroyed and do we really want to create new living organisms from the extinct ones?"

In the Baltic region, the climate of pre-Fennoscandia cooled in time, and movements of the earth's crust resulted in drastic changes in the configuration of landforms. Separation of crustal plates may have resulted in deflection of the Gulf Stream, which caused climatic changes. In Paleogene time, portions of pre-Fennoscandia sank, older islands vanished, and new ones arose from the sea. The ancient amber-producing forest was destroyed, its trees fallen and decayed, and the debris either washed away or was buried in sands, gravels, and

Figure 1.17. Necklace made using natural tear drop shapes of amber being sold on Arbot Street in Moscow, Russia.

TABLE 1.1	Geologic Calendar and Formation of Amber					
Era	Period	Epoch	Time	Characterize by	Episodes in the Formation of Baltic Amber	Examples of Various Resinous Flows
CENOZOIC	Quarternary	Neogene (Recent)	Present to 10,000 years ago	Development of humans	By 3000 B.C. amber was traded as far away from the Baltic region as central Russia	Copal, kauri gums Colombian copal formed
	NEOGENIC	Pleistocene	11,000 years ago to 1.8 million years ago	Wide spread glacial ice	"Young amber" formed in areas not covered by glaciers	From deciduous trees in lake regions of Poland; Tanzania, African resin formation
		Pliocene	1.8 to 5 million years ago	Tertiary Sea invasions; formation of mountains, gradual cooling of the climate in Baltic region	Amber lumps being transported and redeposited	Cotui, Dominican Republic Hymenaea resins formed
		Miocene	5 to 24 million years ago	Grazing animals developing	"Blue earth" formed in Baltic region	From trees producing Simetite, Dominican and Mexican amber. Sarawak (Burmite)
	PALEOGENE	Oligocene	24 to 38 million years ago	Saber-toothed tigers or cats	Drastic changes in topography and climates	From trees producing Rumanite, Bitterfeldite and Baltic succinite Arkansas amber
		Eocene	38 to 54 million years ago	Advent of mammals	Growth of the huge forest with a variety of giant trees in pre-Fennoscandia	Baltic succinite, Rumanite
		Paleocene	54 to 65 million years ago	Advent of birds	Giant conifer trees growing	
MESOZOIC	**CRETACEOUS**	Upper or Late	65 to 98 million years ago	Age of last dinosaurs	Deciduous, Leguminosae, Araucariaceae, oak, cedar, and cypress trees growing	From Siberia, New Jersey, Alaska, Kuji, Japan, Grassy Lake, Alberta, Canada, reassigned Burmite, Taimyr, Russia, Seward, Nebraska, Wyoming, and Kansas, Switzerland
	JURASSIC	Lower or Early	98 to 146 million years ago	Age of dinosaurs		Artic Coastal Plain, USA Cedar Lake, Manitoba, Canada, Isle of Wight, UK Austrian resins Choshi, Japan, Jordan, Lebanon,

clays. These remains formed the clay-rich geologic stratum known in the Baltic as the "blue earth," named by German scientists who studied the coast of the Sambian peninsula, where most of the amber was found in the nineteenth century (see Fig. 1.18).

The blue earth layer, which contains most of the amber, is about 28 to 30 meters beneath the surface and, along the coast, is about 4 meters beneath sea level. Thus, today wave action loosens amber pieces during a storm and washes them ashore.

During the ensuing Great Ice Age, the remnants of the continent were covered by vast, moving sheets of ice that relentlessly plowed through the sedimentary formations laid down by erosion of pre-Fennoscandia. The comparatively soft stratum of blue earth was easily pushed away and incorporated in glacial moraines over much of its extent. It was by this process that Baltic amber, for the most part originally confined to a single stratum, became widely distributed over northern Poland, Sambian peninsula (Kaliningrad Oblast), Lithuania, and other lands adjacent to the Baltic Sea (see Fig. 1.19).

The Great Ice Age occurred during the Pleistocene epoch, approximately a million years ago. The land gradually became free of the icy covering. As water was released, it flowed into depressions carved away by the glaciers. Thus, the Baltic Sea was formed. Northern Europe had now practically assumed its present forms of vegetation and animal life. The action of the waves on the shoreline cliffs of the Baltic loosened quantities of amber. The agitated salt water carried the amber away to sea and deposited it in

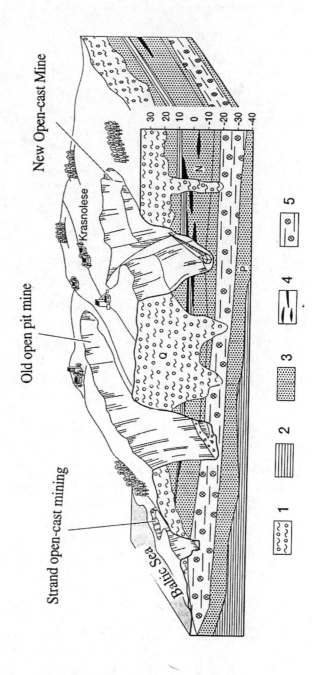

Opposite: Figure 1.18. Block diagram of the geology of amber-bearing deposits in Samland (based on Gurewicz/Kazanow 1976), reprinted with permission from Slotta and Ganzelewski 1996.

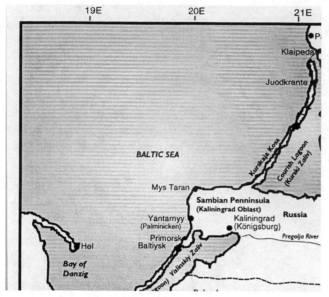

Figure 1.19. Map of Kaliningrad (Sambia) and environs.

the seabed. In the past (and even today), violent storms blowing in the direction of the coast cause lumps of amber to be thrown out along the shore.

In this simplified overview of the geologic history of Baltic amber formation, events are described that began 60 to 40 million years ago, in an era just after the age of dinosaurs and extending to the period when mammals began to roam the Earth. In terms of geologic time, this is a relatively recent development. The vast vegetation of the period that formed amber was the second such period in geologic time—the earlier providing the basis for formation of our coal beds.

BALTIC AMBER COAST TODAY

All amber deposits on the Sambian peninsula were redeposited. In the past, it was thought that the northwest shore of the Bay of Danzig contained the formation where Baltic amber was found in the earth layer in which it was originally formed. The amber deposits are found in varying concentrations in the "green wall," "white wall," and "gray wall" or quicksand, but mainly at an industrial concentration in the "blue earth"—old terms used by Zaddach, the first geologist to study the amber-rich deposits of Sambia. Current geologists have shown that the amber-bearing deposits of Upper Eocene and Lower Oligocene in the Kaliningrad region of Russia were accumulated as individual cycled-bundles. Russian scientists report that the blue earth glauconite, a characteristic mineral of amber-bearing sediments, shows traces of oxidation, which testifies that the glauconite was brought from eroded old sediments. European scientists believe the amber deposits were connected either with the delta of a large river, eroded primary accumulations of fossil resin in soils of the "amber forest," or with concentrations of amber in shallow parts of marine basins. This accounts for the Baltic amber, also called succinite, being found in many regions of northern Europe (Kharin 1995, 49–50).

Succinite, or Baltic amber, is found in areas other than the Baltic seacoast. The southeastern shore of the North Sea, along the coastlines of Kent, Essex, and Suffolk in England, produces small quantities of amber. In the past, the western shore of Denmark and the Frisian Islands produced larger amounts than are produced now. Even today, however, collectors and local "rock hounds" may walk along Danish shores early in the morning after a storm and pick up small pieces of amber. Amber may be found in any area along the

Sambian peninsula, and the German, Polish, and Lithuanian shores of the Baltic Sea (see Fig. 1.6). In spite of this distribution of amber for the past 100 years, roughly 98 percent of all amber has been produced in the Samland promontory, a high ridge of land approximately 20 to 25 miles square, jutting out into the Baltic Sea. The Sambian peninsula, now part of Russia, was formerly the major amber-producing area of East Prussia.

The cliffs on the Sambian peninsula coast rose almost vertically from the sea. Naturalist Willy Ley (1951, 4–5), a native of Samland, wrote in the 1950s describing the land "rising from the sea like the walls of a fortress" with forests on the top of the ridge and at the beach a narrow sandy coast. He described the coast as dotted with countless 200-ton boulders for hundreds of yards from the shore. These were brought there by glaciers of the Pleistocene epoch. The boulders originally were carried from Scandinavian countries as their receding coastlines were washed out.

Since 1992 the Russian Amber Company in the Kaliningrad region has been attempting to bring international awareness to the amber region and is involved in scientific study and dissemination of scientific information. The largest reserve of Baltic amber deposits is located in the northwest region of the Sambian peninsula, Russia, on the Baltic seacoast. Geology reports of 1982 estimated the reserves of explored deposits at Jantanyi (or Palminiken) to be 132,000 tons, with the deposits included in the Prussian suite of upper Eocene. The amber-rich layer extends beyond the limits of the current mined deposits of Primorsk. Russian geologists believe the perspectives for amber occupy approximately 700 square kilometers on the Sambian peninsula (see Fig. 1.20). Recent reports indicate that the layer of amber-bearing blue earth is buried under approximately 25 to 40 meters of soil, depending on the exact locality examined. The blue earth varies from 1 to 17.5 meters in thickness in this region with the "vein" of amber-bearing earth estimated to be over 2 kilometers in length. Lithuanian geologists believe the second largest reserve of amber is located in the Courish Lagoon, a bay previously known as *Kurisches Haff* (German). The bay is separated from the Baltic Sea by a long sandbar, now called *Kuršiu Merija* in Russian. Amber lumps are occasionally found buried in the sands of the dunes on the sandbar. Dredging in the Courish Lagoon has produced vast quantities of amber since the middle of the nineteenth century. Geologic estimates indicate that approximately 100,000 tons of amber remain in the area (Savkevich 1988, 1–2, 5–6).

Figure 1.20. Distribution of amber in Sambian Peninsula (Kaliningrad Oblast) in the "blue earth" of the Upper Eocene deposits. (Data from Kharin 1995, Basharkevitch et all 1983, Srebvrdolsky 1984, Sviridow and Sivkov 1992.)

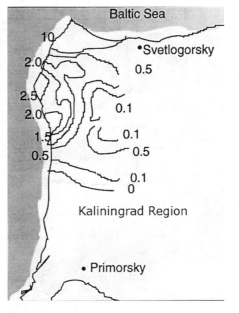

SUMMARY

The Baltic Sea region has been the original source of amber for the world since prehistoric times. Yet, amber is not confined to the Baltic area. Even succinite, a product from the upper Eocene and lower Oligocene epochs, has been found in other parts of Europe and Asia. Amber, or fossil resins, originating from the earlier Cretaceous period to the Quaternary period are found in various parts of the world (Raimundas 1978, 1).

Although still considered a beautiful and rare gemstone, amber is no longer defined as a *mineral* as it was in the past. In 1995 Russian news reports stated that amber was reclassified as an "organic semiprecious gemstone." American and European scientists define amber as fossilized tree resin, the word "fossil" indicating prehistoric origin. Tree resins become fossilized through "various changes due to loss of volatile constituents, processes such as oxidation and polymerization, and lengthy burial in the ground" (Frondel 1968, 381).

PART II

HISTORIC ASPECTS
OF AMBER

1555 Baltic Map - Gdańsk Shoreline as in Olaus Magnus, Historia de gentibus sepentrionalibus, earumque diversis statibus conditionibus , , ,

HISTORY OF BALTIC AMBER

> . . . . *where weep*
> *Even now the sister trees their amber tears*
> *O'er Phaeton Untimely dead.*
> *H.H. Millman,*
> *from the poem "Samor,"*
> *London 1818*

EARLY MAN AND AMBER

It is not known exactly when Baltic amber was first used, but prehistoric amber artifacts found in the Baltic region provide evidence that ancient ancestors of present-day Lithuanians, Latvians, Estonians, Poles, and others living along the Baltic coasts were familiar with amber. It is known that early Stone Age people living near Baltic shores were aware of the beauty of this material and prized it highly (Gimbutas 1963, 19).

Amber's novelty of appearance, its shining luster, and its colors made it a treasure so rare that early chieftains took amber with them to their graves, perhaps in the belief that objects buried with them could be used in another world. Such amber hoards provide us with a means for piecing together the story of early barter and the development of trade routes along which amber and other commodities passed from tribe to tribe and from country to country. Northern Europe had other raw materials to trade with peoples in distant southern lands, but amber was one of the most valued. And, fortunately, amber was imperishable enough to leave a record of this trade.

The Paleolithic Period or Old Stone Age

The Paleolithic Period or Old Stone Age, the most ancient recognized period of human development, is estimated to have lasted over a million years, ending about 8000 B.C. This period is characterized by the use of chipped stone tools. Humans at this time lived in caves. They were nomadic hunters who traveled many miles in warmer summer months following migrating animals. Baltic Paleolithic amber

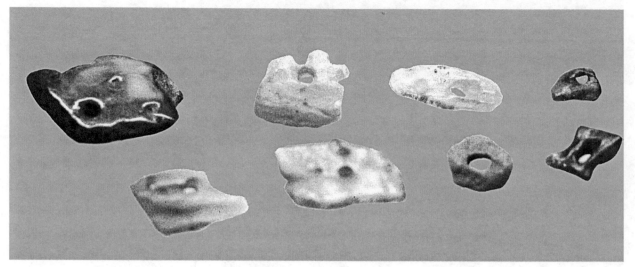

Figure 2.1. Amber pieces with natural holes used as amulets during Old Stone Age. Courtesy Muzeum Ziemi Collection, Warsaw, Poland. Photo by Patty Rice.

was studied by Kwiatkowska (1996, 16) of the Museum of Earth in Warsaw, who reports the oldest amber artifacts found date back as far as about 12,000 B.C. These artifacts included crude amulets in the shape of ornamental pendants, some using natural holes (Fig. 2.1) in the amber piece to attach it to their garments. One such site was Meiendorf near Hamburg, Germany. Another ancient Baltic site, reported by Burdukiewicz (1981, 5–11), was located at Siedlnica in the parish of Wschowa, Poland. This site produced similar amber artifacts made during the Old Stone Age.

Surprisingly, Paleolithic amber in unworked lumps has been found in locations far away from the Baltic. One interesting Paleolithic find with Baltic amber artifacts was in England from Gough's cave in Cheddar and Cresswell Crogs. Since this era was during the last Ice Age, the British Isles were not yet separated from the European Mainland (Tratman 1955). Amber was found in cave dwellings at Grotte d'Aurensan in the Hautes Pyrenes, at Kostelik and Zitmy in Moravia, and at Judenes in Austria. Since these locations are in southern Europe, it is thought that hunters may have found the amber further north and carried it for its magical powers as they traveled south. Traces of settlement in the Baltic region are scarce because during most of the Paleolithic era ice covered much of the area. But, with the summer thaw, hunters moved north following the animals (Fraquet 1987, 6).

The Mesolithic Period

The Mesolithic period, (estimated to have extended from 8000 B.C. to 4000 B.C.) found prehistoric humans using bone, wood, stone, and amber. The climate had changed, the temperature had gradually improved, and forests had begun to grow. Mesolithic amber artifacts have been found preserved in a bog near the Halleby River in

Denmark. These were the work of ancient hunter tribes. One piece was carved with geometric lines and five sticklike figures. The human figures were scratched on the surface of the amber. The etchings were in geometric designs similar to cave drawings. Some believe phases of the moon and seasons for hunting are recorded on some Mesolithic artifacts (Fraquet 1987, 6). The earliest three-dimensional works of art in northern Europe are found in the Denmark region and date from approximately 7000 B.C. These are carvings from large pieces of amber representing animals, including a "bearded" elk, a bear, and a water bird. Because of being preserved in a bog, with oxygen excluded, the amber is still yellow and the surface was prevented from oxidizing (Fraquet 1987, 8).

Ethnographic studies illustrate that early Baltic people from the Mesolithic period were largely a farming and cattle-raising Indo-European tribe. In addition, fierce nomadic warrior tribes, who fought with stone battle-axes, also roamed the area (Grabowska 1987, 11). These warrior tribes were given the label "battle-ax culture" by anthropologists. The amber artifacts shaped into the form of a double-headed axe recovered from graves in the Baltic region probably represented these feared battle-axes, and were perhaps an amulet to protect the wearer from harm. Eventually, the nomadic tribes intermingled with the agricultural group, forming a culture called the "Old Prussians." They are the forebears of the Western Balts and are the same people the Roman writer, Tacitas, later called the "Aisti" (see Fig. 2.2).

Some warrior tribes did not join members of the earlier farming culture. These tribes subsequently became the Stone Age ancestors of the Eastern Balts, today's Latvian and Lithuanian peoples. The Aisti and Eastern Balts picked up amber lumps along the shore of the Baltic Sea and used this resource for their own ornaments and as a trading commodity.

The Neolithic Period

The Neolithic Period, also called the New Stone Age, ended about 3000 B.C.—some say in the Baltic it lasted until about 1800 B.C. During this period evidence of gathering houses, where community activities took place and ceremonies were performed, are found. Finds have produced large ceremonial sacrifices of amber in clay pots. It is believed their purpose was to assist in bringing about a good harvest and an abundance of food. Finds at Mollerup and at Laesten in Denmark have produced large amounts of amber, as many as 4000 and 13,000 beads, respectively. Such large quantities must have been obtained either by collecting throughout the winter or through bartering. The beads are often dark red and some are clear with internal "sun spangles," or discoidal fissures, most often induced by heating.

Figure 2.2. Stone Age amber pendant in shape of double-headed axe from Juodkrante, western Lithuania, dating from about 2500 B.C. Size: 8.89 cm or about 3.5 inches tall. Reprinted from Klebs 1880.

This suggests the pots may have been heated during the "sacrifice" (Fraquet 1987, 8).

Baltic archaeologists find amber used on a far wider scale during the Neolithic period, from about 5000 to 1800 B.C. During this era, decorative amber cylindrical beads, nodular beads, and amber shaped into a double-headed axe were made. Artifacts of this type have been recovered from ancient graves from settlement sites in Poland, such as Zaborze, Nosewo, Wierzbowo, Złota, Rzucewo, and Suchacz. Recent archaeological excavations in the Żuławy region in the lower Vistula valley have found Neolithic "amber workshops" approximately four thousand years old. The Vistula River, supplied by the Baltic Sea, provided an abundance of raw amber. When the surroundings nearby were exhausted of raw amber, the workshops moved along to a new location with better resources (Kwiatkowska 1996, 16–7; see Fig. 2.3).

In the mid 1980s, archaeological digs dating back to 2000 B.C. near Niedźwiedziówka, Poland, found an ancient center for amber carving and sculpting. The location produced raw amber scraps, unfinished carvings, and amber jewelry along with tools made from flint and other stones. It is estimated that about 800 to 900 of these workshops functioned in the region of Niedźwiedziówka (Kosmowska-Ceranowicz and Konart 1989, 200). The reconstruction of these workshops enabled scientists to compare their findings with ethnographic studies to establish the processes and techniques used when producing amber ornaments. Raw amber nodules were cut into halves using a piece of thick thread and a deer-horn edge, holes were drilled with a flint chisel. Then holes were smoothed out and given a final polish using sandstone plates and animal fur. During this period, characterized by refinements of skill in working with stone, better polished stone implements were manufactured. The oldest amber artifacts found in several places in Lithuania date back to this age (Kosmowska-Ceranowicz and Konart 1989, 200).

Conscious of the peculiar beauty and other unique properties of amber, primitive tribes in the Baltic region endowed the material with mystical qualities. Amber disks with designs suggesting a religious meaning were discovered at several archaeological excavations (see Fig. 2.4). In fact, these are the oldest known symbols that indicate that ancient Baltic people were sun worshipers (Kosmowska-Ceranowicz and Konart 1989, 4). The dotted design in the form of a cross radiating away from the perforation in the center was the symbol for the cult of the sun's wheel. The small indentations in the design were filled with resin that provided a decorative accent to the polished surface of the amber disk. These round amulets are linked to

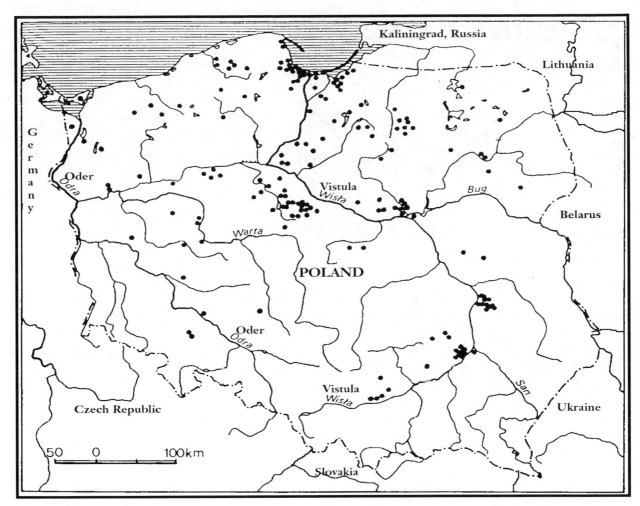

the late Stone Age.

The importance of amber in Baltic culture is further indicated by its influence on pottery designs. Balts in the Vistula Basin developed a form of pottery called "face-urns," vessels that bore human features on the covers or necks. Thus, the culture was labeled the "face-urn" culture. Not only were eyes, nose, and mouth markings included on face urns, but beaded necklaces were depicted around the neck of the urn, and the symbol of the sun was depicted on the lid (Gimbutas 1989, 65). Some urns had brass earring loops strung with glass and amber beads. Often, these large jars were used as incinerary jars for cremation and have been found in burial mound excavations (see Fig. 2.5).

Prehistoric amber artifacts have been found in about 60 localities in Lithuania. Amber articles found in Neolithic grave sites along the Baltic seacoast of Lithuania attest to local preservation of amber as far back as 5000 years ago. In this period, the region from Sambia

Figure 2.3. Map of archeological sites containing amber in Poland. Distribution of excavation sites of Stone Age amber artifacts over the territory of Poland by Mazurowski, 1983, as in Kosmowska-Ceranowicz and Konart, 1989, 11.. Courtesy Muzeum Ziemi, Warsaw, Poland.

Figure 2.4. *Stone Age amber disk from Sambia Promontory, dating from about 2500* B.C. *Disks of this type are the oldest known symbols of sun worship from the ancient Balt culture, Museum of Archeaology, Gdańsk, Poland. (Size range of similar objects: 2.6 cm to 8.5 cm diameter.)*

Figure 2.5. *Incinerary jar with amber and blue glass on loops from an early Balt culture. These artifacts are from Biskupin, Poland.*

and the Courish Lagoon to Palanga in the north was densely populated. Graves of former inhabitants contain numerous individual pieces of amber jewelry, including pendants, beads, buttons, and amulets. Digging for Neolithic artifacts in swamp regions of the Sventoji River near the Baltic Sea, Rimantienė, a Lithuanian archaeologist, found thousands of natural and polished amber articles (Gudynas and Pinkus 1967, 42–4).

However, one of the most famous older finds and larger hoards of amber was found at Juodkrante (Schwarzort [East Prussia] before 1940), Lithuania, a village near the Courish Lagoon, in the heart of the amber-bearing region. It is located on a long sandbar that separates the Courish Lagoon from the Baltic Sea. While dredging and sifting sand from the bottom of the lagoon in search of raw amber for the industry during the 1850s, searchers found a priceless treasure of about 434 Neolithic amber artifacts.

Included among the Juodkrante archaeological finds were five figurines, crudely shaped in stylized human forms (see Fig. 2.6). One portrayed a woman, another only a face. Amber human figurines were found elsewhere, in numerous Stone Age tombs. A site found near Gdynia, Poland, included human figures. Varying in shape and size, they were usually carved from flat pieces of amber and fashioned into small idols or human shapes. Several of these figurines were drilled through, possibly enabling the owner to wear the piece as an amulet to avert disease or evil. Another theory, based on the owl-like facial appearance of most of these figures, is that they represent a deity. During the Neolithic era an owl goddess was worshiped by different peoples. This owl goddess was thought to be feared because she was the messenger of death.

Also, the owl symbol, or hieroglyphic, in ancient Egypt was a symbol for death (Dahlström and Brost 1996, 72).

The largest portion of the Juodkrante treasure consisted of cylindrical beads, amber buttons, small disks, and pendants shaped like stone axes. The amber buttons were lenslike in shape, with cone or V-shaped holes for attaching the buttons to articles of clothing. Such perforations were characteristic of both the Neolithic and the succeeding early Bronze Age. Primitive flint and bone drills, employed in shaping and piercing amber artifacts, left concentric marks in the holes, providing clues for understanding the past. Apparently, the drills were short and required that holes be drilled from both sides for complete perforation. Also, when amber buttons or beads were drilled through, drilling from both sides prevented the amber from chipping as the drill pierced the bottom of the hole (see Fig. 2.7). When attempting to date these amber findings, archaeologists study such drill markings for indications of technique and to determine effects of aging on unpolished portions of amber. But since many surfaces of these primitive amber pendants, buttons, and disks are polished, it is often difficult to learn how they were initially shaped (Gudynas and Pinkus 1967, 42–3).

The Juodkrante amber artifacts were exhibited in the amber collection of the University of Königsberg until the 1940s. It was once believed the entire Juodkrante collection was destroyed by fire in World War II, and these treasures of the past were lost forever. However, as with the Amber Room, various stories exist about the fate of the collection. Fortunately, the collection was photographed before the war, and copies of the original five stylized human figurines were cast in gypsum. Two copies of the figurines were also carved from amber by a Lithuanian amber worker (Encyclopedia Lituanica 1970, s.v. "amber"; see Fig. 2.8).

The collection also included figurines shaped like animals, probably representing a horse or a horse's head. Among other finds from this period are objects in animal shapes, such as a figurine of a wild boar found near Gdańsk and a bear found near Słupsk in Poland. These pieces are of the same type found in "Fino-Ugrian" sites in Lithuania. It is generally thought that these figurines were worn as amulets to provide protection during the hunt, similar to American Indian's use of fetishes. Based on the findings of worked or polished artifacts, Polish archaeologists indicate that in the early Stone Age animal figurines and "worked pieces" of amber were preferred (see Fig. 2.9; Kwiatkowska 1996, 18–9).

Neolithic amber artifacts usually are found in regions near or

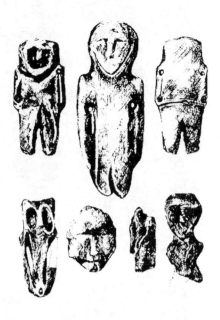

Figure 2.6. Juodkrante figurines. Stone Age amber amulets from Kuršiu Zaliv (Courish Lagoon) near Juodkrante (Schwarzort). Based on Pelka 1920. Size range about: H. 9.5 cm to 7.74 cm, W. 5.7 cm to 3.3 cm, weight about 7.8 g to 36 g. Three pieces from these collections are located in the Institute of Geology and Paleontology and the University Museum in Göttingen, Germany (former Federal Republic of Germany) and in the Amber Museum Palanga, Lithuania.

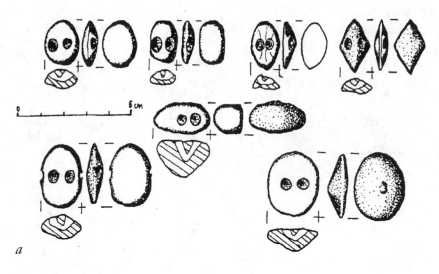

Figure 2.7. a, Diagram of drill holes made using primitive tools. Reprinted from Kwiatkowska 1996; b, Primitive drill displayed in Museum of Archeology in Gdańsk, Poland.

in the area where natural deposits of raw amber occur. These artifacts are found where there were settlements of people such as the Rzucewo group near Elbląg, or the Stegna hoard, which contained 47 amber ornaments with characteristic perforations shaped like the letter V. The beads were similar to those found in the Juodkrante group from the Curonian Lagoon in the mid-nineteenth century as described previously.

According to Polish archeologist Kwiatkowska, the process of working amber remained unaltered from the Neolithic up to the Medieval period. This involved selecting the piece of amber and dividing it into smaller sections, if required, and then roughing each section into the desired shape. Any necessary holes were drilled, then the surface was sanded and polished. Polish archaeologists have found the tools to vary throughout the ages only as new materials were introduced to the culture. During the Stone Age, tools consisted of cutting cords, borers of flint and bone, flint knives and awls, stone grinding disks, and leather. The leather was prepared in a special manner for use in polishing (Kwiatkowska 1996, 16).

Even today, where the sea splashes the Vistula sand bar near Sobieszew, it occasionally tosses up Neolithic oval ornaments with drilled holes, along with natural seastone lumps and throws them upon the strand. It is believed that these artifacts originated in ancient settlements now buried under the sea. What is surprising is that similar Neolithic amber objects were found in distant regions of Poland far from deposits of raw amber—some as far away as Silesia and even into former Czechoslovakia. These ornaments were of the Juodkrante type, but were recovered from a "bell beaker" culture site at Slapanice. These ancient people made pottery that was shaped like a bell, hence the name bell beaker culture. The occurrence of these amber artifacts provides evidence that amber was traded and used as a means of payment even during this early time period (Grabowska 1983, 11).

On the coast of Lithuania, a smaller Neolithic hoard of amber artifacts was found somewhat farther north in Palanga. It included

150 artifacts similar to those found in Juodkrante. These are now on exhibit in the Amber Museum of Palanga, along with photocopies and models of the Juodkrante find. (See Chapter 3 for more about Lithuania.)

The original Palanga collection of amber artifacts belonged to Count Tiskevicius. Some amber articles in the count's collection were found in peat bogs near Palanga by fishermen searching for raw amber material. Other pieces were discovered in old graves in sandy areas. The artifacts as a whole represent different periods. Neolithic amber artifacts in the Palanga collection include pendants, beads, disks, and buttons with cone-shaped perforations. Some are decorated with dots, chiseled lines, and pits. Also in the collection is a flat amber carving in a stylized human shape, with detail showing the waist and a division between the two legs. This unusual piece reveals the skill Stone Age amber carvers developed, despite their crude flint tools. Other articles in the Palanga collection date to the Bronze Age and Iron Age (Gudynas and Pinkus 1967, 44; see Fig. 2.10).

Figure 2.8. a. Amber stylized human figurines. H. 9.57 cm x W. 4.35 cm. Courtesy Muzeum Ziemi Collection, Warsaw, Poland, photographed by author.

b. Author demonstrating how the primitive drill was used (see Fig. 2.7b). Photo by A. Caceras, Michigan Mineralogical Society.

In July 1976 Moscow News Agency reported a prehistoric amber hoard that was discovered during excavations near the fishing village of Ventaga in Lithuania. The find included 24 pieces, with some amber containing insects, flower petals, and leaf imprints (Hunger 1977, 79).

Farther along the Baltic coast, in eastern Latvia, in the region of the Lubana Lake swamps, 2252 amber articles were found by Latvian archaeologist Ilze Lozė (1969a, 124–34). Many of these amber ornaments were not only worn as beads, but were sewn to clothing as decorations and as amulets to impart magical protective powers. Such powers were attributed not only to the amulet shape, but to the amber as well. Amber buttons, again similar to the ones in the Juodkrante and Palanga collections, were found. These Luban buttons were also perforated by primitive flint drills, although the shapes of the perforations were different, since V-shaped drills were used rather than the sharp pointed drills used elsewhere in the Baltic region.

As indicated by such artifacts, Stone Age humans were superstitious and attributed supernatural powers to amber, perhaps because of its power of attraction. Seeing tiny bits of straw and feathers jump to amber lumps after the amber had been rubbed on animal skins must have awed primitive peoples.

One of the largest Neolithic finds was discovered by

Latvian archaeologist Vankina north of Liepaja on the Baltic shore near Sarnate, Latvia (Gudynas and Pinkus 1967, 42–4). The find consisted of a complete Neolithic amber factory including flint and bone tools. There were also splinters of flint that were used by early Balts to cut lumps of amber. The ease with which raw amber can be processed no doubt added to its value for these primitive people. It is no wonder this shining substance was used for amulets and adornment by these northern tribes as early as the Stone Age.

The distribution of amber finds dating from the Neolithic period in Latvia and Estonia was plotted and mapped by Lozė (1969*a*, 124–34; 1969*b*, 1–15). Similarly, Lithuanian archaeologist Rimantiene (1992, 367–76) mapped the spread of archaeological digs yielding amber artifacts in Lithuania. In 1953 Latvian archaeologist Sturms (1953, 167–205) plotted the Neolithic homesteads and burial grounds in which amber was found in Germany and Poland, finding over 100 recognized Neolithic burial sites. Currently, Polish archeologists, such as Kwiatkowska (1996, 16) and others continue this work.

AMBER TRADE DURING THE STONE AGE

Amber was one of the principal commodities for barter in early Europe and the Mediterranean. Archaeologists found amber as far away as central Russia, western Norway, and Finland, which indicates the establishment of trade as early as 3000 B.C. (Childe 1939, 200). Even farther away, some amber ornaments found in Egyptian tombs believed to date back to the Sixth Dynasty (3200 B.C) were identified as Baltic amber in the past, but recent research by Polish scientists indicates that semifossilized resins were used. (Niwinski 1997).

The Aisti and Baltic tribes in the area of northern Poland and Lithuania not only traded their raw amber, but must have traded finished articles as well. Amber beads found in Juodkrante, for example, included those in various stages of completion as well as considerable quantities of finished beads. Furthermore, finished amber found in Finland, Sweden, Norway, northwestern Russia, and the Middle Urals display the same shapes and styles of artifacts found at the source, indicating the pieces were probably worked by early Balts at the places of origin before being exported.

Amber was also traded into the interior of Germany. An archaeological find near

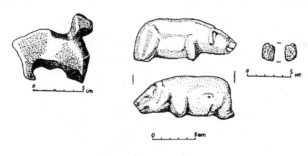

Figure 2.9. Drawings of bear styles from various locations. Drawings from figure found near Slupsk in 1880 and polished amber bear found in Siedlnica, Leszno region. Reprinted from Kwiatkowska 1996.

Landsberg, or present day Gorzow Wkp. Dobiegniew, Poland, produced artifacts among which was a carving of amber representing a wild horse that dated to the Neolithic period (Ley 1951, 6; see Fig. 2.11). Amber ornaments were found with remains of Stone Age lake dwellers of Switzerland and France (Buffum 1897, 32). In fact, almost every European country has produced amber artifact finds. Such general use of amber during the Stone Age indicates widely established trade and furnishes evidence of early esteem for this beautiful golden substance originating in northern regions of Europe.

Away from the European continent, mound tombs (or tumuli) of England, especially in the vicinity of Stonehenge, are archaeological sites that have yielded amber ornaments. The people who built Stonehenge were sun worshipers and the shiny lumps were thought to be a "substance of the sun." The British finds range in time from the new Stone Age to the early Bronze Age (Gimbutas 1963, 55).

The last part of the Neolithic period introduced copper tools and, for this reason, it is sometimes designated as the Chalcolithic period, although it is actually a division of the new Stone Age. Copper was used for both tools and weapons but was replaced by the more suitable bronze, a harder, stronger alloy of tin and copper. The Copper Age was of short duration, whereas the Bronze Age, beginning about 3000 B.C., ended with the advent of the Iron Age (around 1000 B.C.). Fraquet (1987, 8) reports on the distribution of amber in the Jutland area and its gradual spread across Europe from 3000 B.C to 1900 B.C., the transition from the Neolithic to Copper and Bronze ages, pointing out the large size of the hoards found. In Mollerup the amber weighed about 4 kilograms and at Laesten it weighed as much as 8 kilograms. In about 1958 a clay pot containing amber was found at Aaruopgaard, near Horsens. The pot contained 300 to 400 amber beads along with copper rings and twisted coils. Since copper was not native to Denmark, it must have been included in commodities traded.

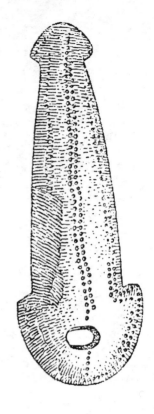

Figure 2.10. Amber axe from Palanga Collection. Size about: H. 9 cm W. 4.5 cm From Mazurowski, 1983, 34.

BRONZE AGE AND IRON AGE AMBER TRADE

The Aisti trade route of the ancient Bronze Age, by which amber was exchanged for metals, passed along the Baltic coast from Klaipeda, Lithuania, to the Elbe River, Germany. Amber gatherers along the coast enjoyed imported copper and bronze, but cultures farther inland were denied access to metal tools and ornaments and continued to use stone hoes with inserted handles for tilling their soil as late as the early Iron Age (Gimbutas 1963, 57). In the Baltic, the early Bronze Age tends to be dated from 1700 B.C. to 1450 B.C., when the Unetician culture covered the region that now forms the country of Poland.

Amber objects dating from the early Bronze Age are rarely found in grave goods because cremation was the predominant burial practice. However, evidence of trading increased and amber played a major role, perhaps even being a means of payment for other commodities. At this time, decorative amber items began to reach southern Europe, indicating the existence of trade contacts. Inhabitants of Pomerania, in northwest Poland, paid for weapons, tools, and ornaments with amber. In Polish regions, such as Radzików and Wojkowice in Silesia, finds provide us with evidence of this trade. Amber beaded necklaces and bronze spirals were among the varied objects found. It appears that amber from Jutland from this time period was reaching Bohemia, being traded via a route running along the Elbe or Vistula rivers toward the Odra (Oder) valley then through the Moravian Gate and Kłodzko vally.

Amber beads, dating from the middle and late Bronze Age (1200–700 B.C.), are found at many sites from the Lusatian culture in Poland. This culture survived unbroken over almost the whole territory of Poland until the early Middle Ages. Amber beads are also found at Lusatian culture sites in Moravia and Slovakia and date from the early Iron Age about 700 B.C. Lusatian culture tribes living in Poland and Silesia acted as intermediaries in the import of amber from East Pomerania. Another group of people, called the Pomeranian culture, had developed from the Lusatian culture and inhabited the lands of East Pomerania. This group developed a ritual burial rite that included use of pottery in the form of facial burial urns (see Fig. 2.12). The face urns had handles, similar to ears, that were adorned with blue glass beads and amber beads. The blue glass, or faience beads, a forerunner of glass from Egypt used in funerary jewelry, indicates that trade and imports from Syria, Egypt, and the Red Sea were reaching the Baltic lands. The presence of faience beads at Pomeranian culture sites and beads and sword pommels made of amber at the cemetery site in Hallstatt bear evidence of trade contacts between people of the Hallstatt culture and those in amber-yielding territories (Kwiatkowska 1996, 17).

Hallstatt Amber

In the region just north of the Alps, a large hoard of amber was found at Hallstatt, Austria, during the 1800s and was dated to the early Iron Age (see Fig. 2.13). It appears that in the Hallstatt culture, amber was widely used by the peasantry. Amber ornaments in these ancient hoards included hundreds of beads, among which was a necklace containing all forms and sizes of amber beads along with 60 blue and green glass beads. Bronze articles found in Hallstatt represented the entire gamut of Etruscan art, from the earliest Assyrian-Phoenician style to the later Celto-Etruscan mixed forms. Included in the hoard was an unusual type of ornament with decorated plates of bone inscribed with circle designs and associated with large oval beads of amber, indicating a relationship to ancient sun-worship symbols (Williamson 1932, 63–7).

AMBER TRADE ROUTES ACROSS EUROPE

Trade probably took place in a number of stages through trading intermediaries. The glass and bronze ornaments were probably exchanged for amber, indicating that amber was reaching highly civilized cultures as far away as Italy and Greek colonies along the Black Sea. Grabowska (1983, 11) reports one of the best archaeological sites from the Pomeranian culture at Gorszewice in in Greater Poland. The objects found included 186 amber beads, as well as bronze and iron weapons, tools, vessels, blue glass ornaments, and beads. It appears that these objects originated in the Hallstatt cultural zone and from Italy and the Pannonian Plain inhabited by the Thracian people.

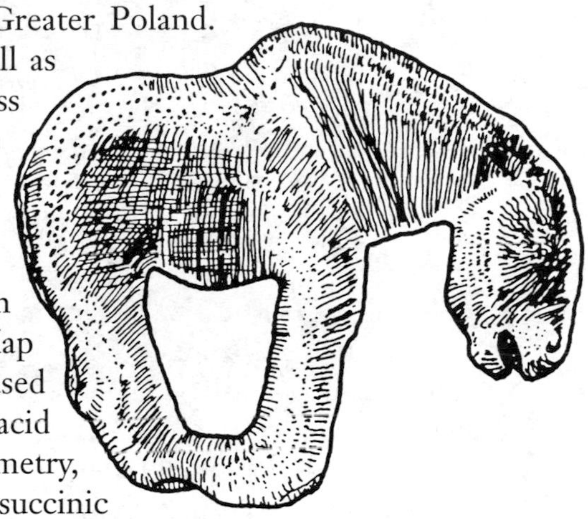

Ancient amber trade routes were traced in 1925, as shown on the old Amber Trade Route Map (Gudynas and Pinkus 1967, 36). These routes were based on amber locations where amber containing succinic acid was found. However, after using infrared (IR) spectrometry, scientists found amber other than Baltic contained succinic acid. Currently, Eastern European scientists are revamping the old trade routes using new analysis of the amber samples. The fact that trade existed is a way to trace ancient history. Historian Buffum (1896, 32) stated: "The amber trade routes are the original trade route along which luxuries of life went out in search of the necessities."

Some historians believe amber was known to Assyrians even in the days of Ninevah (ca. 2000 B.C.). A broken obelisk in the form of a tapered four-sided shaft of stone inscribed with Assyrian cuneiform was found in the nineteenth century. The cuneiform writing, presumably by a king of Ninevah, was translated in about 1876 by Oppert (1876, 1–15), a noted Assyriologist. Oppert believed the inscription indicated the early existence of commerce between northern Europe and Assyria. His translation follows:

In the sea of changeable winds [indicating the Persian Gulf]
His merchants fished for pearls;
In the sea where the North Star culminates [indicating the Baltic Sea]
They fish for yellow amber.

In 1925 DeNavarro (1925, 481–507), a Baltic historian, described early trade routes as indicated by finds of ancient amber artifacts. He determined that the tomb amber was Baltic amber

Figure 2.11. Neolithic amber horse figurine, Woldenberg/Dobiegniew, Poland. c. 2000 B.C. The figure is 4 inches (10 cm) from head to tail. Redrawn from Pelka 1920, Spekke 1956. This horse is so famous among archaeologists that it was used as a symbol for the 1997 Baltic Amber Symposium in Gdańsk.

because of its succinite content and excluded simetite amber of Sicily, which does not contain succinic acid. By plotting locations where Bronze Age tomb amber had been discovered, he delineated routes traveled by early merchants. Latvian archaeologist Sturms added to the early work of DeNavarro by knowledge gained in later discoveries at Juodkrante. This entire project is being restudied by the Polish Academy of Science and Italian scientists to verify the amber source. New information will be available in the near future.

Research shows that Baltic tribes traded amber mainly along European rivers. During the middle Bronze Age (1800–1200 B.C.), a central route began in Jutland, Denmark, and passed along the Elbe River southward to what is now Hamburg, Germany. An important hoard of amber artifacts was found at Dieskau, where the route separates, with the main branch following the Saale River. From the Saale, amber finds indicate the route continues south along the Naab to the Danube, then onto the Inn River at Passau along the present border of Germany and Austria. Crossing over the Alps through the Brenner Pass, the route follows the Eisock River until it joins the Adige River flowing into Italy.

Early research on amber trade routes pointed to a second branch following the Elbe River into former Czechoslovakia. Another branch route turned off at Passau and followed the Danube to Linz and upper Austria. The hoard of amber found at Hallstatt, Austria, proved that another branch of the Bronze Age amber route followed the Slazach River (Spekke 1957, 48–61).

Other amber finds indicate that a western route diverged from the main central route on the Saale, then passed westward just north of the river Main to connect with the Rhine, then traveled south along the Rhine into the Aare River watershed.

Apparently, the Baltic Bronze Age and the transcontinental amber trade for tin and copper began simultaneously. It is known that amber was traded to central European cultures, who then traded it to the Mycenaeans and early Greeks. Amber from the source area of Sambia was shipped to the mouth of the Vistula River, where it was combined with amber from Jutland and then transported southward along the Bronze Age central trading route. Based on finds of amber in Mycenaean tombs, Gimbutas (1963, 57), a Lithuanian anthropologist, believes this trade began around 1600 B.C.

In pre-Greek cultures, amber played an important role as a luxury and treasure of the educated and ruling classes. Amber beads, pendants, amulets, and hairpins with amber finials were found in tombs of the Mycenaean kings of Crete. An enormous quantity of flattened spherical amber beads was found among a treasure of golden ornaments in the Mycenaean acropolis, providing evidence that amber was considered a magnificent substance. While excavating the acropolis, Heinrich Schliemann (1878) found 400 beads that proved to be of Baltic origin. In 1885 Helm (1885, 234–9), a German scientist, chemically tested Mycenaean amber and identified it as succinite. Using the modern IR photography techniques in 1963, Beck (1971, 234–6) compared the spectra of Mycenaean amber with the spectra of Baltic amber and found them identical, furnishing strong evidence that the Cretan amber was indeed Baltic in origin.

In later periods, amber was exported on a more extensive scale. For example, an eastern route, beginning at Gdańsk, was established during the early part of the Iron Age (between 1000–500 B.C.). Danzig, now Gdańsk, in the heart of the Baltic amber region, was a natural starting point for expeditions across Europe to exchange amber for iron, bronze, and other commodities. The route followed the Vistula River toward Nakło; then, crossing lakes between the Vistula and the Warthe (Warta), it continued down the Warthe River into the Oder River near Głogów. From the Oder, the route passed to Wrocław, Poland, and over the Carpathian Mountains by way of Glatz Pass, then followed the March River, with the main branch continuing through Carniola and entering the Adriatic Sea near Trieste, Italy. This route appears to be of great significance during the first and second centuries, indicated by numerous finds of Roman coins and objects originating in Italy or Roman provinces dating from this period. The Roman artifacts are now found in three Polish territories located along this trade route's course. A trade center site containing stores of amber was discovered at Wrocław, Poland. Further evidence is indicated by the presence of amber workshop sites located at Jacewo in the Kujawy region; Regowo by Skierniewice; a site near Warsaw; and one in Świlcza in the Rzeszów region of Poland (Kwiatkowska 1996, 17).

Additional routes branched off to Czech Republic and Vienna, Austria. An eastern route began at Klaipeda and followed the Dnieper River, continuing on to the Black Sea region. It will be interesting to see if modern analysis of amber hoards using newer scientific techniques will revamp the plotting of ancient trade routes.

The amber trade originating in the eastern Baltic region fluctuated as Mediterranean cultures developed and declined. After the destruction of the Mycenaean culture of Central Europe, for example,

Figure 2.12. Face urn cinerary jar with amber beads on ear loops, La Tene period, from Biskupin, Poland.

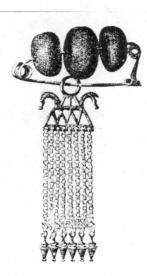

Figure 2.13. Bronze chain and fibulae with amber beads from Hallstatt. Reprinted from Pelka 1920.

amber supplied to Crete sharply diminished in quantity. The early Balts, however, found new outlets for their amber, which had become known as the "Gold of the North" in the Near East.

Near East trade existed as early as 900 B.C. This is evidenced by an amber statuette, about 20 centimeters high, representing Assurnasupol, King of Assyria (885–860 B.C.), found in archaeological excavations on the banks of the Tigris River. Chemical tests suggest it is of Baltic origin. The Assyrians progressed westward toward the Mediterranean and are reported to have demanded tribute from the rich merchants, the Phoenicians. Fraquet (1987, 13) reports that this contact was the origin of the amber statue. She also reports a second statue that mysteriously appeared on the U.S. market, thus casting doubt as to the statue's authenticity. IR spectroscopy by Curt Beck (Fraquet 1987, 13), proved one of the statues to be Baltic amber and the other a resin from the Middle East. The exact route amber traveled to reach Assyria is a matter of conjecture, but it may have been traded along routes through Russia and the Caucasus or through Central Europe via Phoenician trade routes in the Mediterranean (Spekke 1957, 61–2).

The main Eastern Amber Route through Central Europe during the early Iron Age continued to be used despite invasions by hostile tribes. In the hostilities, the Lusatians of Central Europe were mediators between the Hallstatt culture in the eastern Alpine area and the amber gatherers. By the beginning of the seventh century, Lusatian intermediaries provided amber to the Etruscans and transported metallic objects from the more highly developed cultures along the amber routes to supply the Old Prussians on the Baltic coast. Such objects included bronze horse trappings as well as jewelry. The hoards and graves richest in bronze ornaments in the source area for amber were concentrated along the lower Vistula and on the Sambian Peninsula. During the period between 800 B.C. and 700 B.C., trade extended into the central Caucasus region. Finds of bronze in new forms, such as belt hooks, spiral plates, and wheel-shaped pendants, appear. These are in the style of similar objects found in the Caucasus (Gimbutas 1963, 83).

PHOENICIAN TRADE IN AMBER

Sun worship was widespread, if not universal, during ancient times. Some historians maintain that the Phoenicians brought sun worship with them from the Orient when they came to the Mediterranean, and temples built to Baal were temples to the sun. Since Greeks and Romans also were involved in sun worship, one cannot doubt that amber, which was more sunlike than any other gem in nature, excited

the admiration of sun-worshiping people. Thus, it was a valued commodity for trade.

Ancient Phoenicians, a hardy seafaring and commercial race, were probably the first sailors to trade amber among the Mediterranean countries, as well as to pioneer sea routes to the Atlantic shores of northern Europe to obtain amber and exchange it for bronze between the thirteenth and sixth centuries B.C. However, amber was probably traded to Phoenicians through middlemen with little knowledge of its place of origin.

In the region between Etruria in Italy and Massilia (the present-day city of Marseilles in France) dwelt the Ligurians. Their land, located at the mouth of the Rhone River, included the Mediterranean terminal of one of the direct routes from the Baltic. It has been suggested that the Phoenicians obtained amber from the Ligurians, a supposition based on the fact that at one time amber was called "Ligurian." The Phoenicians were clever traders and went to great lengths to protect their trade secrets. To obscure their sources of foreign goods, they fabricated tales of sea monsters and other dangers encountered on their sea voyages.

Now that the Phoenicians had seen amber gathered from the sea, they were determined to keep the secret for themselves and, thus, guard their lucrative trade. When their fleets returned to Syria, many tales were told of perils to the north, of lodestones that would draw ships to destruction on hidden reefs, of whirlpools that would suck them down to the bottom of the ocean, of witches who enchanted men by turning them into beasts, or terrible sea serpents and awesome monsters (McDonald 1940, 155).

By the time the Roman Empire was established, earlier Phoenician sources for amber were forgotten. Some Greek and Roman writers reported amber as being dug in Liguria, but Pliny, the Roman naturalist (A.D. 23–79), declared this to be false. He scoffed at Demonstratus, who called amber *lyncurium*, believing it to form from urine of the wild beast known as the lynx—red amber from the male and white from the female. Demonstratus also informed his countrymen that in Italy there were other wild beasts known as *languri* (thus, some people called amber *langurium*). Zenothemis called these wild beasts *langes*, stating that they inhabited the banks of the river Po. Based perhaps on the discovery of insect or plant inclusions, or amber's piney scent when burned, another historian, Sudines, suggested amber was in reality produced by the *lynx-tree* that grew in Liguria.

Fact and fancy intertwined to explain the mysterious characteristics of this highly desirable golden ornamental substance.

Factually established is that, with the travels of the Phoenician traders, amber spread to many countries of the Mediterranean, where it sold at the price of gold. As amber was traded for tin in England, the Frisian Islands, and the Denmark coast, it would have been within easy access. Thus, it was possible for the Phoenicians to trade directly with amber gatherers, eliminating the middlemen as time passed.

The name "Gold of the North," then commonly used for Baltic amber, is still used among those dealing in amber when referring to the finest amber variety.

GREEK CULTURES AND AMBER

Despite continual invasion by barbarian tribes, amber remained an important commodity in Greek trade from early Greek cultures onward. Homer, in his *Odyssey* (1000 B.C.), portrayed a "cunning trader" enticing female servants to his ship by dangling amber beads before their eyes. In another section, Homer describes echoing halls "gleaming with amber." According to the *Odyssey*, among the jewels offered to the queen of Syria by Phoenician traders was "a golden necklace hung with pieces of amber." In describing his adventures in Syria, the character Emmaeus remarks (Book XV):

> Thither came the Phoenicians, mariners renowned, greedy merchant men, with all manner of goods, in the black ship . . . there came a man, versed in craft, to my father's house, with a gold chain strung here and there with amber beads, and the maidens in the hall and my lady mother handled the chain and gazed on it, offering him their price.

The sunlike color of amber was of special significance to the Greeks. Homer, for example, described Penelope's necklace as "golden, set with amber, like the radiant sun!" Another description states (Book XVIII):

> For his henchman bare a broidered robe, great and very fair, wherein were golden brooches, twelve in all fitted with bent clasps. And the henchman straightaway bare Eurymachuss a golden chain of curious work, strung with amber beads shining like the sun.

In his descriptions of ancient art objects and wares, Homer mentions only this gemstone.

Among the Greeks, amber was desired mainly for its decorative qualities. Accordingly, it was used in inlays together with ivory. In jewelry, it was used as parts of necklaces and also in pins or *fibulae*. The latter were either made of bronze or gold. Often they were dec-

orated with amber in a manner similar to that illustrated in the Hallstatt object shown in Figure 2.13. Among the more exotic uses of amber was its use in inlays decorating harps.

The remarkable property of amber attracting small bits of lint, pith, or other light objects after it had been vigorously rubbed was first discovered among the Greeks by the philosopher, Thales, about 600 B.C. Since the Greek word for amber was *elektron*, it was from this early discovery of the curious attractive properties of amber our modern word electron and its derivatives come.

The source of amber, however, was shrouded in mystery, and a popular legend proposed that amber was the hardened teardrops of the Heliades, the sisters of Phaëthon, that were shed over a mysterious river called the Eridanus somewhere in the frigid northern vastness of Europe. Herodotus (ca. 484–425 B.C.), in his characteristically skeptical manner, not only doubted the existence of any such river, but stated that he had not found a single person who could say, from his own experience, that there was such a river or there was even a fabled sea into which this river was supposed to run.

The actual existence of the Baltic Sea did not appear in historical records until about 46 B.C. Pomponius Mela, a Greek known as "Spaniard Mela," made a first obscure mention of the Baltic. He described a bay above the Elbe River full of large and small islands.

In this bay . . . is Scandinavia, inhabited up to now by the Teutons, and it is prominent among the other islands because of its greater size and the fertility of its lands (Spekke 1957, 77).

After about 600 B.C., fashion for amber declined in Greece proper, but continued for about 200 more years in Greek colonies located on the Italian peninsula. Between the sixth and fifth centuries B.C., the Etruscan people on the Italian peninsula used amber as inlays, for beads in fibulae, scarabs, and small figure type pendants. This area called Etruria was the first part of the Italian peninsula to work amber extensively. Therefore, ancient Italian amber artifacts are often labeled as Etruscan. However, many Italian amber artifacts were produced in settlements such as Apulia, Calabria, Campania, Lucania, or Latium. One type of amber head pendants carved in this area has characteristic large almond eyes and hair depicted simply by incising a group of parallel wavy lines. (see Fig. 2.14). Commonly found designs were fertility amulets carved in the form of rams' heads, cowrie shells, bulla bottle pendants, and other animal figures. Monkey figures and lions carved from amber are also found. Since this was a pre-Christian era, many of the carvings represent Greek and Roman myths and legends (Fraquet 1987, 16).

Figure 2.14. Amber artifacts in shape of head with typical almond-shaped eyes. Roman style eyes were more open. Size above about 4.5 cm, below 4.2 cm. about 3 cm x 2.5 cm (From Pelka 1920, Stronge, Catalogue of Amber, 1957, as in Reineking von Bock 1981, as redrawn by Fraquet 1987.)

The settlement of Picenum on the Italian peninsula probably produced the greatest number of finds of ancient amber in Italy. It is thought that amber found here was not worked locally but was imported. Among the ancient artifacts found at Picenum was a carving of a youth and a person reclining on a couch (see Fig. 2.15; Fraquet 1987, 16). This piece is now located in the Metropolitan Museum of Art in New York and was on exhibit at the American Museum of Natural History amber exhibition in the spring of 1996. Another amber artifact from Italy, which was also on display in the 1996 amber exhibition, was a carving of a winged deity with an Etruscan youth sitting on its lap (Grimaldi 1996a, 151).

Interestingly, records indicate that Etruscan actors were sent to Rome to placate the gods when there was a plague. During a later period, amber objects carved in images of Etruscan actors have been found in Roman regions. There were many identical items produced, which indicates there may have been mass production of this theme (Fraquet 1987, 16). Perhaps the amber carvings were thought to have the same protective powers that the earlier Etruscan actors had during the plague!

During the La Tene and Roman periods from about 400 B.C. to A.D. 400, development in amber-working techniques took place. To refine their work, craftsmen began using primitive lathes and, thus, trade was expanded. Amber continued to be used by the inhabitants of Amberland in the north. Grave goods, including amber artifacts, were found in richly furnished graves. At Wrocław, Poland, for example, a prince's grave contained an amber necklace mixed with glass and metal beads (Kwiatkowska 1996, 17).

ROMAN ERA—"GOLDEN AGE OF AMBER"

For several centuries before and after the birth of Christ, inhabitants of the Baltic region did not transport amber along trade routes to the south, trading it instead to Germanic tribes, their immediate neighbors east of the Vistula River. A trade center developed in Sambia at the mouth of the Nemunas River in the area of the present Klaipeda, and smaller centers developed in Glindia and Sudovia, as indicated by large quantities of Roman coins and artifacts found in these areas (see Fig. 2.16). The task of maintaining these long, tenuous trade routes across Germany was a constant struggle because of tolls levied by tribes along the routes.

Both Pliny, the Elder (A.D. 23–79) and Tacitus (A.D. 55–120), Roman historians, described the influence of amber during the "Golden Age of the Roman Empire." Pliny wrote the *Natural History*, a 37-volume encyclopedia. In Book 37 entitled, *The Natural History of Precious Stones*, he expounded on amber, which he considered to be an object of luxury popular mostly among women, who had no real justification for its use.

Emperor Nero regarded amber so highly that he described his beloved Poppea's hair as having the color of amber. Other ladies of Nero's court dyed their hair to match the color of this gem (and perhaps the color of the hair of the emperor's favorite). Since yellow was held in esteem as an imperial color, amber's value increased so much that Pliny tells us "the price of a figurine in amber, however small, exceeded that of a living slave."

Figure 2.15. Roman artifact of woman and youth reclining on couch. Courtesy Metropolitan Museum of Art, New York.

Pliny describes several kinds of amber obtained from the "Aisti" amber-gathering tribes (1857, 402–3):

> There are several kinds of amber. The white is the one that has the finest odor, but neither this nor the wax-colored amber is held in very high esteem. The red amber is more highly valued; and still more so, when it is transparent, without presenting too brilliant and igneous an appearance. For amber to be of a high quality, should present a brightness like that of fire, and not flakes resembling those of flame. The most highly esteemed amber is that known as the "Falernian," from its resemblance to the color of Falernian wine; it is perfectly transparent, and has a softened, transparent brightness. Other kinds, again, are valued for their mellow tints, like the color of boiled honey in appearance. It ought to be known, however, that any color can be imparted to amber that may be desired; it is sometimes stained with kid-suet and root of alkanet; indeed, at the present day, amber is even dyed purple. When a vivifying heat has been imparted to it by rubbing it between the fingers, amber will attract chaff, dry leaves, and thin barks, just in the same way as the magnet attracts iron. Pieces of amber, steeped in oil, burn with a more lasting flame than pith or flax.

At the beginning of Nero's reign, demand for amber was so great that to obtain a supply for gladiatorial exhibitions, a Roman knight, Pytheas, was sent to the north in search of the actual source. Some historians believe this to be one of the most significant historical events of the Roman era because it opened direct trade with Baltic cultures near the Vistula River. The manager of the gladiatorial spectacles, Julianus, commissioned the knight to cross barbarian territories, a feat never before attempted by a Roman.

To obtain amber from the northern territory for Rome, the knight first had to travel across Roman territory, across the Alps, and down the Danube to Carnuntum, a fortress and the main base for the

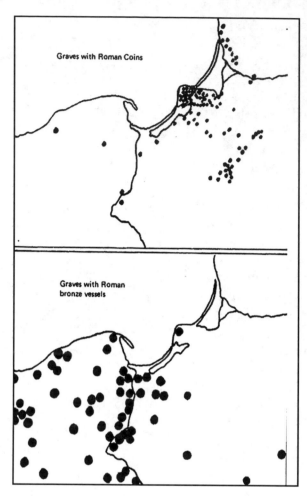

Figure 2.16. Sites where Roman artifacts have been found in the Baltic region. Based on Wheeler 1955.

Roman Danube fleet. The ruins of this ancient city are located near Hainburg, Germany. Lying about 600 miles (600,000 paces) north of Carnuntum, the amber coast was eventually reached, but no exact description of the knight's route is recorded. Past historians assumed that he followed the amber route along the Marsch River to the Danube, crossing overland to the Vistula River, which empties into the Bay of Danzig. Thus, the existence of a main route for procuring amber for Rome, bypassing middlemen, was established (Ley 1951, 12).

According to Kosmowska-Ceranowicz and Konart (1989, 20):

Polish historians have located a little known manuscript from a sixteenth century chronicler describing the history of Prussia and includes a description of Pompey's expedition one hundred years earlier than Pytheas. Pliny, the Elder, described Pytheas' expedition in 320 B.C. in a similar manner as the Polish manuscript, which indicates the travel across Roman territory to Massalia (now Marseilles) to the north in the quest of tin and amber and having reached an island that previously was not precisely located. The Polish manuscript appears to locate the island as well as the river Eridanus being rich in amber. (See Fig. 2.17).

Pliny describes the knight's influence on the use of amber in Rome after his successful expedition (Pliny 1962, 199).

This knight traversed both the trade route and the coasts, and brought back so plentiful a supply that the nets used for keeping the beasts away from the parapet of the amphitheater were knotted with pieces of amber. Moreover, the arms, biers and all the equipment used on one day, the display on each day being varied, had amber fittings. The heaviest lump brought by the knight to Rome weighed 13 pounds [about 10 pounds U.S. weight].

In all, about 13,000 pounds of amber were brought back as a gift to Nero from the German or "Aisti" kin. Nero not only adorned the circus with amber, but made it available to gladiators to wear as amulets or charms on their breasts to assure their victory.

Amber pieces were occasionally studded decoratively, like sequins, over the entire garment of a gladiator. One gladiator's amulet of amber was found with the words "We will conquer" carved on it (Szejnert 1977).

The journey following the land route described above was probably the cause for shifting the major amber trade from the Elbe River, or central route, to the Vistula, or eastern route. About this time, the northwestern shore of Denmark and the Frisian Islands diminished in importance in the amber trade, and Sambia became the center of production and trade. The extent of Roman trade is shown in the Sambian area by discoveries of Roman coins and vases in ancient tombs. Roman coins excavated in the vicinity of the Gulf of Danzig date mainly from A.D. 138 to 180, suggesting that this was the peak period of Roman trade to that region (Strong 1966, 9–10).

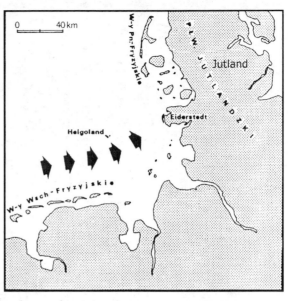

Figure 2.17. Presumed location of the island reached by Pytheas in his search for source of amber according to study by J. Kolendy, 1985. Reprinted with permission from Kosmowska-Ceranowicz and Konart 1989.

In Italy, Aquileia became the main Roman center for manufacture and importation of amber, since it was located at the end of the route to the amber coast. Aquileian workshops manufactured decorative carvings in amber, representing bunches of fruit, animal figures, and vegetables. It was customary to give small amber carvings of ears of corn and fruit as New Year's presents. The significance of such gifts is related to the magical properties of amber as well as to the carved forms. Small pots for cosmetics were elaborately decorated in relief, but on occasion large vases also were produced. Amber carvings from Aquileian workshops are of high quality and are truly exquisite objects of art. So abundant had amber become that Emperor Elagabalus, in an extravagant display of luxury, paved a portico of his palace with pulverized amber (Strong 1966, 11–2).

The last Roman to write about amber was Tacitus (A.D. 98). The eastern trade route, shown earlier as the main source of amber prior to Tacitus's time, had its origin along the western coast of Denmark near the Elbe River and reached the Mediterranean near Marseilles. However, when Tacitus was living, the center of trade had shifted to Samland. In his *Germania*, he describes the tideless Baltic as a sea beyond Sweden in an easterly direction "sluggish and almost without movement; which seems to surround the whole of earth because the last rays of the setting sun, until sunrise of the next day, keep the sky so bright that it darkens the stars."

The eastern Baltic Sea in midsummer does not darken at night because of its far north latitude. Tacitus was no doubt attempting to

Cultures	Year	Baltic Culture	Baltic Amber Artifacts
Fall of Caesar in Western Rome Territories	V 476 AD IV		Amber hoards beads & raw amber found in Poland at Basonia, Vistula valley, Lublin region,
Table 2.1 Culture, Facts, and People Joined in History, Showing Growth of Knowledge Revealed by Amber from Paleolithic to the Fall of Caesar in Western Roman Territories.* (Part 1)	III II I 0		Amber workshops in Świlcza near Rzeszów, Artifacts from Cecele, Szurpily& Osowa (NE Poland) Amber discs with banded edges from Pomerania, Warmia, & Mazuria, Amber workshop from Jacewo, Kujawy region. Amber hoard with glass & metal beads from Partynice (suburb of Wrocław), Poland
	I--BC	*Przeworska* Culture	
La Tene Period	II--BC		
	III-		
Hallstatt Culture	IV- V-	*Mound* Culture of East Baltic	Artifacts from Komorów & Gorszewice, Poland
Wessex Culture Bell Beaker phase 700 - 410 BC	VI-	Grave Casket Culture	Amber artifacts from Radszików & Wojkowice in Silesia, Poland
	VII-	Face Urn Culture	
	VIII- IX-	(Lusatians)	Double-headed axe artifacts from Zaborze, Nosewo, Wierzbowo, Zlota, Rzucewo &
Mycenean Culture (*Achajow*)	X- 1000 BC	*Unetician* Culture	Suchacz, Poland
	1700 BC	*Rzucewo* Culture	Artifacts of Gdańsk area, Poland 900 amber workshops from Niedźwiedziówka Żórawy, Poland
	2000 BC		
	3000 BC	Bell-Beaker Culture	Amber human figures in Lithuania Ornaments of Joudkrante (old Schwarzort) 2252 amber articles from Luban Lake, Latvia Amber pieces from Sarnate, Latvia
Narva Culture 3400-3100 BC		Funnel-Beaker Culture	850 ornaments from Sventoji, Lithuania
	4000 BC		
Meglemose Culture	**Early Neolithic**		
	5000 BC	Battle-Axe Culture	Amber bear from Gdańsk, Poland.
	Mesolithic	Eastern Baltic	Amber horse of Dobiegniew, Poland (old Woldenberg) (some date as 2000 BC)
	8000 BC		
	Paleolithic		Amber discs of Siedlnica, Poland

Vertical labels: Iron Age / Bronze Age / Stone Age (Neolithic) — Ages of Antiquity

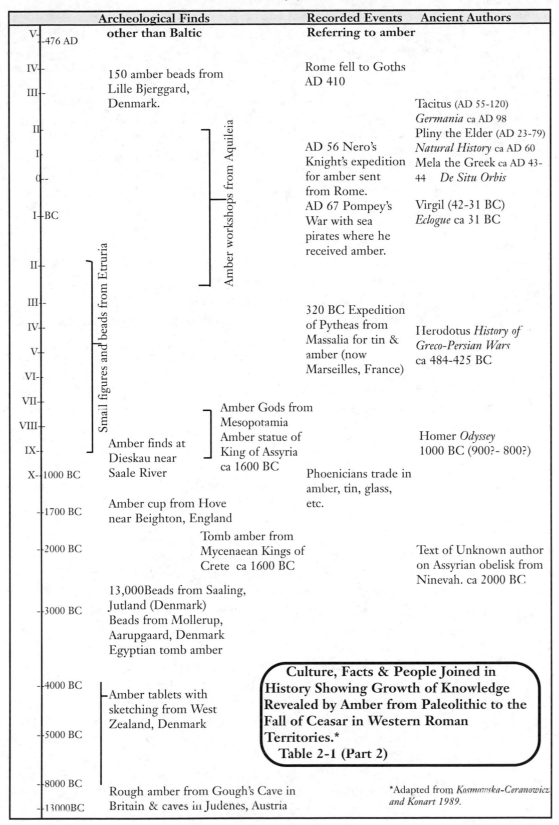

	Archeological Finds	Recorded Events	Ancient Authors
	other than Baltic	Referring to amber	
V-476 AD			
IV-	150 amber beads from Lille Bjerggard, Denmark.	Rome fell to Goths AD 410	
III-			Tacitus (AD 55-120) *Germania* ca AD 98
II-			Pliny the Elder (AD 23-79)
I-		AD 56 Nero's Knight's expedition for amber sent from Rome.	*Natural History* ca AD 60 Mela the Greek ca AD 43-44 *De Situ Orbis*
0-			
I-BC		AD 67 Pompey's War with sea pirates where he received amber.	Virgil (42-31 BC) *Eclogue* ca 31 BC
II-			
III-		320 BC Expedition of Pytheas from Massalia for tin & amber (now Marseilles, France)	Herodotus *History of Greco-Persian Wars* ca 484-425 BC
IV-			
V-			
VI-			
VII-		Amber Gods from Mesopotamia	
VIII-		Amber statue of King of Assyria ca 1600 BC	Homer *Odyssey* 1000 BC (900?- 800?)
IX-	Amber finds at Dieskau near Saale River		
X-1000 BC		Phoenicians trade in amber, tin, glass, etc.	
-1700 BC	Amber cup from Hove near Beighton, England		
-2000 BC	Tomb amber from Mycenaean Kings of Crete ca 1600 BC		Text of Unknown author on Assyrian obelisk from Ninevah. ca 2000 BC
-3000 BC	13,000 Beads from Saaling, Jutland (Denmark) Beads from Mollerup, Aarupgaard, Denmark Egyptian tomb amber		
-4000 BC			
-5000 BC	Amber tablets with sketching from West Zealand, Denmark		
-8000 BC			
-13000 BC	Rough amber from Gough's Cave in Britain & caves in Judenes, Austria		

Vertical label (left, spanning II-BC to IX): Small figures and beads from Etruria

Vertical label (spanning II to I-BC): Amber workshops from Aquileia

Culture, Facts & People Joined in History Showing Growth of Knowledge Revealed by Amber from Paleolithic to the Fall of Ceasar in Western Roman Territories.*
Table 2-1 (Part 2)

*Adapted from Kosmowska-Ceranowicz and Konart 1989.

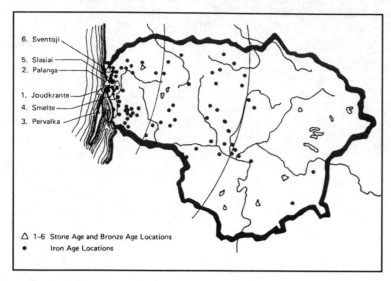

6. Sventoji
5. Slasiai
2. Palanga
1. Joudkrante
4. Smelte
3. Pervalka

△ 1–6 Stone Age and Bronze Age Locations
●   Iron Age Locations

Figure 2.18. Amber artifact finds from Lithuania. As in Gudynas and Pinkus 1967.

describe these "white nights." In another description, he probably refers to what is now familiar as the aurora borealis, which can often be seen from the Sambian coast: "Imagination adds that the gods and the flaming coronets on their heads become visible" (Ley 1951, 12).

Tacitus believed there were countries and islands in the northern sea where dense forests grew and, stimulated by the sun, produced amber resin in profusion, and amber fell into the sea and was washed upon the German shore. This idea is amazingly close to the truth regarding amber's origin, except for Tacitus's belief that these trees were growing during his own time, when in reality they had long ago been swept into oblivion.

At its peak, the Roman Empire expanded as far as England, Albania, Croatia, and over Eastern Europe. Amber artifacts from the first millennium after Christ mark this influence of Roman soldiers traveling as far away as the British Isles. For example, in a Dorchester cemetery, an amber bottle-shaped pendant was found along with a coin of Constantine dated A.D. 300. The late Roman graves in England frequently produced amber beads in pairs, a phenomenon based on amuletic properties amber was believed to possess. During the fourth century Rome fell to the Goths (see Table 2.1).

AMBERLAND OF THE EARLY BALTS AND ITS ARCHAEOLOGY

While the story of amber has been preserved in literature of the Mediterranean cultures, virtually nothing except the evidence of exhumed artifacts tells us about the inhabitants of the northern amber-producing region. What were the early Balts like? How did they regard and use amber?

The answers to these questions are gradually being found by Polish, Lithuanian, and Latvian archaeologists as they search grave sites dating from A.D. 100 to 1300. A Great Migration period from A.D. 400 to 700 took place in the Baltic region after the fall of the western half of the Roman empire. This brought about a collapse of earlier established trade routes with European civilizations around the Mediterranean. Archaeological finds dating from this period prove that the amber trade route then moved farther to the east. Grabowska

(1983, 12) reports a cache of amber found at Basonia in the Lublin region that dates from the beginning of the fifth century and contained 300 kilograms of raw amber and 39 kilograms of amber beads. This site indicates a route that led from Gdańsk area in Pomerania upstream along the Vistula and along the San and Dnestr rivers through Poland to the Ukraine on to the Black Sea. The cache included amber disks with a banded edge; similar shaped amber disks have been found in sites in Pomerania, Warmia and Mazuria. They appear to have been produced on a mass scale because similar ones have been found in Scandinavia, indicating exportation to that area. During the Great Migration, there was great upheaval in the culture with a mass exodus of people leaving behind settlement voids or "ghost towns" in the region of Poland. This led to a lack of interest in amber as a trade commodity until the early Middle Ages when the Slavs adopted it as their favorite raw material for use in the production of jewelry and ornaments.

Study of the mounds suggests that ancient Lithuanians progressed through three distinct stages of cultural development during this period. A decline of family-oriented systems took place between A.D. 100 and 400, followed by the evolution of Baltic "class-oriented" communities from A.D. 500 to 800 and, finally, by the development of an early feudal period between A.D. 800 and 1200 (Gudynas and Pinkus 1967, 44).

Numerous amber artifacts found in graves and homesteads along the Baltic seashore in present-day Lithuania date to the Iron Age; their sources are shown on the map of Lithuanian archaeological sites (see Fig. 2.18). Lithuanian archaeologists Gudynas and Pinkus located 104 graves or homesteads, 61 of which were located in the seashore zone within 100 kilometers of the sea. A second zone, 100 to 200 kilometers from the sea, contains only 33 sites. The distant zone, over 200 kilometers from the sea, has only 10 sites (Raimundas 1978, 1).

The pattern of these excavations attests to the important role of amber in the lives of Baltic inhabitants. The number of Roman coins and vessels found in excavations suggests that trade with the Romans was most brisk along the shores, where the largest quantities of amber were recovered. It appears that while coastal inhabitants enjoyed an abundance of amber along with the luxuries produced by its trade, amber was scarce even in Amberland for peasants living farther inland. Yet, it is well attested to by the shape and quantity of amber artifacts that amber was revered by Balts not only for its beauty, but also for its magical properties. The Eastern Balt culture, between A.D. 100 and 400, was rich in superstition and beliefs in

Figure 2.19. Bronze and Iron Age amber beads and spindle whorls, glass beads, bronze spirals and buckles from first millennium of our era. Courtesy Palanga Museum, Lithuania.

powers inherent in natural objects. For example, spindle whorls, generally found in "flat-mound" graves of females, were made of amber and, thus, empowered to prevent evil spirits from "hexing," or tangling, yarn as it was being spun.

Burial mounds of this period in the Shveksne region contained necklaces of amber beads alternated with small blue glass and enamel beads and separated by bronze spirals. Similar amber- and glass-beaded necklaces dating from this era were excavated in mounds near Palanga (see Fig. 2.19). The glass beads appear to be Roman imports, since the largest quantities are found in sites where many Roman coins were found (Gudynas and Pinkus 1957, 44).

A remarkable concentration of blue and green glass beads was found in Massuria and the Lower Nemunas basin. Presumably, the demand for bright-colored beads increased as the region developed into a center for fabricating primitive jewelry. However, about the third century A.D., a glass factory began producing not only blue hemispherical beads, but also bronze spirals so commonly found in early Baltic jewelry. Some authorities state that amber was so common among amber-gatherers that they desired something more exotic, such as these glass and bronze ornaments. Samland homestead sites, yielding finished amber jewelry along with pieces still in rough or half-finished condition, prove that amber beads continued to be worked in Amberland, but, as Gimbutas (1967, 129) points out, there was a scarcity of finished amber in relation to the amount of raw material available.

Despite a changing culture from A.D. 500 to 800, when class systems developed in the Baltic society, the Balts continued to wear amber jewelry. Based on the belief in amber's efficacy as a personal protective material or amulet, amber was given the name *gintaras*, meaning *protector*. This protective power kept amber from being usurped by other gemstones when wealthier persons developed a taste for more elaborate forms of jewelry. Graves from the well-to-do sections of the community contain much jewelry, including silver fibulae, gold plaques and bronze ornaments. A common adornment for women of this period was a collar made from bronze or silver. The construction of these collars portrays the advancement in workmanship and technology of Baltic craftsmen of the era.

New styling is also found in amber jewelry from this period. Biconical amber beads were now combined with characteristic bronze spirals and glass beads. A grave site in Rùdaičiai (ca. A.D. 500–800) produced a necklace containing 27 small amber beads cut as truncated bicones. A grave of the same period in Lazdininkai contained a necklace made with 13 biconical amber beads, 12 small glass beads, and 2 bronze spirals. Flat amber beads were also commonly used.

The style of combining amber beads with small blue glass beads and bronze spirals was still prevalent during the late Iron Age or early feudal system period of the Balts between A.D. 900 and 1200. Although new jewelry materials were coming into vogue, amber did not fall into disuse. Strings of small amber beads were widely used because of the persistent belief in the magical powers of the gemstone. Also, in the region that now comprises Poland a new form of amber artifact in the shape of a small cross begins to appear in large quantities. Historically, Poland officially converted to Christianity in A.D 966. Amber goods dating from this period are found throughout Poland. Amber workshop sites have been discovered at the coastal towns of Wolin, Kołobrzeg, and Gdańsk (Wapińska 1967, 17; see Fig. 2.20).

Figure 2.20 Amber artifacts in shape of cross, Malbork Amber Museum. Size about 2.5 cm x 4 cm. From Gdańsk and Warsaw Museums, Photograph by P. Rice.

By A.D. 1000 to 1200, graduated amber beads came into use, with the largest bead in the center and the others diminishing in size toward the ends. Examples of this style were found in graves in Nikėlai, Paulaičiai, and Švekšna in Lithuania. Bronze collars, first used in an earlier period, were still very much the fashion for neck adornment in the late Iron Age. Some collars were made of a bronze chain with a single small amber bead in the center.

Surprisingly, graves in eastern regions of Lithuania rarely contain amber articles. Puzzled archaeologists offered several explanations, the most convincing of which related to burial practices. From about A.D. 600 to 1300, inhabitants living east of the Sventoji River began cremating their deceased. Therefore, any amber placed with the dead was completely destroyed. It is assumed that Eastern Balts used amber to as great an extent as did their amber-gathering neighbors to the west, since, prior to the introduction of cremation, Eastern Balts buried their dead in flat mound graves that contained amber articles. Furthermore, raw amber found in alluvial deposits in moraines in eastern Lithuania, along the banks of the Nemunas River and in numerous other localities, provided readily available raw mate-

Figure 2.21. a, Baltic sun god pin and pendent with hanging chains, L. 10 cm. Patty Rice Collection.

rial. This leaves little doubt that eastern Lithuanians used amber widely. Some archaeologists believe amber was prized more highly in eastern Lithuania than elsewhere and was saved for exchange for imported bronze and iron tools, rather than manufactured locally into articles of ornament.

Further insight into the culture of this later Balt period, from A.D. 900 to 1200, is provided by grave sites in the Klaipeda and Palanga regions, where more than 300 amber beads were found. Apparently, both women and men adorned themselves with amber, the latter wearing comb-shaped amber pendants on their belts, while spears of warriors and bridles for horses were decorated with unpolished lumps of amber. The remarkable custom of placing amber in graves of horses might have been based on a belief in the magical powers of amber. A related custom of braiding one or two amber beads into a horse's mane was probably designed to confer the same protection as the wearing of amber on the person of his owner.

When a woman was buried, round polished amber disks were wrapped in her headdress and placed under her head. Women's graves also contain polished spindles of amber and, occasionally, a small pot containing a natural piece of amber. The significance of these finds, probably owing their origin to the supposedly protective properties of amber, remains unexplained.

Turning to another area of Europe, it is know that from A.D. 800 to 1000, sea trade was dominated by Vikings sailing unchallenged on the Baltic. A sea-borne trade to Swedish and Danish colonies thrived, as is shown by amber ornaments found in burial mounds on the Swedish islands of Gotland and Örland. One of the Baltic tribes, the Curonians, inhabited the region near the Lithuanian and Latvian borders. They grew powerful and began competing with the Vikings for sea trade. The Curonians were a fierce tribe that not only traded amber, but led a war of piracy against Scandinavian countries, as described by Gimbutas (1963, 156):

Trade and wars of piracy between the Baltic and Scandinavian Vikings continued intermittently throughout the tenth and eleventh centuries. Rich and well-settled Curonia attracted the rapacious Vikings from Sweden, Denmark, and even from Iceland, but they, in turn, were decoyed by the Curonians who plundered their coasts. Thus, the powers were balanced by piratical raids on both sides.

Figure 2.21.b, Baltic necklaces with hanging chains and "popcorn" style made by Lithuanian artists; Patty Rice Collection.

The Curonians became the Baltic Vikings, and they were the most restless and richest of all the Balts during this period. Curonian graves are rich in bronze and amber ornaments. Their jewelry forms followed geometric patterns with hanging attachments. Several chains were secured to large brooches. Today, this style of jewelry is often copied by Baltic artists (see Fig. 2.21). Curonian ornaments are found all across the Baltic region, attesting to the important role amber played in the life of the forefathers of today's Lithuanians during this little-known era.

Old woodcut depicting "The Tree that Exudes Amber."

BALTIC AMBER GATHERERS-
GATHERING/MINING/PRODUCTION

Amber which is thrown up by the Baltic breakers is collected on the sea coast, and they are so greedy for it that they search for it even in the water and in the sand on the sea bed. . . . As piety has declined among the people, so has the price of amber fallen, for previously people were wont to pay a great deal for holy figures or beads made of amber, and women often wore their rosaries all the time. But today they make little boxes, little spoons, little vases from amber, even bird cages—and all this because of its fragility, rather for the eye than for use.
—Giovanni Francesco Commendone,
Papal Legate at time of Sigismund Augustus (1548–1572)

AMBER MONOPOLY AND GUILDS
OF THE MIDDLE AGES

Following the cessation of trade with the Mediterranean because of the fall of the Roman Empire, little was recorded as to the status of Baltic amber and the cultures living along the shores of the Baltic Sea. Polish archaeologists make reference to amber being mentioned in Diocletian's edict of A.D. 301. As emperor of Rome, Diocletian's rigid and oppressive regulations fixed maximum prices of commodities throughout the empire. Amber was included in the regulated items. During the sixth century, the Asti, or ancient inhabitants on the Sambian peninsula, sent a quantity of amber as a present to Theodoric the Great, the Ostrogoth king and founder of the Ostrogothic Kingdom in Italy. Theodoric not only acknowledged the gift, but developed a special interest in amber. He sent a mission to distant Baltic shores in quest of the "Gold of the North." Surprisingly, Theodoric was fortunate enough to secure an amber lump weighing from 7 to 8 pounds (3.18–3.7 kg)! (Kolendo 1990, 17).

During the tenth century, Boleslav, Duke of Poland, succeeded in making Christians out of the pagan Old Balts at the point of a sword. His father, Mieszko, actually led the Poles into Christianity and Poland was officially converted in A.D. 966. This was most likely an attempt to appease the crusading and marauding Saxons. During his son Boleslav's reign (A.D. 992–1025), the Christian church was firmly established. Boleslav III (who reigned from 1102 to 1138), conquered Pomerania, defeated the pagan Prussians, and defended Silesia against the Holy Roman Emperor. In A.D. 1161, the Western Balts, or the *Borussians* as they were called, succeeded in casting off Christianity and Polish rule (*Infopedia* 1995).

In the next century, inroads made by these pagans upon lands of neighboring Christians and their advance into Pomerania in A.D. 1229 induced Konrad, Duke of Masovia, to call upon the Knights of the Teutonic Order to conquer these unruly barbarians. Having originated during the Crusades in 1198, the knights were proficient warriors and entered the region of East Prussia, where they fought for 50 years, virtually exterminating the Baltic Old Prussians and restoring Christianity to the region. During this period German colonists began to settle in the region.

In earlier times, amber was the absolute property of the finder, but by the time the knights were called upon to conquer the Balts, the dukes of Pomerania had claimed for themselves all amber found as far as the coast of Danzig (now Gdańsk). The control of amber was the reward given the Teutonic Order for their victory over the conquered Prussians. The amber monopoly, taken over by the order, was extended along the coast of both West and East Prussia or the region that now comprises Poland and the Sambian Peninsula of Russia. By 1283 the Teutonic knights were absolute rulers of Prussia. Baltic amber was a lucrative trade material desired by neighboring lands and by nobles whom the knights needed to keep as allies during this "illegal tyrannical control" over their newly claimed territory (Kwiatkowska 1996, 17). Many amber gifts were given to nobility in nearby territories. The knights claimed every piece of amber found and finders were obligated to give up their amber, but received very little payment in return. An edict was issued forbidding amber working in the knights' territories. This monopolistic law had far-reaching effects, influencing all changes in the production and sale of amber during the last 770 years! Today, newly uncovered raw amber is still regarded as the property of the state in Poland, Lithuania, and Russia (see Fig. 3.1).

In A.D. 1274, an early Latin document from the German Grand Marshal of the Knights of the Teutonic Order granted the right of gathering amber to the bishop of Sambia. The bishop was allowed to fish for amber in a certain locality near Lockstadt. Later, in 1312, the same right was granted to Gdańsk fishermen off the Bay of Danzig near Gdańsk. Between 1340 and 1342, Grand Master Dietrich von Altenburg made an agreement giving rights to fish for amber to the Monastery of Oliva (Grabowska 1983, 15). The only way to receive such rights was through the Grand Marshal of the Teutonic Order. As amber increased in value, being the choice material esteemed for rosary beads, the order rescinded these treaties. By the middle of the fifteenth century, the order had again gained complete control, earning for themselves the title of *Amber Lords.*

In an attempt to organize the amber rosary trade to yield a greater profit, the grand marshal issued an order in 1394 forbidding anyone to possess raw amber. All collecting of amber was henceforth to be done under the supervision of a *beach master*, and all amber was to be delivered to the amber lords. This order was brutally enforced. Any unauthorized person observed picking up amber was speedily hanged from the nearest tree. To intimidate local residents, gibbets were erected along the Sambian coast. Numerous local legends tell of cruel beach masters relentlessly carrying out these orders. The tales describe how ghosts of the beach masters were forced in the afterlife to haunt scenes of their multiple crimes. One such legend tells of ghost riders roaming the shores, inciting fishermen to revolt against the oppressive amber laws. Another story portrays the fate of one particularly heartless judge, Anselmas of Lozenstein, a man who settled every incident of amber pilferage, no matter how small, with instant hanging. In the afterworld, as punishment, legend has it the cruel judge was condemned to eternally wander. His weary cry could be heard across the Baltic coast on stormy nights calling out, "Oh, my god, free amber! Free amber!" (Spekke 1957, 10: see Fig. 3.2).

Making sure every piece of amber was reserved for the order was only the beginning of organizing the

Figure 3.1. Amber figure of Teutonic knight in full armor. This finely detailed sculpture is constructed from two natural lumps of amber. The frontal carving is from "goose fat" butterscotch-hued amber with mottled translucent orange areas. The back of the knight has natural markings. The base is opaque amber with natural flaky dark reddish crust similar to pieces found in sand dune areas of northeast Poland. Size: Total 569 g, overall 20 cm x 15 cm x 11.1 cm. Purchased in Poland ca. 1965. Courtesy Burgess-Jastak Collection.

Figure 3.2. Engraving of amber fishermen and gibbets on the amber coast. Reprinted from Wagner 1774.

amber rosary bead trade. After raw amber was collected, it was stored in warehouses in Bruges, Lubeck, Augburg, and Venice to be sorted and worked. The order prevented establishment of any independent amber works in its own country near the amber source, believing there would be no way to prevent workmen from obtaining amber illegally (Williamson 1932, 9).

In 1302 amber turners in Bruges were organized into the first of the amber guilds. This was soon followed by formation of a guild in Lubeck (perhaps as early as 1310), first mentioned in the Burgher's Register in 1317, with the oldest scroll from the guild dated 1360. Since the knights shipped amber to guild members to be made into rosaries, the workers called themselves *Paternostermachers* or "makers of Lord's prayer beads" (rosaries). The guilds were assigned their own church for holding divine services and had their own patron saint, *St. Adalbert*, the first missionary in the amber region. St. Adalbert was martyred in A.D. 997 by a non-Christian priest and is now the patron saint of Poland, with his own feast day on April 23. During the early 1300s, the Bruges and Lubeck guilds produced rosaries for the entire Christian church (see Fig. 3.3).

Records of the order show that, in 1399, the grand marshal kept his own amber-worker at his castle at Marienburg on the Vistula River in the heart of amber country not far from Gdańsk. This castle, now called Malbork Castle, has been converted to a museum, and it contains a room exclusively dedicated to an exhibition of medieval amber treasures collected from all over Poland. Master Johan, the grand marshal's cutter, made objects such as rosaries, artistic reliefs for altars, and mosaics of amber, all of which were entirely for the order's own needs and not for sale. According to a record of expenditures from 1399 to 1400 in the accounts of the treasurer of the order, "Jan," the carver, received 472 marks for six months' work on amber objects, and in the next year payment for a rosary and two "pictures" in amber (Grabowska 1983, 15).

The influence of the Teutonic knights weakened after the battle at Tannenberg in 1410 when they were defeated by the Poles and the Lithuanians. The Prussian cities gradually began allying with

Poland and the king of Poland gained greater power.

In 1418 Grand Marshal Michael Kumeister von Steinberg wrote a letter to the queen of Hungary, the wife of Sigismund of Luxembourg, mentioning that he intended to send her a rosary made of white amber. According to Polish historians (Grabowska 1983, 15):

> The political situation in the Teutonic Knights' state—formed on foreign land and breaking all laws—required far-reaching connections with the rulers of contemporary Europe. And so to these they gave gifts, in keeping with the current custom, and among these gifts were amber objects. . . . In an inventory of the personal property of Charles V, King of France, there is a statuette of John the Baptist; a register of the treasures of the Louvre in 1418 specifies an amber crucifix; Charlotte of Savoy boasted that she owned a bas-relief in a gilded frame showing St. Veronica. These references may be to works of art made in the territories of the Teutonic Knights.

Figure 3.3. 1698 Engraving of "Paternostermachers" workshop. Reprinted from Weigel 1698.

Flourishing and growing in power, by 1420 the Bruges Guild was composed of 70 masters and over 300 helpers and apprentices. Each master was allowed two apprentices and three journeymen for training. There was turmoil in the guilds right from the beginning. Fifteenth century records of the Bruges Guild include correspondence with the Teutonic knights regarding disputes over raw materials. Appeals were made to the duke of Burgundy and to the Hanseatic League. Despite disagreements, by the late 1400s, the Pomeranian towns of Kolberg (now Kołobrzeg, Poland), Köslin (Koszalin) and Stolp (Słupsk) also had established amber guilds. Records indicate that these guilds were in existence in 1480 and had their headquarters at Stolp (Słupsk, Poland).

In 1466 the City of Danzig (Gdańsk) was taken from the Teutonic knights and placed under the sovereignty of the king of Poland under terms of the Second Peace of Thorn. The knights were left in possession of the eastern part of Prussia. Gdańsk grew and industry thrived. Dutch merchants, encouraged earlier by the Teutonic knights, had settled at this prosperous seaport. The knights tried in vain to keep amber workers from settling in Gdańsk. By 1477, despite the efforts of the order, there were enough craftsmen in the city to form a guild. The city council confirmed the statute of the *Brotherhood of Amber Craftsmen* on 31 March 1477, eleven years after the Peace of Thorn.

King Casimir, the Jagiellonian, again made amber free, granting to Gdańsk the right to control the seashores in Gdańsk and Pomerania. Along with many other privileges, Casimir increased the number of localities subject to the city by adding amber-yielding Hel peninsula, on the western end of the lagoon to their number. From then on the Gdańsk Guild made efforts to increase amber supplies and condemned "botchers." Thus, traditions of amber craftsmanship became strongly developed as the Gdańsk Guild grew. It was much later before guilds were formed in other cities such as Elbląg in 1529 and Königsberg in 1641 (Grabowska 1983, 15).

Although the order had previously objected to the king's actions, in 1483 the

Figure 3.4. Rosaries made of amber. Length 20 to 26 inches or 50.8 cm to 66 cm. Patty Rice Collection.

knights agreed to include Gdańsk in their trading organization and to supply the Gdańsk Guild with raw amber. The order, however, provided competition to Gdańsk by establishing an amber turner named Hilger in Königsberg (Kaliningrad). Hilger produced amber products for the order from 1499 to 1510 and appears to have worked exclusively for the order. Although the knights supplied much amber to Hilger, most of it was sent to guilds in Pomerania and Gdańsk (Ley 1951, 15–6). Trade from the guilds sent rosaries to all Christian regions. Each religion, Catholic, Islamic, and Buddhist, had its own style of prayer bead. The rosaries for the Catholic religion were made of 55 beads with larger beads spaced in between a series of smaller beads. This was to assist in counting the number of *Ave Maria*'s said for each small Maria bead, and the Lord's Prayer said for each of the larger Paternoster beads. The smaller rosaries generally consisted of 33 beads and larger rosaries consisted of 165 beads (see Fig. 3.4).

During the early part of the sixteenth century, the Teutonic

Order, now under the feudal rule of the Polish monarchy, regarded Poland as its enemy. In 1511 the order, in an attempt to improve its position, appointed Margrave Albert of Ansbach and Bayreuth, a Hohenzollern and relative of the king of Poland, as grand marshal. In 1525 Grand Marshal Albert became a Lutheran and proclaimed himself Duke Albert of Prussia, possessing all the rights of the old Teutonic Order, including the amber monopoly. Continual friction existed between the guilds and the dukes of Prussia, primarily because of the conversion of German principalities to Lutheranism after the Reformation. Thus, leaving the Catholic church resulted in a lessened demand for rosaries from populations that were once Catholic. With their traditional customers gone, the rosary-producing guilds found it difficult to absorb the raw material that they were obligated to buy.

The Gdańsk amber-turners and several other guilds complained to the German emperor, requesting mediation between producers and buyers. In 1526 the Gdańsk Guild had 46 members and was constantly growing in number. This disturbed the city council, thus a decree of March 2, 1549 limited the number of amber workshops to 40.

In Duke Albert's attempts to again make his amber monopoly profitable, he requested Aurifaber and Gobel, ducal court physicians, to investigate the medicinal uses of amber. Their investigations produced the earliest scientific tracts designed to promote amber as a remedy for sundry illnesses (see Chapter 4 for medicinal uses of amber).

Production records for this period are meager. In the early 1500s in a volume called the *Succini Historia*, by Andreas Aurifaber, court physician, it was reported that the average yield of amber from the Gdańsk area was about 110 kegs with "one year bringing more than another." An old map, dated 1539, portrayed the amber seacoast of Vistula Split with amber packed in kegs

Figure 3.5. Map of Vistula Split called amber coast, dating from 1539, showing amber being collected and stored in barrels for shipping.

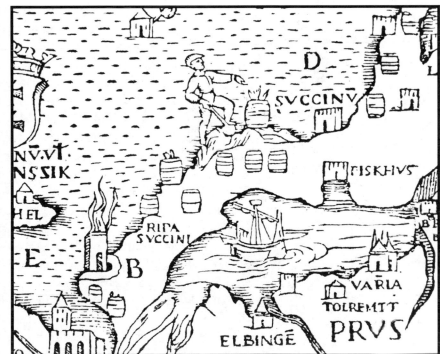

Figure 3.6. Moslem prayer beads. Size: 10 inches or 25 cm. Yellow opaque amber.

or barrels ready for shipping (see Fig. 3.5).

According to Grabowska (1983, 16), Aurifaber, recorded in his *Succini Historia*, stated:

> I saw lumps of amber the size of a man's head, from which His Highness the Prince ordered that bowls and goblets should be made. . . . Amber is sent to Gdańsk where Mr. Pawel Jeske [Paul Jaski] and his family, amber merchants, take it. They put out the amber for lathing and initial working, and sort it according to color and variety. The amber is then sent by ship to Antwerp, and thence, not without great profits, to Italy, France, Spain, Turkey, and the pagan lands where it is used, so they tell me, for ornaments to show magnificence and in burials. . . . Mr. Jeske recognizes where a piece of amber comes from in the same way that we recognize Hungarians, Italians or Scots.

To offset the loss of the rosary business, Duke Albert encouraged the use of amber in new artistic works. He commissioned many large pieces to be carved into beautiful articles. Amber was often used in combination with other gemstones or combined with ivory and tortoiseshell in making chests, altar pieces, mirror frames, and even tables. Many elaborate carvings of religious figures were made.

Finding sales of raw amber too great a burden, the duke decided to put the entire trade into the hands of an agent. In 1533 the head of a prominent merchant family in Gdańsk (Danzig), Paul Koehn von Jaski, signed an agreement transferring the amber monopoly to his firm, but Duke Albert reserved the right to maintain his own amber-turner to produce amber products for his own needs. White amber was also exempted from this treaty because of the outstanding medicinal properties it was believed to possess.

At this time, Duke Albert was annually receiving the equivalent of 15,000 gold dollars for amber, mostly from the Near East trade, since trade with France, Italy, and Spain had collapsed. As trade in rosaries for Christian regions declined, Paternostermachers made prayer beads for Buddhist countries. The Buddhist prayer strand, and Hindu rosary from which it is derived, are usually round similar-sized beads that have 108 beads, corresponding to the number of Brahmin present at the Buddha's birth. The new monopoly exploited the Near East markets even further by focusing on the production of Muslim prayer beads, an Islamic counterpart of the rosary, creating an increased demand for amber. The Islamic prayer strands have 99 beads with the hundredth bead believed to be in paradise. art-region traders came to the amber country to trade silk carpets ornamented with gold thread in exchange for amber goods. Elaborate amber stat-

ues, carved by the Gdańsk Guild for temples of the Middle East, brought new wealth to the guilds (see Fig. 3.6).

Despite new managers for the raw amber industry, fighting continued between producers and craftsmen. One well-documented dispute that lasted over a period of years was with the Gdańsk Guild. Finding it difficult to consume all the raw material they were required to buy because of declining trade in nearby regions no longer requiring rosaries, a few master craftsmen complained to the Council of Gdańsk. On hearing of this, Paul Jaski, though he had recently leased the amber monopoly, refused to supply any amber to the Gdańsk Guild. At the thought of losing their entire source of livelihood, the guild came back begging for the old terms, with a promise to discipline the masters who had complained.

Discovering that the guild was at his mercy, Paul Jaski drew up the petition of 1538, in which he agreed to supply the guild with amber as long as he was allowed to sell finished pieces produced by the guild. He then proceeded to give the guild only inferior quality amber. After continued complaints from the guild and pressure from the council, a new agreement was made with Jaski around 1546 or 1548. Once again, he promised to sell finished amber, but this time he demanded that the guild's finished goods be sold to him exclusively. Without much foresight, the guild masters promised to "police their own ranks" to guard against craftsmen selling their finished work on their own. The major problem overlooked by the guild was that Jaski had not promised to buy a fixed quantity of goods. Again the guild found itself with more raw material than it could consume (Ley 1957, 15–6).

Complaining to the council, the guild stated that they were "as severely oppressed by Paul Jaski as the children of Israel had been by King Pharaoh" (Ley 1957, 15–6). On behalf of the Gdańsk Guild, the council requested Duke Albert to sell one-third of his amber directly to the Gdańsk Guild, bypassing his amber agent. However, placing the troublesome amber monopoly in the hands of his lessee had relieved Duke Albert of many problems. Therefore, wishing to keep peace with Jaski, the duke ruled in his favor.

The dispute with Gdańsk craftsmen continued. In both 1552 and 1555, the people of Gdańsk appealed to Duke Albert. Finally, Jaski lost his contract with the amber-turners, but Duke Albert did not sell to them either. This caused problems for the craftsmen, but records do not indicate where or how they acquired their raw materials thereafter. Other sources must have been found, as evidenced by artistic work produced by the guild. Although the House of Jaski con-

Figure 3.7. Figure of Ares displayed with cylindrical beads—decoration for men's clothing, Gdańsk, ca. 1600. Figure is 13 cm. Necklace has 49 cylindrical beads each, about 1.5 cm to 2.2 cm diameter by 3 cm to 4.4 cm long. Total length: 82 cm, weight: 380 g. Photograph taken in Malbork Castle Museum by author.

trolled the Pomeranian inland amber monopoly, it was limited to East Prussia and the Pomeranian diocese. Other inland areas did not have such rigid restrictions on amber as did Prussian regions.

Under Polish law, for example, inhabitants of amber-bearing regions could obtain a license to extract amber from their own land, whereas along the shore from near Gdańsk to Polsk, amber was the property of the city of Gdańsk. Therefore, the Gdańsk Guild was able to independently obtain a supply of raw amber. But the principal supply remained in the hands of the amber monopoly controlled by the House of Jaski. Later in 1636, King Wladislaus IV of Poland finally commanded Jaski to again sell amber to the Gdańsk Guild at the same price obtained elsewhere.

As a sign of the increasing prestige and power of the amber guilds, the Stolp Amber Guild became sufficiently affluent in 1534 to ask for and receive permission from Duke Barnim the Elder to maintain its own brewery. All guild masters were held in high regard, and in 1555 the princes of Pomerania referred to the "united, dutiful, time-honored Guild Masters and mutual brothers of the amber-working district of Stolp, Kolbrzeg and Köslin" (Williamson 1932, 104). By 1575 further recognition was given to amber-workers by raising them to the level of merchants. Duke Johannes Friedrich confirmed their rights as trading merchants by edicts dated May 20, 1574 and March 24, 1575.

Guild craftsmen gained rights to socially interact and trade with other merchants, receiving the title *Merchants and Amber Trading Corporation of Stolp*. In 1583 the societies of Kolbrzeg, Stolp (Słupsk), Köslin, Elbling (Elbląg), and Gdańsk united, with Gdańsk becoming the headquarters of the guild union.

In Königsberg (now Kaliningrad,) fine quality amber sculptures were produced by individual artists for European courts before the formation of the Königsberg amber worker's guild. It was not until 1641 that the Königsberg Guild was officially formed. As evidence of the skill of its members, it is recorded that a master craftsman of Königsberg, Christian Porschin, invented an amber lens to use as a "burning-glass" to start fires. It was said to be "much quicker in action than were the glass lenses." Having developed a special method to produce transparent amber, Porschin also attempted to make amber "eyeglasses for spectacles" (Williamson 1932, 104).

Similar ingenuity was shown by workers in the Gdańsk Guild, who, upon request, produced a crown carved from a single piece of amber. The crown was presented to King John Sobieski of Poland by the citizens of the city. Unfortunately, this crown no longer exists. However, beautifully preserved pieces from the 1600s in the Malbork Castle Museum, Poland, are a figure of Ares and amber beads. Ares is a fully three-dimensional sculpture and is exhibited along with a large strand of cylindrical beads with relief carved floral designs over each bead (see Fig. 3.7). The beads were decoration for men's clothing in Gdańsk.

Figure 3.8. Solitaire game casket, Poland, first half of eighteenth century. Size: 24 cm x 24 cm, H. 6 cm. Photograph taken in Malbork Castle Museum by author.

In 1611 David Konarski, Abbot of Oliva, a Cistercian monastery, offered an amber statuette of the Madonna and Child to the Jasna Góra Monastery in Czestochowa. This treasure still survives today, though many of Poland's medieval amber pieces were destroyed. The amber is worked in the Gothic style rather than the style of Madonna carvings from the seventeenth century. Therefore, Polish art historians believe this statuette was created by an earlier master craftsman in the monastery's amber workshop during the mid-sixteenth century. This is entirely possible because Oliva Monastery was one of the amber workshops that began as early as 1340. Another amber piece from Oliva monastery, remaining today, is a medallion found in the sarcophagus of Anna the Jagiellonian, in the Wawel Cathedral. This is a royal medallion with a portrait of King Stephen Bathory. Medallions were a new trend in amber art during this period and were modeled after coin designs. Pieces of this type were often treated as amulets or talismans (Grabowska 1983, 16).

Figure 3.9. Elaborate altar constructed with amber, ivory, and horn, with medallions of transparent amber etched on reverse side in the "verre églomisé" style. ca. seventeenth century Size: H. 85.5 cm, W. 36 cm. Photograph taken in Malbork Castle Museum by author.

One amber-turner of Gdańsk, described by Otto Pelka (1920) in his book on the artistic history of amber, was Michael Redlin. Pelka included a drawing made by Redlin for an elaborate two-tiered jewelry cabinet inlaid with amber and ivory. This drawing is included in many current books on amber. The amber cabinet was carved and engraved with landscapes and historical scenes.

Gdańsk workers used the natural colors of amber, ranging from bone to shades of yellow and brown, in mosaic fashion to produce multicolored works such as game boards, chessboards, and decorative jewelry caskets (see Fig. 3.8). Redlin also made a chessboard of multicolored amber, complete with matching chesspieces. Another masterpiece by this artist was a magnificent twelve-branched chandelier that took two years to finish and that combined amber, ivory, and inlaid portraits of Roman and German emperors under clear amber overlays. All three objects were given to the tsars, Ivan and Peter, by the Elector of Prussia, Frederick III, but were among the objects lost in Moscow in the destruction of wars. In 1681 a chair of amber was presented to the tsar of Russia. This piece was supposed to be one of the most unusual objects ever made of amber!

Clear amber was often engraved or incised with decorations on its reverse side, a technique adapted from glass-etching technology. In producing coat-of-arms designs, artisans painted parts of the miniature engraving in color, using an art technique similar to that known as *verre églomisé*. This specialized work was often used on amber in constructing the elaborate pieces of the seventeenth century (see Fig. 3.9).

The etching technique was often done in the following manner. The artist placed a sheet of gold foil on the bottom of a transparent piece of amber, then applied a layer of lacquer. The etching was done in reverse on this surface and was followed by a thick layer of lacquer, applied to cover the back of the amber piece and protect the design. The translucent glowing warmth of amber, enhanced by placing gold leaf under clear pieces, imparted a bright glow to amber altars designed for small chapels. This glowing quality lent itself well to amber crucifixes as well as other ritual pieces. Religious altar sets, including chalices, ewers, water sprinkles, and candlesticks, were sometimes fashioned of amber. Many of these works of art were destroyed

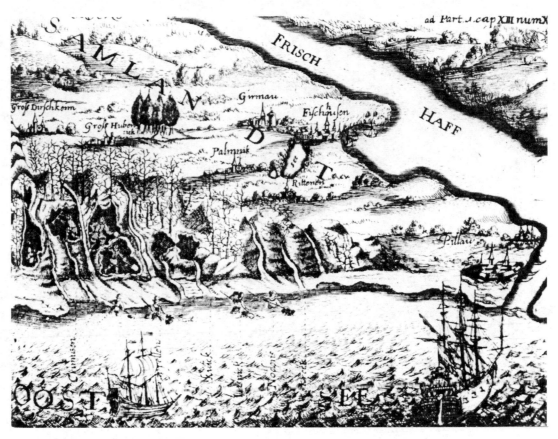

in European wars, but a few examples of these priceless works can still be found in museums in England, Copenhagen, Stockholm, Germany, and Italy. The largest collection today is housed in the Museum of Silver, in the Pitti Palace in Florence, Italy.

Königsberg and Gdańsk, were centers of amber craftsmanship during the fifteenth century, since both were located in amber-bearing regions. It appears that Königsberg treasures were sold to Western Europe, whereas Gdańsk products tended to remain in Poland, some in the royal treasury, others with ecclesiastic and nobility patrons. Thus many of these amber artistic works fell victim to Polish-Swedish wars. Grabowska, the leading authority of amber in art and historic amber-working techniques, reports that in the "Register of things remaining after the Swedes and after the flight of the Swedes, listed in 1661, on 1 December at Wiśnicz," the only amber objects mentioned were:

> "A carved amber casket with St. George on the lid; a pair of carved amber candlesticks; two amber statuettes, the Blessed Virgin and St. John; an amber beaker; a tiny amber box; two pieces of amber with embedded flies; a small amber box con-

Figure 3.10. Map of Samland from 1634 showing amber fishing and mining regions. Reprinted from Hartknoch 1674. Courtesy Malbork Castle Museum.

Figure 3.11. Lid to seventeenth-century casket made in Gdańsk as displayed with back lighting in Malbork Castle Museum. L. 24 cm W. 19 cm

Figure 3.12. Seventeenth-century casket with various shades of amber and ivory. Malbork Castle Museum Collection. L. 13.3 cm W. 10 cm H. 12.5 cm.

taining a relic of St. Joseph; four pieces of unpolished amber; amber dice." Two hundred years later a Swedish diplomat acknowledged "The Swedish Marshal Charles Gustav Wrangel, a great knight in the European field of battle, looted Poland mercilessly. He collected enormous treasures and built himself a castle near Malar to house his loot." He described amber pieces of furniture, especially Gdańsk cupboards of a color which is not found in Sweden (Grabowska 1983, 16–8).

The development of the skilled amber craft was encouraged by nobility and ruling classes. Not only did they commission amber artistic works, but they would employ talented artists in their courts. Their workshops had the finest equipment so their own amber craftsmen could work with the most modern equipment of the time.

King Sigismund III Vasa, of Poland (1587–1632) and of Sweden (1592–1599) was a great patron of the amber craftsmen of his day. He also worked amber, making many exquisite pieces, some of which remain today and are attributed to his craftsmanship. It is said that because of his political difficulties, being obsessed with the idea of winning the Swedish crown, he would escape from his problems by participating in the arts. This king painted, etched, carved silver and gold figurines and a church plate set with pearls, golden goblets, candlesticks and lamps for the shrines at Czestochowa and Loretto (Kulicka 1996a, 22).

One item produced by Sigismund still remains in the treasury of Wawel Cathedral. This is an amber goblet known as the *hobbler* because it has no stem to stand on. The cup is crafted from one lump of amber and at the bottom is an etching, in the *verre églomisé* style, of the likeness of Sigismund III. The king often used the etched portrait of himself as his signature for his own work (Grabowska 1983, 18).

The amber monopoly remained with the House of Koehn von Jaski until 1642. At this time, Frederick William, the Great Elector, succeeded in retrieving amber rights by paying the enormous sum of

400,000 thalers (approximately $28,000), thus, returning the possession of amber to the state (Ley 1957, 18).

New laws were enacted to protect the state's interest. Not only was possession of raw amber once again forbidden, but a special permit was needed merely to walk along the beach where amber might be found. However, the former severity of the amber laws was lessened by making punishment dependent on the quantity of amber stolen, but punishments were still very harsh. Possession of 2 pounds of amber merited death by hanging. If more than 2 pounds were unlawfully taken or found in possession, the miscreant was put to death by breaking on a wheel (Gudynas and Pinkus 1967, 94; (see Fig. 3.10).

Every third year, fishermen living along the shore had to swear the "Amber Oath," requiring them to promise to "denounce any smuggler, even if he should be the closest blood relative" (Williamson 1932, 108). The result was that the fishing community became riddled with distrust and suspicion, and families were ruined by denunciations, merited or otherwise. An "Amber Court," with a special amber oath for priests who administered the oath, dealt with smugglers to further insure that illegal gathering would be held to a minimum. In still another measure, "beach riders" were sent out after storms to prevent local inhabitants from being tempted to pick up amber that washed up on the beaches. Ironically, the local inhabitants were not only required to turn in all amber found accidentally, but were compelled to go out in search of it. As an incentive, they received salt for amber, weight for weight, as pay.

Figure 3.13. Amber and silver loving cup by Jerzy Skryba's workshop. Photograph taken in Malbork Castle Museum by author.

Figure 3.14. Baroque cupboard-style cabinet from King Stanislaus Augustus Poniatowski, ca. eighteenth century. Size: L. 26 cm W. 13 cm H. 35.5 cm. Made of amber, ivory, horn, metal, and mirrors. Photograph taken in Malbork Castle Museum by author.

Though little expense was involved in obtaining raw amber, the monopoly was again losing money. Trade diminished steadily. The controllers of the beaches, including beach riders, supervisors, paymasters, and amber judges, required salaries. The guilds were becoming impoverished. In 1690 approval for closing the Elbling (Elbląg) Guild was requested of the council. Yet guildsmen continued to make elaborate amber pieces for gentry and royal patrons (see Fig. 3.11).

Not much information is available for sixteenth, seventeenth, and eighteenth century guild activities, but artistic works, along with the production of rosaries, continued to provide outlets for the guilds' raw materials (see Fig. 3.12). Lubeck Guild records of 1692 describe two elegant crucifixes and a cabinet encrusted with amber made by Johann Segebad and Niklas Steding. The *incrustation* process was used to completely cover, or form a crust, over a base of wood with amber. This was actually a veneering process used extensively from the seventeenth to the nineteenth centuries. Elaborate chessboards were constructed by this means, with the natural colors and textures of exotic varieties of amber fitted together into mosaics. Chess sets were often made of amber in both Königsberg and in Gdańsk. One chessboard given to King Sigismund III had amber plates on the border surrounding the checkered squares, which was decorated with ornamentation attributed to master craftsman, George Schreiber, known as Jerzy Skryba in Poland, of Königsberg, who was active from 1610 to 1643. Other pieces by George Schreiber, such as the loving cup made with silver and amber plates decorated with a relief design, can be found in Malbork Castle Museum, Poland (see Fig. 3.13).

Additional artistic amber objects made by George Schreiber can be found today in the Darmstadt Museum, Germany, and other European collections. One such work, signed by Schreiber, is a lidded cup. The sides of the vessel are made from richly carved amber plates. The amber plates are bound together with engraved silver bands. According to Grabowska (1983, 18–9):

> Schreiber carved and engraved a grotesque-cum-floral ornament, copied from Netherlandish patterns, with technical facility onto the polished surfaces of the smooth plates. His decorative forms were linked with the shape of the object. . . . He never exceeded the boundaries imposed by the function of the walls of the vessel.

During the seventeenth century, craftsmen in Gdańsk tended to have one common characteristic that art historians use to identify their work. Grabowska (1983, 18–9) describes it as:

Amber . . . was treated here like any other raw material, for example, metal or bone. It was only the Gdańsk amber craftsmen who made full use of the diversity of the natural properties of this raw material—which occurs in a range of colors from white, through all shades of yellow to brown, red and also green and azure. It is in fact this which makes it possible to pick out the group of objects made in Gdańsk in the 17th century.

The Gdańsk Guild had difficulties with craftsmen who were not members of the guild, as well as with suppliers of raw amber. Such persons, working without authority, carved and sold amber goods, and the guild claimed great damage to its members thereby. This damage may have been real, since some independent carvers produced fine quality sculptures and were even commissioned by the Council of Gdańsk for artistic amber work. One such freelance sculptor was Christoph Maucher, who was regarded so highly by the guild in 1705 that they opened their ranks to aid him and gave him instructions for some of his artistic masterpieces (see Fig. 3.14).

In the early 1700s the Stolp Corporation became part of the Prussian state, at which time it attempted to gain cheaper rates for raw amber by joining with the Königsberg Corporation. On November 3, 1702, the two guilds amalgamated. But the union was not always calm, because on at least three occasions, the Königsberg Guild tried to obtain all raw amber from the Königsberg Chamber of Amber for themselves. Frederick I and Frederick William I, of Prussia, upheld the rights of Stolp to share in amber production.

The Amber Room—A Magnificent Work and Wonder

In 1713 Prussian King Frederick William I exhibited the most spectacular of all works of amber art—an entire room, including walls, ceiling, and doors, covered with mosaics of amber pieces of varying shades and hues. This magnificent room was furnished with amber vases, dishes, candlesticks, snuff bottles, powder boxes, and cutlery. It had taken 12 years and the work of several architects and craftsmen to complete. This was a room of glistening amber, a material inaccessible to any other sovereign, even though many nobles along the Baltic treasured amber to the point of employing their own amber turners.

The room was originally designed, in approximately 1701, by Andreas Schlüter, a noted architect, sculptor, and eminent representative of classical baroque art. Polish art historians point out that the role of the designer should not be exaggerated because it was the expert work of the master craftsmen who made the magnificent design glow. It was the amber that made the room famous!

Figure 3.15.a, Amber room: Top, Amber room as it appeared in 1930s-40s (source unknown) b, center, amber disc with letter R, representing the initial of Peter the Great reconstructed by Russian artists; c, left, sample panel as displayed in Kaliningrad Amber Museum, Russia.. Photo by author in 1995.

Gottfried Wolffram, court amber master for Danish King Frederick IV, was recommended by the monarch to Frederick I, king in Prussia, to carry out the work for making a complete room of amber. By 1707 only one wall was completed. After a dispute over payment for work completed, Wolffram was dismissed. King Frederick I then entrusted the task of continuing construction of the amber panels to two Gdańsk masters, Gottfried Turau and Ernest Schacht, for a fee less than Wolffram's. These two craftsmen had mastered all the secrets of the art of working this raw material that "so deceptively appeared so easy."

The room was built entirely from a mosaic of smoothly polished plates of amber in a multicolored arrangement. There were delicate engravings on the back of transparent disks of amber. Scenes engraved on the amber showed the life of fishermen. Ornaments carved in bas-relief presented fully flowing figure sculptures. In 1711 the rest of the room was finished. Thereupon, the workmen gave the king a bill that, because of the previous difficulties, compared their fee for two walls with the fee charged by Wolffram for the first wall completed, ensuring that the sovereign would see that their work was not as expensive as his former craftsman's and no dispute would develop.

The glittering amber panels were then installed on the walls of a corner room on the third floor in the Berlin castle in Prussia's capital. Eye witnesses reported that "when the sun shone through the windows, it was like standing in an open jewel box" (Williamson, 1932, 108). Upon seeing the room in 1913, Peter the Great, the Russian Tsar, extravagantly admired its beauty.

In 1716, after Tsar Peter's famous victory over the Swedish armies at Poltava,

King Frederick William I, son of Frederick I, presented the entire room to the tsar. Some reports say this gift was in exchange for tall Russian soldiers, rather than simply a generous gift. Nevertheless, the amber panels of the room were carefully dismantled, packed in boxes, and taken by sleigh to St. Petersburg. It was first installed in the tsar's winter palace.

In 1755 Tsarina Elizabeth transferred the entire room to Tsarskoye Selo (Tsar's Village), where Catherine the Great's summer palace was located. The tsarina added many previous gifts of amber given to the Russian royalty by Frederick the Great. Among these treasures was a splendid mirror with the frame carved of amber. The carving depicted the Imperial Russian crown held up by two armed men at the top. The pedestals were carved into representations of the goddess of war on one side, and, on the other, the goddess of peace. Beneath were figures of Neptune and a dolphin, intended to represent Russia's power at sea. At the foot appeared carvings of armor, soldiers, and arms, representing Russia's land power.

In 1760 several amber carvers from the guilds in Königsberg were called to Russia to complete the carving of the room under the instruction of the imperial architect, Carlo Rastrelli the Younger, son of Bartholommeo Francesco Rastrelli, an Italian architect who had designed many monumental architectural structures in St. Petersburg. Some reports say this work was directed by him. However, we definitely know much designing was done to install the room. Italian Rococo supplements were added to the chamber to accommodate differences in the size of the new location. The original room had only a 16 foot ceiling, whereas the new summer palace had a 30 foot ceiling. At least five amber craftsmen, namely, Friedrich Roggenbuch, Johann Roggenbuch, Johann Welpendorf, Clemens Frick, and Heinrich Wilhelm Frick, from Königsberg continued to work on the amber room until about 1763. Venetian mirrors in gilded frames, white door posts with gilded ornamentation, and candelabra added a subtle sensitivity to the ambiance of the room and in no way detracted from the overall color scheme or architectural concept.

Many other amber art objects were included in the furnishings of the room. A large decorative curio cabinet, called a *Kunstkammer*, common among the elite of the region, featured numerous amber artifacts. It is reported that by 1765 there were over 70 amber items on the inventory. These included small chests, or sarcophagi, snuffboxes, candlesticks, crucifixes, plates, knives and forks, and altar pieces. Most impressive was a small corner table made of encrusted amber that was added in 1780 (see Fig. 3.15).

A German press attaché in the Soviet Union wrote of this

splendid room before it was removed from Catherine's summer palace. Following is this interwar account, as quoted by Grabowska (1983, 26):

> The style of the amber room at Tsarskoe Selo is a mixture of baroque and rococo and is a perfect miracle, not only because of the value of the material, the artistic value of the sculpture and the lightness of the forms, but mainly because of the beauty, the warm shades, of the amber—here darker, here lighter—which gives inimitable charm to the whole room. All the walls are covered with a mosaic made from polished pieces of amber of uneven shape and different sizes, but of a fairly uniform golden-brown color. The walls are divided into panels by frames covered with amber reliefs, and the panels are filled with four Roman landscapes executed in a mosaic of various-colored stones, showing allegories of the five senses also in frames with amber reliefs. What labor was required to create this unique masterpiece! The fantastically rich baroque style used in decorating the room only made the task more difficult. It was done perfectly, regardless of the technical difficulties. From this difficult, breakable material, they made baroque forms of ornamentation and the other decorations, small busts, various figures, coats of arms, trophies etc., frames and panneaux.
>
> The whole decoration is equally delightful in daylight and in artificial light. There is nothing ostentatious, nothing showy here and the decorations are simple and harmonious, so that some people may pass through the room not realizing from what material the wall coverings, window and door frames and wall ornaments are made. The amber covering most closely resembles marble, but it never evokes the impression of cold splendor which is so typical of marble, and at the same time it is much more beautiful than paneling of the most costly wood. The amber room is lighted on two sides by three French windows looking out onto the palace courtyard. There are mirrors on the window columns, with gilded and carved frames down to the floor. Between the windows, there are glass cases containing various small objects, like chess men, snuffboxes, amber caskets etc.
>
> On one of the walls, the dates 1709 and 1760 are engraved—the most important dates in the history of the amber room.

This unique masterpiece of amber artistry remained in Tsarskoe Selo for 178 years! It was preserved in magnificent condition

until the outbreak of World War II. Russian museum curators tried to hide the room, covering it with blankets and cotton, for safekeeping from the Nazis as the frontline approached St. Petersburg and Tsarskoye Selo (now Pushkin). When the Nazis occupied St. Petersburg, General Erich Koch, the German officer in charge of the art treasures of Europe, ordered the room to be dismantled, crated, and shipped to the Prussian Fine Arts Museum in Königsberg. It took six men about 36 hours to dismantle the room. The panels were packed into 27 crates before they razed the palace. The room is reported to have been on exhibition in Königsberg in 1942 (Rice 2004, 305)

When Allied air raids began on Königsberg, the Prussian Fine Arts Museum's director, Dr. Alfred Rohde, despite eminent danger to the chamber, put off dismantling the amber panels. However, by April 1945 it was in cases and seen in the castle by impartial witnesses! According to Grabowska (1983, 27), "This is the last authenticated information on the fate of the amber room during the war."

Despite Boris Yeltsin's 1991 announcement to the press that the amber room was hidden in Germany, no one knows the room's exact whereabouts. However, speculation abounds. Some fear it was destroyed by fire during the bombing raids toward the end of the war. Another story reports the panels being crated up and placed on a German passenger ship leaving Gadynia in 1945. Unfortunately, the ship was torpedoed while still on the Baltic Sea and sank (Patterson 2005, 9). Divers have attempted deep sea diving to recover the amber panels but none have yet been found. Treasure hunters have speculated that it was stored in underground tunnels in Weimar, Germany, or even in bunkers of the brewery frequented by Koch, the Nazi S.S. commander who was last in charge of the panels. In 1992 *People Magazine* provided details on searching for the amber room in an article entitled "A Treasure in Amber." Other teams of divers and spelunkers have investigated the authenticity of these claims. As yet no reports have come forth on the room's location (Bishop 2003).

Yet another curious account from Berlin is discredited by Polish art historian Grabowska (1983, 27) in the following manner:

> There is one more clue in a letter from Frau Liesel Amm from Berlin, published by Ryszard Badowski in his book, *Tajemnica Bursztynowej Komnaty* (The Secret of the Amber Room). However, the information given here does not fit in with our knowledge of the properties of amber. Frau Amm writes that after she went down in one of the cellars of the Königsberg castle she saw "a mass resembling melted honey, in which there were charred fragments of wood." [If only one] could get a mass

Figure 3.16. Amber Room in 1992 with no panels and painted simulations of amber lining the ceiling. Photo by author.

resembling melted honey from pieces of amber. . . . Let us add that at the heart of a fire, the temperature reaches 1,200°C and already at 300°C amber ceases to exist. Frau Amm, if she had seen the smoldering remains, would not even have guessed that from them the amber room had been burnt. The only thing which might have betrayed the place after the fire would have been the scent! A scent stronger than the bitter smell of burning, aromatic, incense-like which would have shown the fate of the amber room, especially to those who knew of its existence but were not witnesses of the drama. . . . And if it was saved, hidden in some underground bunker, then it will still not have been spared by fate.

The Sunday, March 19, 1967, issue of the *San Diego Union* included an Associated Press news report from Warsaw, Poland, that read:
San Diego Union, B-6 Sunday, March 19,1967
Kings Amber Claimed Buried

Warsaw, Poland AP

Erich Koch, imprisoned former Nazi ruler of East Prussia, was quoted in a Polish newspaper yesterday as saying a valuable "amber chamber" made for King Frederick the Great is buried somewhere in Kaliningrad, Russia, formerly East Prussia's Königsberg.

Koch died in 1986 and, since he was the officer in charge of the panels in 1945, he would have been the only person who might have actually known of the panels' hiding place, if any. Koch never revealed their location. Curator Alfred Rohde never disclosed the secret of where the room was stored either.

Many interesting theories relating to the room's disappearance, increasingly bordering between fact and fancy, continue as speculation as to the whereabouts of the amber room is contemplated. Some believe the room was stored in an underground cellar, while others say even searching in these would be futile. What is known is that when the Russians reached Königsberg in April 1945, the amber room had disappeared. No one knows if it had been destroyed or whether its hiding place may still be discovered.

While the search for the original amber room continues, amber artists from the People's Master Artists of Applied Art in Latvia in 1975 took on the seemingly impossible task of restoring the room. With the help of art critics, librarians, archivists, and archaeologists, Boris Nikolayevich Blinov, his wife Antonina Georgiyevna, and their son Boris collected and thoroughly studied all available records, including those from the former lapidary factory at St. Petersburg, whose artisans restored the amber room at the end of the nineteenth century. To assure an exact replica was constructed, preserved photos of the lost amber room were enlarged to actual size, making all details down to the smallest piece visible. Colors of amber were matched and sketches made for patterns so mosaic designs would be accurately shaped. For one panel alone, over 3000 patterns were necessary.

To provide an estimate of the amount of amber required for completion of the full-sized room, careful mathematical calculations were made. A model of the room was constructed at one-fifth its actual size. The model required only 230 kilograms of amber, and thus, verification was provided that the contemplated work was feasible (Suprichev 1978, 20–2).

Despite discouraging opinions of skeptics, the Blinovs completed the first portions of the replica of the original masterpiece; a large amber disk carved with the seal of King Frederick William I, and the upper wreath design of the basic panel. These pieces were presented to specialists of the State Committee for Protection of Cultural and National Monuments in Leningrad (now St. Petersburg) and were enthusiastically received. As a testimonial to the work of the artists, the former curator of Katherine the Great's Palace, and now an Honored Scholar of Culture of Russia, A. Kuchumov (Suprichev

Figure 3.17. Author in restored Amber Room July 2003. The mosaic and carved amber panels are separated by tall narrow mirrors and florentine gilded frames and cherubs. The floor is constructed using wood parquetry.

1978, 20–2), stated:

The work of B. Blinov in his restoration of articles of the amber room, . . . I believe . . . [in] principle and general direction of the reconstruction have been correctly undertaken. The fine details of the trim—the carved wreaths, the medallions and other features have been done at a high level of artistry, true to character. The minutest details not only concede nothing to the original, but in some cases surpass the thoroughness of detail. The result can stand on its own as a work of art.

In 1979, the Soviet Government began funding the reconstruction of the Amber Room. A sample of an amber disk with a crown and a letter *R*, to represent the seal of Peter the Great, in the mosaic as well as two mosaic amber panels are on display in the Amber Museum in Kaliningrad along with a large color photo of the original amber room (see Fig. 3.15a, b, c).

By the summer of 1992 only about 30 to 40 percent of the

Figure 3.18 Old Guild Halls in Gdańsk in 1998 as viewed by Sean Odenwalder. Photo by author.

panels for the amber room were completed. The work was being continued by architect A. Kendrinsky, Master Craftsmen A. Zhuravlow, and A. Krylov. Even this work was not on display for visitors to the palace in 1992. At that time only the gilded friezes were restored in place, but painted simulations of amber panels line the ceiling of the room (see Fig. 3.16). By 1993 many of the panels had been reconstructed and about 40 percent were installed in the palace. The two best Russian carvers from Pushkin, Alexander Zhuravlow and

Alexander Krylov, have worked since 1986 developing techniques to reproduce exact copies of the panels using the same techniques as the seventeenth and eighteenth century amber masters. Alexander Krylov presented information on restoration and preservation of amber art at the 1997 Baltic Amber Symposium in Gdańsk.

In February 1996, the American Museum of Natural History presented an amber exhibition and featured the Russian amber workers who were restoring the amber panels. A portion of the exhibition recreated a corner of the restoration for view by visitors. Two completed amber panels and the amber corner table were installed. The detail of the mosaic panels was magnificent. The craftsmen set up their workshop in the museum and were working on a partially finished panel. It was interesting to see the craftsmen using a handheld hair dryer to soften beeswax to adhere amber platelets to the wooden base of the panel. I was told the workers had examined incrustation work done by medieval guild amber workers to learn their secrets for adhesives and how to prepare amber by "clarification" and by dying the amber. The Russian craftsmen could be seen carefully cutting each amber piece to the exact shape and thickness, then polishing and fitting the mosaic into place.

The work of completing this project progressed slowly because of the breakup of the Soviet Union into independent countries. Now workers at St. Petersburg must purchase amber at a much higher price than before, when the centralized Soviet government provided the amber to the museum practically free. In 1996, amber craftsmen had to pay about $250/kilogram for raw amber. The project required a tremendous amount of amber—several tons–that had to be imported from amber mines in the Kaliningrad region of Russia. Only about 20 percent of the raw material was usable in the completed panels, with 80 percent being waste material as a result of the crust, oxidation, and breakage. The Russian carvers needed other jobs to support their families as they often were paid four or five months late. Even though the carvers were trying to revive the old techniques, they still used current techniques to produce transparent amber. Ground amber can be melted under very high pressure to produce a high grade of transparent amber (Shedrinsky 1996, 23–6).

In April 2000, one of the original Amber room's Florentine mosaic pictures was found in an antique auction and seized by German authorities. A Nazi soldier had taken it as a war prize when dismantling the room in 1941. The dealer was fined and the art was returned to Catherine's Palace in Russia in solemn ceremony with Russian and German officials present (Sautov, et al. 2003).

By 2003, the restoration of the amber room was completed

Figure 3.19. Engraving of amber fishermen on Baltic coast. Reprinted from Treptow 1900.

and the reconstructed panels were in place and on exhibition in the Tsarskoye Selo Palace in time for the *300 Year Jubilee of St. Petersburg* (see Fig. 3.17). The project had taken six tons of high quality amber from the Baltic Sea. The total cost of restoration came to $11.35 million. The German company, Ruhrgas, donated in 1999, $3.5 million to purchase amber and pay the restorers. Russian President Vladimir Putin and German Chancellor Gerhard Schroeder inaugurated the rebuilt room at Tsarskoye Selo during the St. Petersburg celebrations at the end of May 2003 (see Fig. 3.17).

In February 2004, The American Amber Association sponsored an Organic Gem Show in Tucson where fragments from the Amber Room were exhibited and Boris Igdalov, the director artist-restorer from Tsarskoye Selo's unique restoration studio, demonstrated amber carving techniques. Restorer-artists were able to replicate any artworks, both existing and lost, by examining old documents.

Eighteenth Century and the Collapse of the Amber Guilds

During the eighteenth century, the demand for artistic work in amber began to decline with the development of much bickering among the guilds. By 1755 the Königsberg Guild included only 68 master craftsmen merely the membership the Bruges Guild had attained as early as 1420. The support and interest in amber art by the elector lessened. As the Napoleonic wars developed, it became increasingly difficult to sell amber objects and to exact payment after they were sold, all of which resulted in the decline in commissions for large art objects. (See Fig. 3.18 old Gdańsk Guild Halls

Curator Alfred Rohde (1942, 88–91), when describing the

eighteenth century, reminds:

> Attitudes toward amber were always a question of faith in hidden powers and inner warmth emanated by this unusual raw material. The age of enlightenment, however, lost its appreciation of this intuitive aspect, and saw in amber only an easily broken material.

Grabowska (1983, 28) points out that large size jewelry boxes, or "caskets," were produced during the second half of the eighteenth century. However, these no longer were decorated in carved amber nor supplemented with ivory as was the case in work by the seventeenth century guild masters. They were usually simple in form, but were still amber platelets encrusted over a wooden base.

In 1811 the Prussian government decided to withdraw from control of the amber industry and again placed the raw amber trade in the hands of a private agent. In the same year, Königsberg Guild closed and several other guilds dissolved as a result of the collapse of trade. However, the Stolp (Słupsk) Corporation continued making amber necklaces for peasants until 1883.

Polish archives reveal the names of several master craftsmen from Słupsk and there are a few surviving amber objects. For example, Frederick Wilhelm Arnold of Słupsk produced empire medallions with allegorical figures etched over the smooth surfaces of the polished golden amber. These pieces are considered to be exceptional for the period. It was the development of industrialization in production that nearly meant the demise of artistic hand-produced amber articles. Truly artistic craft work was devalued. The demand for lathe-turned beads influenced amber workers to produce these on a mass scale. Grabowska (1983, 28) reports:

> As early as the end of the 18th century, 25,000 lbs. of these were sent from East Prussia to Russia alone: that is, 222 barrels of amber beads for 300,000 guldens. Beads—honey-yellow, transparent or opaque, spindle-shaped or round—were the amber products typical of the 19th century. . . . Amber treasures have been preserved in natural history rooms among geological collections, as interesting specimens of this rather rare mineral [sic].

Figure 3.20. Plate from first book on amber, illustrating amber fishermen. Reprinted from Hartmann 1677.

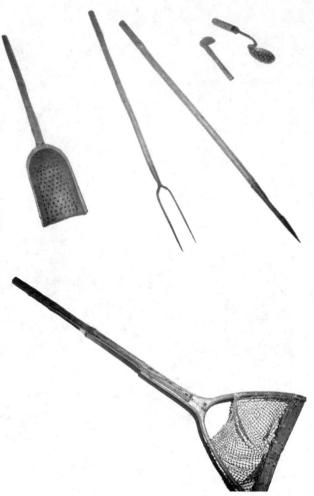

HISTORICAL METHODS OF GATHERING AMBER

Tennyson's *Lover's Tale* mentions lovingly:

> *The loud stream,*
> *forth issuing from his portals in the crag*
> *(A visible link unto the home of my heart)*
> *Ran amber toward the west.*

Since prehistoric man picked up the first piece of amber from the Baltic shore, natives of the Baltic coastal region have collected pieces washed up from amber-bearing strata that lie submerged in the sea. During storms with strong onshore winds, large quantities of amber are loosened from the seafloor. Only slightly less dense than seawater, amber chunks are dislodged easily from floor deposits and tossed ashore, sometimes becoming entangled with seaweed and flotsam. Thus, amber is left high and dry upon the beaches with the ebb of the tide.

Some amber is still found on the beaches and is called "sea-amber" or "seastone." Characteristically, it is semipolished, with only traces of the weathered crust found on buried amber. Furthermore, it is usually free of its fissures because of having been subjected to the violence of the churning sea, as well as to abrasion from tumbling along the bottom of the sea. Thus, masses of amber that originally contained cracks or fissures tend to separate into smaller lumps, free from such flaws.

Along the southern coast of Sweden, the western coast of Denmark, and the Frisian Islands, native collectors from local villages still gather amber exactly as early humans first collected it. After a storm, the beach is full of warmly dressed amber-collectors bucking bitter sea winds as they walk the beach in search of small lumps of amber. Such specimens are generally quite small, and they are much more scarce than they were in the past. Amber gathered by combing the beach is sometimes called "strand-amber" (see Fig. 3.19).

To improve the yield and prevent amber from being washed back to sea, special "amber-catchers," called *Kascher* in German, were constructed with nets fitted to the ends of poles about 20 to 30 feet long. Exactly when this method was introduced is not known, but the

Figure 3.21. a, Prongs and forks for dislodging amber lumps form the sea bottom; Size of spear and fork is abt. 3.5 ft. to 4 ft. or 1.06 m to 1.22 m, b, Net for raising amber from the sea bottom. Size from end of handle to net board is abt. 4.5 ft. or 1.37 m. Archeology Museum Collection in Gdańsk, Poland.

first book written on the amber industry, published by Philip Hartmann in 1677, described the procedure, which he called *Schöppen* (see Fig. 3.20).

Hartmann pictured an amber fisherman equipped with a long-handled net and clad in protective leather clothing with a pouch or "cuirass" for stowing amber attached to his chest. This gear was worn by fishermen who waded waist-deep into the waves, thrust their long-poled nets into the oncoming surf, and scooped up the masses of floating seaweed

Figure 3.22. Engraving from 1868 of amber poking from a boat done on calm days. Reprinted from Runge 1868.

that perhaps held entangled amber. The contents of the nets were dragged ashore and sorted by women and children. Amber collected in this manner was referred to as "drawn-amber," or, by the local fishermen, as "scoopstone" (Gudynas and Pinkus 1967, 36).

The Baltic Sea was cold in all seasons, but especially so in November and December, when most storms occurred. Gear became frozen in the icy waters and huge bonfires were kept burning to thaw out the cuirasses so they could be used again. The force of the stormy sea often swept fishermen from their feet, and drowning, especially in such clumsy gear, was common. For protection, some fishermen connected themselves together by ropes, while others dug long poles deep into the sand and attached themselves to the poles when the force of the water pulled them seaward (Gimbutas 1963, 57).

During Hartmann's time, 20 to 30 bushels of amber could be harvested in 3 or 4 hours when amber fishing was most favorable. A bushel of amber weighs about 70 pounds (31.75 kg), and even in the late 1800s when scoop fishing was still a commercial method of obtaining amber, such a rich haul was exceedingly rare.

In marshy areas, the amber search was carried on by men on

Figure 3.23. 0. Helmet divers searching for amber offshore from Brüster Ort about 1869. (After Klebs.)

horseback, called "amber-riders," who rode into the water at low tide, searching for amber left behind by the receding tide. Hartmann described amber divers whose only equipment was a wooden spade carried to loosen the amber from the sea bottom, in warm seasons of the year when the waters were calmer and less cold. Near sandbank areas, amber divers descended to the seafloor in the lagoons, which were separated from the Baltic Sea by long strips of land called "Nehrung" or "sandbars." Using their paddles, divers stirred up the sandy bottom to dislodge the slightly buoyant amber.

Other methods were devised to collect the material from the floors of the lagoons in the sheltered harbor area from Gdańsk, (now Poland) to Memel (now Klaipeda, Lithuania). Using long-handled heavy spading forks with prongs bent in a sharp curve, amber fishermen in broad-beamed rowboats raked the bottom to dislodge amber lumps. These were then scooped up by another fisherman equipped with a small net similar to those used by open sea amber catchers (see Fig. 3.21).

This procedure was described in detail by Haddow in 1891 (34–7):

> The men engaged in this put out in boats, each of which has four or five occupants. The work can only be carried on in a clear, calm sea, as the amber has to be fished up from the bottom, and a sharp and practiced eye is needed to distinguish it even in the smoothest sea One boatman loosens the amber with a particular sort of spear, while another holds his kascher, or net, in readiness to catch it. The length of the kascher poles and spear poles varies from 10 to 30 feet. The iron spear head is a plate of iron, the shape of a half-moon, or triangle, three or four inches in length and the same in width. The net is six or eight inches in diameter. When large blocks of stone have to be moved to set the amber free, crooked forks and prongs, sometimes eighteen inches long and twelve inches apart, are employed. During operations, the boat leans over and the gunwale is brought nearly to the surface of the water.

In the area of Brüster Ort (now Mys Taran, Russia), a somewhat different method was used. Large boulders of the seafloor were hauled up to be used as building stones, and large quantities of amber were often found wedged between the boulders. Amber lumps were raked up with dragnets, the latter fitted with sharp rims to dislodge the small stones and pieces of amber. Hooks were used for loosening the large stones, which were then raised to the surface between strong tongs. Amber was obtained in this manner only on the northwest corner of the Sambian peninsula. This tedious method of raking amber from the bottom was called "Bernsteinstechen," or "amber-poking" (Williamson 1932, 60). The methods of "catching" and "poking" for amber may have appeared during the period when the House of Jaski of Gdańsk (1533–1642) controlled the amber monopoly. The increase in supply from these more productive techniques would have enabled Jaski to pay the revenue to the former grand marshal of the Teutonic Knights (Oppert 1876, 1–15; see Fig. 3.22).

Figure 3.24. Engraving from 1869 of dunes at Schwarzort, now called Juodkrante, Lithuania (north of the region of current Kaliningrad Oblast, Russia). Reprinted from Berendt 1869.

In 1725 a new method for obtaining amber from beneath the sea was attempted. The government hired two professional divers from the city of Halle to dive for amber too deep to be reached with forks and probes, but the experiment ended in financial failure. Years later, in 1869, another attempt at diving was more successful, yielding a rich harvest. The firm of Stantien and Becker provided divers with modern equipment to collect amber lying on the seafloor, from Brüster Ort (Mys Taran) in the west to near Gross-Dirschkeim in the east, (former East Prussia), until the supply was exhausted (see map Fig. 1.19, 22).

An interesting account of diving off the amber reef at Brüster Ort describes the costume (see Fig. 3.23) of the divers (Diving for Amber ca. 1890):

First a woolen garment covered the diver's entire body. Over this was worn a one piece "India-rubber" suit. The helmet was designed with a small air chest which was connected to an air pump in the boat above by a 40 foot [12.2 meters] long rubber tubing. Another tube attached to a mouthpiece which was held between the diver's teeth. The helmet had three small glass openings to enable the diver to look to the front and to each side. The helmet was screwed on to the diver, a rope was tied around his waist and lead weights attached to his feet, shoulders and helmet. The diver was then ready for his plunge deep into the amber world. For periods up to five hours, the diver walked the sea floor, hooking, dragging, tearing amber from its bed with his heavy two-pronged fork. The diving boats were manned by seven men each—two divers, two pairs of men who worked the air pumps alternately, and an overseer. The overseer stood in the boat to receive amber from the divers' pockets and to see that no pilferage took place.

Strict regulations on the gathering of amber by local fishermen were still enforced. In Königsberg in 1826, for example, a permanent executioner was retained to put to death people willfully gathering amber without a lease from the government. In 1811 the government leased the amber rights to the Douglas Consortium for 25 years. However, the enterprise was not successful, and the lease was surrendered before its expiration. After this, local inhabitants were allowed to lease either as individuals or as communities and were then allowed to collect amber by any method they chose. An account of the bustling of new activity was described by naturalist Ley (1951, 26):

> For the first time the fishermen worked with real enthusiasm. Travelers who visited East Prussia in that period from 1837 to about 1860 remarked on the very large number of small craft engaged in amber poking; especially off Brüsterort, the "cape" of the Samland, hundreds of vessels could be seen on clear days. For the first time in history, there was no forced labor and no smuggling. And since the beach was now accessible to anybody, the fishing villages could start another type of business; they became resorts. And for the first time in history, the Department of Internal Revenue experienced a succession of years of profit from amber.

Now geologists were able to study the beach areas and lagoons to the north and south of the peninsula of Sambia. Their findings indicated that the sandbars that protected the lagoons near Gdańsk (Vistula Split or *Frische Nehrung* [German name]) and the Courish sandbar (*Kurische Nehrung* [German name])to the north of

Samland, now called *Kuršiu Merija* [Russian]; or *Kuršiu Nerija* [Lithuanian]) were of relatively recent origin and were still shifting (being blown by the winds) in the direction of the lagoons (see Fig. 1.19, 22). Therefore, to keep the Courish Lagoon, or *Kuršiu Zaliv*, navigable for shipping goods from Königsberg (Kaliningrad, Russia) to Memel (Klaipeda, Lithuania), it was necessary to dredge the channel, removing sand blown in from the dunes on the sandbars. Several towns along the Courish sandbar were important in this amber shipping route. Goods were brought from Königsberg to Cranzbeek, at the southern end of the lagoon, loaded on barges, and then transported along the inside of the sandbar, stopping at ports along the way until reaching Memel. Toward the northern end of the sandbar was Schwarzort (now Juodkrante, Lithuania), where amber amulets and other artifacts from the later Stone Age were found (see Fig. 3.24).

Figure 3.25. Amber mining at Sasau, 1868. Reprinted from Runge 1868.

The amber stratum in the Courish Lagoon is of recent formation, with the amber resting on sea bottom sand among organic debris. At times, amber artifacts similar to those found in Old Prussian (Asti) graves were found in company with the raw amber. The presence of these artifacts in the Courish Lagoon, taken with drawings of old maps, supports the view that the lagoon was formerly connected to the Baltic. Repeated inundation of the shores probably washed the artifacts from graves into the lagoon (Haddow 1891, 35).

The possibilities for obtaining rich amber yields by dredging encouraged Wilhelm Stantien, an innkeeper of Memel (Klaipeda, then East Prussia), to secure a lease from the government in 1854 to produce amber using this method. With only a fishing boat equipped with dredging equipment, he began a successful venture. Near Schwarzort (Juodkrante) in 1855, approximately 1000 pounds (453.59 kg) of amber were obtained from the lagoon. Two years later, when dredging was repeated, considerable amounts of amber were again obtained (Ley 1951, 17).

In 1869 Stantien obtained financial support by joining merchant Mority Becker, forming the firm of Stantien and Becker. They began work with small manpower dredges, but later employed a steam-powered dredge. In 1861 they were allowed six dredges, paying the state 30 marks per day for dredging privileges. In addition, they were keeping the lagoon free of sand and were required to work at least 30 days from May to September and to pay a minimum of 900 marks per year.

Six years later, Stantien and Becker had 11 large steam dredges operating. Sand at the bottom of the lagoon was removed to a depth of 35 feet (10.9 m), dumped on the sandbar dune, and searched for amber. Working time per year, because of severe winters, was about 30 weeks. During the summer, while the weather was good, up to a thousand people were employed. In 1868 the dredging operation produced 185,000 pounds (83,895.65 kg) of amber. On average, 165,000 pounds (74,842.35 kg) of amber were produced per year (Williamson 1932, 127–8).

A new contract was later signed, governing amber production from 1882 through 1900, with the payment to the state totaling 200,000 marks per year. In the year 1889 less amber was produced than in previous dredging, and in the following year even less was produced. It became apparent that the lagoon was becoming depleted. The amber deposits no longer yielded enough to support the dredging expenses. However, dredging in the Courish lagoon from 1860 to 1890 had recovered approximately 2250 tons of amber! Because of the drop in yield, in 1890 Stantien and Becker halted operations in the lagoon but not their search for amber. Although dredging had been a success, the firm also experimented in land-mining operations.

Previous attempts by the government to obtain amber from the land simply took the form of raking or scraping, with overseers on horseback guarding the coast and watching for any "suspicious unauthorized digging." Now that geologists were able to examine the Sambian coast in detail, the knowledge gained was applied to delineating the geologic strata along the seashore. This led to the recognition of the *blue earth* as the amber-rich stratum. Geologic investigations of the Sambian region by German geologists introduced names such as "wild earth," "blue earth," and "green wall" to identify strata of different layers of Tertiary deposits that produced amber. Blue earth is in reality a gray-greenish colored glauconite clay composed of ferric-iron silicate mineral. Glauconite is formed by precipitation from seawater and contains potassium, which yields radiometric ages that measure the time of sedimentation. Today the sediment can be analyzed for radiogenic argon producing a rough age of the stratum (Poinar 1992, 17).

Pit-mining now appeared likely to assure reliable supplies of amber. Mining attempts had begun as early as 1662, but had not reached the blue earth amber-yielding stratum. Most amber had been found in secondary alluvial deposits or as small "nests" of nodules probably washed ashore after storms during earlier geologic periods and incorporated in sediments containing silt, clay, and sand. Such nests were small, sporadically encountered, and quickly exhausted. Occurrences of nests were discovered near Palmnicken (now Yantarnyy, Kaliningrad), in Pomerania and elsewhere along the Baltic coast (Ley 1951, 33; see Fig. 3.25).

Glacial debris containing amber has been found inland in the northern German lowlands, Pomerania, Poland, Sambian Peninsula (now Kaliningrad Oblast), Lithuania, and Denmark. All of it appears to be derived from the blue earth, which had been excavated by glaciers of the Pleistocene epoch and carried inland. Such unimportant secondary deposits were generally found accidentally during operations requiring digging, such as excavations for gravel and building. Some of these deposits were mined, but even systematic exploitation tended to be a hit-or-miss venture. Glacial deposits in East and West Prussia, Poland, and Lithuania tended to be the richest. Near Gdańsk,

Figure 3.26. Old Anna shaft mine workings of the firm Stantien and Becker near Palmnicken, about 1890. Reprinted from Andrée 1924.

at Klukowo (former Gluckaw) a glacial deposit was worked for at least 170 years. An unusually large piece of amber, weighing 11 pounds 13 ounces (5.34 kg), was found there as late as 1858 (Bauer 1904, 543).

By the end of the eighteenth century, mining was conducted by the Prussian government in the striped or banded sands of a lignite formation containing some amber just above the blue earth (refer to Fig. 1.18, 27), but was abandoned after 14 years. Such formations were no longer systematically worked because of the expense ofK the process compared to the yield (Bauer 1904, 544).

Earlier, in 1850, a geologic survey of the Prussian province of Samland by Zaddach (1867, 85–197) contributed much to scientific understanding of amber and its natural distribution. It was he who first stated that all amber originally came from the blue earth layer and not from the striped sands. Encouraged by information from Zaddach's report, the firm of Stantien and Becker signed a contract with the Prussian government on January 19, 1870, to mine amber. In return, Stantien and Becker paid the equivalent of $1500 per year per acre of land used. They received exclusive rights to obtain amber by mining, since prior contracts with various communities had expired around 1867 and had not been renewed, as underfinanced efforts of mining by the communities had ruined much of the shoreline and

Figure 3.27. Transported from the mine in electric rail cars, blue earth was dumped into "wash houses," where the amber was extracted from the soil. Reprinted from Bolsche 1927.

destroyed otherwise productive land. Near Kraxtepellen on the Sambian coast, for example, improper techniques of an earlier mining attempt resulted in a long stretch of the coast collapsing into the underground tunnel. As a result of these difficulties, only shore collecting, catching, and poking contracts were given to communities (Ley 1951, 34).

Stantien and Becker's lease allowed them to mine for amber at Palmnicken (Yantarnyy) between Pillau and Brüster Ort on the west coast of Samland. Meeting with great success, they established a large open-pit operation on the shore of Kraxtepellen. The yield from this mine increased steadily for five years, but because the blue earth amber-bearing stratum in this area was only 6 to 8 meters below sea level, wooden dams were required to keep seawater from entering the diggings. This posed such a problem that in 1879, when the Royal Prussian Mining Administration began working inland from the north coast of Samland at Nortycken, the operation was forced to halt, as water continued to enter the mine pit from the water-bearing sands above the blue earth. Despite drawbacks, Stantien and Becker were more successful in their mining attempts than the government had been. Between 1870 and 1875 they mined about 10,000 pounds (4,535.9 kg) of amber per year at a yearly expense of 300,000 marks, or at about half the sale value of the amber.

In 1875 Stantien and Becker contracted to build deeper and larger underground workings, the Anna Mine (see Fig. 3.26), near Palmnicken, driving entirely through the blue earth until non-amber-bearing rock was reached (Ley 1951, 34). The whole of the amber-bearing stratum could then be removed via shafts, tunnels, and galleries, and work did not have to stop during the cold season of the year as it did during open-pit operations. A further benefit was that the land above was not destroyed in the process. In the first year, the operation yielded 450,000 pounds (204,115.5 kg) of amber, with a similar yield obtained the next year, followed by another 600,000 pounds (272,154 kg) in 1877. There seems to be some confusion about the actual date of the opening of the Anna pit. Kosmowska-Ceranowicz (1996*a*, 4) reports that the mine was in operation from 1883 to 1922 and was later succeeded by quarrying.

An eye-witness description, by Haddow (1891, 37), of the Palmnicken mine portrays the hardships workers endured to obtain amber:

> At Palmnicken we visited the diggings in which, about thirty paces from the domain of the waves, the sea-gold is sought. It is an amazing sight! In the downs, shafts and galleries are made.

The fresh water is pumped out. Forty feet under the sea level the pits are dug, and the perpendicular boring reaches a depth of fifty feet. The workmen stand in three parallel rows, knocking to pieces every clod of the blue earth, the stratum in which amber is oftenest found. A group of six or eight men is placed under each overseer. While he stands watching, that which is found is thrown into a vessel of water. The men grouped nearest the sea when they have examined the blue earth, throw it with large shovels from the lowest floor of the pit to the higher platform, which is reached by long, narrow ladders. Here the refuse material is taken in charge by a group of men and women, and flung from shovels to the third or upper most platform, whence it is carted away. All the operations accord with the rhythm of a slow and monotonous melody which the overseers sing. This regularity of movements is intended partly to prevent pilfering which, however, cannot be altogether prevented, although the miners are carefully searched before leaving the pit after the day's work. It is not astonishing that in the whole range of diggings not less than twenty hundred-weights [100 kilograms] are raised on many a day. Men, women and children in all imaginable costumes, in the oddest of attires, shielding themselves against the sharp, whistling winds, digging vigorously or swinging their shovels to the languid strain of the somber melody;—what a singular spectacle is this!

Though tunnels extended underground, mining was also continued in open pits, where workers were constantly exposed to Baltic winds. Underground mining employed innovative systems for pumping out water, carrying the soil away to be washed, and separating the amber.

By 1885 the mines of Stantien and Becker were producing over 900,000 pounds of amber per year. For the following ten years, the yield fluctuated between 600,000 pounds (272,154 kg) and 850,000 pounds (385,551.5 kg). Of that figure, 10,000 pounds (4535.9 kg) to 12,000 pounds (5443.09 kg) per year were produced by shore-gathering and catching. A record amount of 1,200,000 pounds (54430.8 kg) was produced in 1895. Encouraged by this success, the state in 1899 once again took over the amber mines, with Stantien and Becker receiving a compensation of 9,700,000 German marks, at that time equal to almost $2,500,000 (Ley 1951, 35).

Thus the old Stantien and Becker Anna Mine became one of the government mines. The state established the *Royal Amber Works: Königsberg* to engage in a variety of amber activities in addition to

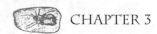

mining. It opened a factory for making amber jewelry and cigar holders and chemical plants to produce amber oil and amber varnish. The mine at Palmnicken (Yantarnyy) was near exhaustion, but the state continued its operation and opened a new mine nearby. All amber again became state property and had to be delivered to the government works, with cash reimbursement being given by collecting agencies that bought amber from the fishermen. The regulations were so strictly enforced that other firms did not dare to buy directly from the fishermen (Jakubowski 1914, 14).

During Stantien and Becker's monopoly, they were threatened by a new invention in 1880 that they worked hard to discredit. It had been discovered that small pieces of amber could be consolidated into large masses by heating to 160°C under high pressure. Given the trade name ambroid, it was declared an imitation by the *Royal Amber Works: Königsberg*. Nevertheless, the invention was purchased by the state works at a high price. Ambroid was difficult to distinguish from natural amber, and experimentation proved it to be harder than natural amber and more suitable for smoker's accessories, particularly mouthpieces. Thus the Royal Amber Works also began using small pieces of amber to make ambroid, and from then on, most cigarette holders and pipe stems were made of this material (Ley 1951, 35).

By the turn of the century, nearly one-half of the total production of amber was devoted to manufacturing smoking articles. Ornamental objects in great varieties were also manufactured, most commonly round or faceted beads for necklaces, bracelets, or rosaries. At the outset, workmen determined how each lump of amber could best be adapted to eliminate as much waste as possible. Rough pieces were sorted according to color, form, and dimensions of the following classifications (Bauer 1904, 550):

Large, flat and mostly cloudy "tiles" and "plates" for smoking articles.

Round, opaque "tears," sometimes flattened at the bottom, for smoking and gem or art articles.

Fine and clear "pouches" or shelly, transparent amber, which frequently contained insect and plant inclusions, for gem and devotional articles, or for preservation as specimens for museums because of their insect inclusions.

The main finished products were divided into four principal categories (Bauer 1904, 550):

Gems: Necklaces, bracelets, brooches, earrings, pendants, fin-

Figure 3.28. The Anna Mine as it looked when it closed in 1925. Reprinted from Bolsche 1927.

ger rings, cufflinks, teething rings for children, etc.

Smoking articles: Cigar and cigarette holders, mouthpieces for pipes, etc.

Objects of art: carvings, jewelry boxes, cups and dishes, writing utensils, ornaments, mosaic pictures, etc.

Devotional articles: Catholic, Muslim and Buddhist rosaries, sacred figures, amulets, etc.

Such articles were widely exported, with each recipient area having its own preference in regard to the variety of amber used. Beads of the purest, fatty, yellow varieties were most popular with Oriental and English connoisseurs; bone or whitish kinds were preferred by inhabitants of West and East Africa; light, clear amber went to the United States; and the finest water-clear specimens were preferred in France. Opaque, inferior varieties were used in Russia and the interior of Africa.

The bulk of the amber was still produced from the government Anna Mine. The mines were protected against sea invasions by

stone dams. Several shafts extended as far down as the blue earth, which occurs in a layer from 13 to 24 feet (4 to 8 m) thick at this place (Bellmann 1913*a*, 129).

In about 1910 electricity was introduced into the mining works, and its use resulted in increased production. Amber-bearing earth was transported by electrically powered trolleys to cube-shaped washing pits. There the cars were tipped into a box holding about 150 cubic meters of amber-bearing soil (see Fig. 3.27). Next the amber

Figure 3.29a. Open-pit operation for extraction of Sambian amber, 1894. Reprinted from Treptow 1900.

Figure 3.29b. Blue earth scooped up by mechanized bucketed conveyers and deposited in rail cars. Open-pit operations in Samland produced over a million pounds of amber in a single year. Reprinted from Ley 1938.

was sprayed to remove most of the soil. The floating crude amber was then moved to a simple washing plant, where it was washed for 10 hours in a revolving drum with clean sand and water to remove dirt and part of the outer crust.

The cleaned amber was sent to the Königsberg works, which consisted of sales offices, stores, sorting plants, and the amber compressing factory. Small grain-size amber lumps (1 to 12 mm) underwent further treatment at the state works, rather than going to the hand-workers, and passed through a cleaning apparatus similar in every detail to cereal grain cleaners. The best pieces were then sorted out by hand and either passed on to the sales office or reserved for the compressing factory. The remaining pieces were subjected to dry distillation in the chemical plant to make amber oil, amber acid, and amber varnish or colophony (Bellmann 1913*b*, 722).

In 1911 production at the Anna Mine amounted to a total output of 389,000 tons of blue earth, containing 382,772 kilograms of amber. An additional 15,000 kilograms of amber were contributed by fishermen. The rough amber and by-product sales in 1911 were as follows:

Raw amber, 66,700 kilograms
Compressed amber, 23,500 kilograms
Colophony, 158,200 kilograms
Acid, 1,300 kilograms
Oil, 31,700 kilograms

The average amber content for the blue earth at Anna Mine reported in 1913 was about six kilos per cubic meter.

By 1914 hand work had been largely replaced by machine work. Electric lathes were used to turn amber mouthpieces and automatic buffing machines were used to polish amber pieces destined for jewelry. Girls strung beads for necklaces and made bracelets, with women taking over most finishing operations, while men worked in the mines (Jakubowski 1914, 14).

No less than 500 women worked at home on cutting and shaping rough amber into jewelry. A weighed quantity of raw amber was given to each woman to be worked in her home by all the females of the household. The finished products, together with parings, were returned to the factory, where the materials were again weighed to guarantee all had been returned. Scrapings, dust, and chips were sent to the chemical plant for making varnish.

The state amber factories employed about 1000 male and female workers, in addition to about 500 female home workers. There were 350 miners permanently engaged in the mines.

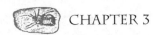

As underground mining costs began to increase, and decreasing amounts of amber were found, it became evident that open-pit mining would be more profitable and had to be adopted in the near future, especially since the demand for raw amber was growing and stocks were nearly exhausted. In 1912 production of blue earth was raised 24,150 tons over that of 1911, with a corresponding increase in raw amber amounting to 12,400 kilograms. Production was maintained at 453,592 kilograms per year until the First World War broke out in 1914 (Petar 1934, 3).

Before World War I, the Lithuanian cities of Klaipeda and Palanga also were commercial centers for producing amber. Approximately 500 workers were engaged in the amber industry at Palanga, producing nearly 20,000 kilograms of raw amber per year, mainly from sea fishing and shore-gathering. Similar production was maintained at Klaipeda, with the addition of amber obtained from dredging.

During the war (1914–1918), the Lithuanian amber industry was almost destroyed, but between 1918 and 1933, when Lithuania became independent, it gradually revived. Though no major mines were ever established in Lithuania, the seacoast adjoining the rich amber beds of the Sambian peninsula furnished an average of about 1 metric ton of amber annually, with the industry controlled by the government. Finders of amber were obligated to deliver it to the state or face a charge of embezzlement. Commerce reports of 1935 indicate that about 5 metric tons were imported annually from Prussia to augment the small domestic supply.

Lithuanian amber products were manufactured at 10 artisan shops or factories in Palanga, Klaipeda, and Kretinga, in all employing 112 persons. About 5000 kilograms of waste, too small and too dirty for pressed amber, was exported to chemical plants at Palmnicken.

One of the products of the Palmnicken chemical plant was amber varnish, which dried to a hard finish but was very dark in color. It was used on ship decks as well as on Stradivarius violins. Besides its dark hue, the expense of the varnish was considered a drawback. Today practically all varnish is made with synthetic resins, which are more plentiful, more uniform in quality, and produce a tougher, longer-lasting varnish.

Amber acid, or succinic acid, another product extracted from Baltic amber at the chemical plant, was used by iron foundries "as a scum producer for the wet dressings of mineral coal and ores" (Weinstein 1958, 228). In the Soviet republics, amber acid was used in the production of soap, bath salts, and pharmaceutical preparations, as well as in the manufacture of rhodamine dyes, photographic chemicals, and artificial leather (Petar 1934, 6).

Approximately 15 percent of the products produced by the chemical plants was amber oil, which was used in wood preservatives, insecticides, and metal casting and flotation (Kraus 1939, 239–40). Raw amber that was unsuitable even for chemical processing was, and may still be, used in Baltic countries to make an ornamental concrete of greenish and light brown colors (Suprichev 1978, 22).

Preceding World War II, only about 25 to 30 percent of the material produced from the mines operating in the Baltic area was of the quality to be used as gem amber and for manufacturing into pressed amber or ambroid. Gem quality material was worked at Königsberg, while the other material was heated in large retorts in the chemical plant at Palmnicken. In 1978 only one-tenth of the amber production was suitable for jewelry. However, modern processing techniques enable the plants to clarify and heat raw amber to enhance the color and clarity, thus enabling more of the amber to be used in the jewelry industry.

After World War I, the Sambian amber-producing mines came under the control of the German government. Demand for Baltic amber increased during the 1920s, with amber being second only to diamonds in U.S. gem imports. Commercial reports show 6809.75 kilograms of raw amber were imported at a value of $41,566 in 1920 alone (Petar 1934, 8).

At this time, underground mining met with a number of difficulties that incurred increasing expense. The porosity of the overburden, as well as the presence of quicksand just above the amber-bearing stratum, resulted in a change to open-pit mining operations in 1923 and the closing of the old Anna Mine in 1925 (see Fig. 3.28). By now surface mines were removing approximately 1,400,000 cubic meters per year of barren earth overlying the stratum of blue earth. Diesel shovels cut away all the overlying deposits down to the amber-bearing layer. Other specially designed shovels, using buckets on an endless belt, scooped up the amber-bearing soil and deposited it in conveyors that carried it to the washing sluices (Baer 1937, 82–6).

Electric locomotives were now used to carry the blue earth from the mines to the Palmnicken (now Yantarnyy) factory. Reporting on the status of mining, Prockat (1932, 305–7) described the procedure for recovery of amber:

> The cars are emptied over a grate in the so-called spray house, and the blue earth is pulped with sea water under a pressure of six atmospheres, by the flushing methods familiar in the extrac-

tion of gold from sands.

The large amber pieces are collected by hand; the smaller ones flow on with the clay slurry, on which they float because of their low specific gravity—under 1.05. It is thus possible to remove the heavy material from the channel, while the amber-bearing slurry flows on to a classifying drum, where the amber is segregated into different sizes. The larger pieces are cleaned of their crust in special wooden drums to which sand and soda solution are added. At the sorting, which is done at Königsberg, these are again beaten by hand, so that only clean amber will be sent to the market.

Modernization of mining and recovery techniques was followed by a record production in 1925 of 547 metric tons of crude amber. Open-pit mining or quarrying cleared the overlying glacial deposits and those from the Upper Tertiary period. Mechanical digging machines were used to extract the blue earth amber-bearing sediments from not only Palmnicken but also Primorskoye and coastal open-pit mines (see Figs. 3.29a and b). In some mines a conveyor belt transported the amber-bearing earth to the amber-processing factory to be sifted. At the factory raw amber was recovered in a suspension of water and loam. The stripped earth was then washed back out to sea. This process would alter the contours of the coastline in the mining areas (Petar 1934, 7; Kosmowska-Ceranowicz 1996*a*, 5).

The Free City of Danzig (now Gdańsk, Poland) was also an important manufacturing center for amber products. In 1925 about 600 workers were employed (Kemp 1925, 212). From 12 to 20 independent manufacturers purchased raw amber supplies from sources not controlled by the Prussian state monopoly, as well as from the monopoly itself, and sold finished amber at their own prices.

In March 1927 a manufacturing and sales agency called Staatlich Bernstein Manufaktur GMBH was organized by leading German amber factories with headquarters in Königsberg. The Prussian state agency, Preussische Bergwerks und Hütten-Aktiengesellschaft, was represented in this sales agency. By 1930 the Prussian state works at Palmnicken employed 600 to 700 men in mining and preliminary sorting and treating of amber, with an additional 150 men at the factory at Königsberg.

A facetor, generally a woman, was required to work as an apprentice for four years before she was allowed to produce faceted amber beads for the market. About 24 percent of the articles manufactured from amber in Germany were sold in domestic markets. Since Muslim prayer beads were required to be of amber, substantial markets were established in the Muslim world. However, after 1930 production dropped because of unstable commercial and political conditions, and the decline in sales resulted in an accumulation of stocks sufficient to meet the demand for two or three years. In 1932 production was down to only 59 metric tons (Petar 1934, 8).

Because of the political unrest in 1933, when Hitler became chancellor of Germany, amber production dropped to zero. The National Socialist Party publicity department attempted to aid the industry and bolster demand for amber by requiring all sporting societies to use amber prizes and decorations for sports awards in place of the usual gold and silver. The resulting improved trade conditions enabled German amber mines to resume operations by March 1934, and 104,332.7 kilograms were produced that year. In 1935—the last year for which figures were published—99,790 kilograms were produced. During this period, approximately 375 men were employed in the amber industry in Palmnicken (Ley 1957, 35).

Historically, amber jewelry has been subject to waves of fashion, being in vogue for a period of 2 to 5 years, then followed at about a 15-year interval characterized by less demand. During 1934 for example, amber again came into vogue and readily found markets around the world.

By 1939 amber was no longer controlled by the German government, and fishermen of Sambia and Lithuania were free to collect amber as they pleased. This, plus modern methods of production and marketing, resulted in overproduction. At the same time, synthetic imitations, produced in large quantities, inevitably led to less demand for genuine amber and a drop in price. As with diamonds and other precious gems, rarity is one of the chief factors in determining demand and price for objects of luxury (Weinstein 1958, 225). This overproduction, however, was of short duration, and by the middle of the twentieth century, supplies of Baltic amber decreased and were made less accessible to the western world. After World War II, production areas were controlled by the Soviet government.

Few specific facts are known about the amber industry during World War II, but it is known to have been nearly destroyed for a period of time. Ley reports from personal correspondence that in January 1945 the Königsberg mine was active, but the Russians made sporadic air attacks on it. Later in that year, the amber mine flooded and remained so for several years. In 1948 the Russians drained the mine. During that summer, they produced 90.7 kilograms of amber daily. According to Ley (1951, 36):

A newspaper article published in March 1949 in the American Zone of Germany stated that most of the professional amber

workers and enough raw amber to last for 20 years were brought to western Germany in 1944 and that the Königsberg Amber Works are now split into two: one in Hamburg (British Zone) and one in Tubingen (American Zone), where amber rosaries (including Mohammedan rosaries) and amber jewelry are being made to satisfy orders from the Benelux nations and from Great Britain. East Prussian amber workers, hearing this, fled through the Iron Curtain.

After the war, amber-processing shops gradually reopened in Palanga, Klaipeda, Kaunas, Plunge, and Vilnius in Lithuania. Hundreds of craftsmen were engaged in various phases of the work, processing up to 10,000 kilograms of raw amber annually. Because of political changes in 1963, the amber mines and pits of historic Samland were incorporated into the Lithuanian amber industry. Lithuanian sources report that these mines produced about 500,000 kilograms of raw amber annually, or about 90 percent of the world's amber production at that time (Encyclopedia Lithuanica 1970, s.v. "amber").

In 1999, two-thirds of the world's amber comes from the Kaliningrad Oblast's (Samland) concentration of amber-bearing earth, but mining in this region is entirely open-pit, with the mine spread over 1200 meters or four-fifths of a mile. The amber is found at a depth of 36 to 40 meters, with the blue earth amber-bearing layer between 5 and 7 meters thick (Almazjuvelirexport 1978, 1). Kosmowska-Ceranowicz (1996*a*, 5) reported in her description of Baltic amber deposits that "in 1948 and from 1951 to 1986 the amount recovered totaled 17,704.5 tons."

Since 1994 amber workers have held a yearly International Amber Fair called "Amberif," organized by Gdańsk International Fair Co., International Amber Association, and the Museum of the Earth of the Polish Academy of Sciences. The Fair has been devoted to educating participants about the current research, preservation, amber testing, manufacturing, and artistic production of amber products. Geological and archaeological methods for testing and determining amber deposits are presented in seminars and poster presentations. Scientists from Germany's Bitterfeld amber region, such as Gunter Krumbiegel, and Poland's Academy of Science, Barbara Kosmowska-Ceranowicz, keep the industry updated (see Fig. Photo documentation for mining on Baltic strand pg, 147-152.).

LITHUANIAN / LATVIAN AMBER WORKINGS

Figure 3.30a.
Lithuanian amber art on black wood to represent On my palm I hold a piece of Lithuania. Thomas Lunas Collection.
Size of wooden background is 12" by 18" or 30 cm. by 45 cm.

Amber, the Gem of Lithuania
How beautiful our little country
As the clear drop of amber.
For a long time I love it—
In the patterns of the weavings
And the songs of my native village.

On my palm I bring you
The gentle name of Lithuania
Splendid like the sun, like
      A pale piece of amber
Drop of the Baltic—
    —Salomeja Neris

Since ancient times early Balts living along Baltic Sea coast called amber "sun-stone" and "pearl of the Baltic Sea." In Lithuania and Latvia artistic amber processing is evident from Neolithic times. Not only did the inhabitants of the Baltic coast gather amber and barter it away along the ancient trade routes, but they also used it in their everyday life. Pieces of amber were used as amulets, some being carved in the shape of animals and others were threaded on strings forming the first necklaces. This is evidenced in archaeological finds and particularly the Juodkrante, Lithuania, collection.

Although the Juodkrante Neolithic amber artifacts, along with the largest collection of choice specimens of amber containing inclusions, disappeared from Kaliningrad during World War II, numerous specimens from the amber mines continue to be housed in the Amber Museum of Palanga. This museum, which opened in 1963, presents a unique collection of cultural materials, including the Palanga amber collection, consisting of 100 artifacts similar to those in the Juodkrante hoard (see Chapter 2). The Amber Museum of Palanga is housed in the mansion that belonged to the late Count Tiskevicius, an amber connoisseur himself. Previous to the museum's opening, Count Tiskevicius displayed the Palanga amber artifacts in the 1950's at a Paris exposition.

In 1936 his collection, which consisted of 153 articles of different ages from Neolithic to Bronze to early Iron Age, was placed in the Kretinga Museum. It was finally transferred to the Palanga Museum in 1960. In that same year, amber from collections housed in various museums of history, ethnic studies, and art, in Kaunas, Klaipeda, Telsiai, Silute, Siouliai, Vilnius, and elsewhere were also transferred to Palanga, Lithuania. By 1963 choice amber collections throughout Lithuania were gathered together and placed in this museum to make it the largest amber museum in the world. The largest and most unusual specimen in the museum weighs 4.28 kilograms and has delicately interwoven streaks of as many as 50 different shades of color.

Twentieth-century Lithuanian artists developed a tradition of making mosaic pictures by using various features of amber—color variations and textures—adhering the design to a black wooden board. This amber art technique is unique to Lithuanian folk art. One such interesting piece, a mosaic of bits of white and blue amber portraying the atomic icebreaker, *Lenin*, stuck in the ice, was on exhibit in the Palanga Museum in 1976. (Suprichev 1978, 22; Kostiaszowa 1997, 45).

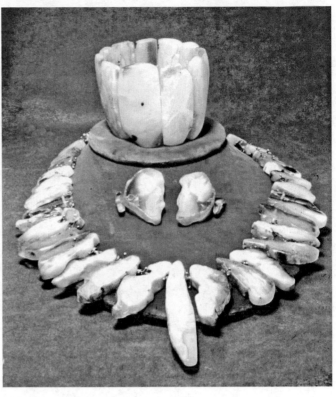

Figure 3.30b. Lithuanian necklace and bracelet of polished slabs of amber. The necklace slabs retain the outer crust along the edges. Pieces range from transparent to fatty to opaque mottled amber. Each slab is separated with a silver twisted wire spacer. Made in Lithuania. Size: Longest slab approximately 64 mm or 2.5 inches, total wt. 96 g. Patty Rice Collection.

Among the exhibits are those explaining the formation of amber, as well as illustrations of the traditional applications of amber throughout the span of Lithuanian history. During the period when this region was under USSR control, visitors from abroad were not allowed visas to Palanga, thus few tourists had access to the museum. At the present time, visitors may travel throughout Lithuania and no visa is required. Lithuania became part of the Soviet Union between 1962 and 1994. During this period, it gradually exported less amber than in the past because less raw material was available and most amber gathered by Baltic coast settlers was sent to shops in Russian cities to supply ornaments for purchase by tourists. The official Russian wholesale outlet Almazjuvelirexport controlled the sale of amber and exported natural, pressed, and heat-treated amber articles. Amber's popularity continued in Lithuania SSR and Latvia SSR along the Baltic coast during this period. In the late 1950's, the world's first instructional center for artistic amber-processing was established as the Secondary School of Applied Arts in Liepaja, Latvia (Anusulis 1979, 126) Graduates of this school worked in plants in Latvia and Lithuania, where amber continues to be processed for domestic sale and export. While under the control of the USSR, only finished amber pieces were exported by the Russian company. West Germany and were the

*Figure 3.30. c,
a, Natural amber brooches
constructed from two
lumps of amber linked
together with chains;
b, Transparent amber
brooches with silver filigree metal work frames.*

main purchasers of amber from this source. Currently artistic works in amber, the Gem of Lithuania, are treasured throughout the world.

In May 1973 a month-long Baltic Amber Fair, exhibiting the most interesting specimens with inclusions, such as necklaces, bracelets, pendants, and brooches made by Baltic craftsmen (particularly from Latvia and Lithuania), was held. The exhibition illustrated the ancient Baltic traditions of working with amber. Similar fairs were held in former Czechoslovakia and former East Germany.

Soviet scientists continued to research various uses of amber and, interestingly enough, found clear amber could even be used in laser technology (Suprichev 1978, 22). However, specific details of its use were not given. Some of the most complete postwar documents written about the Sambian deposits were authored by a Lithuanian scientist and curator at the Palanga Museum, V. Katinas (1971), and by a Russian scientist, S. S. Savkevich (1970). Information from their research is included in Chapter 6.

Lithuanian artists typically use early Baltic motifs in their jewelry even though the significance of their symbolism and powers may have been long forgotten. Amber is most often combined with silver wire and chain. Lumps of semipolished amber are sometimes strung with silver spirals separating the pieces. One naturalistic style involves slicing a lump of amber into slabs about 5 centimeters thick, then polishing both sides of each slab, and leaving the natural crust markings on the edges around the slabs. These are strung with the longest one in the center and identical slabs across from each other, often spaced with ornamental, twisted silver wires (see Fig. 3.30b). Very small chips of amber are also strung on slender wires attached together to form the so-called popcorn clusters, which then are hung on elaborate chains.

Another popular style of Lithuanian and Latvian artists uses natural lumps of amber shaped into large free-form pendants commonly set with silver filigree mountings and several silver chains hanging below the pendant. Brooches often are constructed from two lumps of amber connected by links of chain. Several chains are secured to large brooches. Such styles are based on designs of the ancient Balts and preserve national ethnic styles (see Fig. 3.30c. a and b).

Because Lithuanian people work so intensely with amber, Lithuanian culture, art, customs, and mode of life grew together, incorporating the mysterious material from the sea. Local artists, who

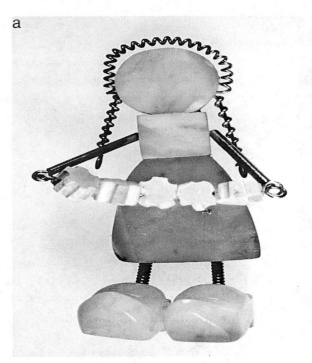

may have been fishermen, became masters in working amber. Fishermen and peasants along the coast carved amber lumps to pass away the dreary winter months, producing numerous ornaments called *Kniepkiniai* or knickknacks in Lithuanian. A typical sculpture may be the amber figure of a folk dancer or amber gatherers with a basket of small amber pieces. Another sculpture of this type depicts clusters of growing mushrooms, a great delicacy among Lithuanians since mushrooms grow abundantly in wooded areas of the countryside. Mushrooms have long been regarded as a symbol of fertility throughout Europe, so once again we see the connection with the mystical qualities of amber (see Fig. 3.31). Not all work was so simple, however, and animal figures or other objects from nature are common subjects in amber sculptures and carvings.

Because of their strong love of nature, local artisans felt amber was most beautiful when the least shaping took place, and every effort was made to retain natural shapes of large rough amber cobblestones. Thus, large lumps of amber were often given just enough polish to enhance their natural beauty, then simply mounted on wooden bases. Large lumps may be slightly polished and incorporated into a design using copper or brass, such as the sunburst sculpture, reflecting the close ties with sun worship between amber and the ancient Balts (see Fig. 3.32).

The School of Lithuanian Art emphasized simplicity of

Figure 3.31. Kniepkiniai (knickknacks): a, Folk dancer with chain of amber flowers, made of opaque yellow amber, single pieces for head, body, skirt, and shoes. Girl carries white bone amber flowers, joined together with twisted wire and springs. Figure is free-standing and three-dimensional. Size: 8.5 cm tall
Patty Rice Collection. b, Amber mushrooms mounted on a base;

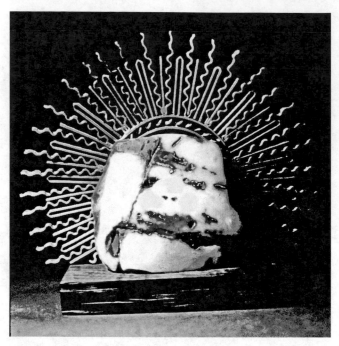

Figure 3.32. Baltic amber sculpture with sunburst design. Brass wires radiate from a brass arc to portray the sunburst rays extending behind the large amber lump, which is the sun disk. The amber lump is mottled yellow-to-brown opaque amber, polished with natural crust remaining. The sculpture is mounted on a black wooden base. Size: Amber wt. 8 kg, overall 5 in. x 5 in. x 3 in. Made in Lithuania.

design and the inherent beauty of amber as seen in ancient amulets and other primitive ornaments. Some amber art was made to replicate the early artifacts and reproduce classical designs found in Roman and Greek tomb amber. Special treatment of the amber to age the sculptures was used to try to replicate the Roman artifacts of the Etruscans and objects found in the excavation of Pompeii. Amber can be heated to make it a darker color, dipped in oils or even a "flambé" with a quick flame over the carved piece to give it an ancient look.

Devotional objects, such as rosaries, medallions, crucifixes, and religious art, were and continue to be made of amber. During the period when Lithuania was under USSR control, amber art of this type became difficult to locate, since carvers were discouraged from producing religious artifacts because the soviet was officially an atheistic culture. Typical Lithuanian mosaics were constructed with pieces of amber, using variations in colors and shapes, to depict scenes related to local myths and religious monuments. One such mosaic depicts the Palanga shore on the Baltic, representing a famous coastline and tall trees of Palanga (see Fig. 3.33*a*). Another example is an amber mosaic of the *Miraculous Mother of the Gate of Dawn*, representing an icon found in a chapel in Vilnius. Lithuanians go to this special chapel to place symbols representing their illnesses or their problems, believing the miracle-working Holy Mother will then cure them or solve their problems. Many gifts of amber necklaces, bracelets, and pins are placed around the icon of the Holy Mother (see Fig. 3.33*b*).

Lithuanian peasants, living in the coastal regions, were familiar with amber and carving was done in their homes using primitive tools. Amber in several strands was worn with the women's traditional dress accompanied by embroidered aprons and blouses. Pieces of amber, shaped similar to the ancient amulets with natural holes, were crocheted together to form a collar of amber. Each polished flat teardrop-shaped amber piece is crocheted directly into the bib just as the ancients would sew amber to their clothing (see Fig. 3.34). Also amber chunks of varying sizes were polished, drilled, and then strung with wire spacers and hanging chains, reflecting the old Curonian jewelry styles. As mentioned previously, the Curonians inhabited the

Figure 3.33a. Lithuanian amber mosaic. The scene on the black wooden background is made entirely of natural Baltic amber and depicts a chapel on the Baltic coast near Palanga. Some of the amber is cut and polished while other lumps are left unpolished. Crustal pieces are used to vary the texture. Size: 25.4 cm x 52.6 cm. Patty Rice Collection.

Figure 3.33b. Amber mosaic of the "Miraculous Mother of the Gate of Dawn," modeled after one in Vilnius, Lithuania. Amber variations are used to depict stars, crown, aura, clothing, and body. Amber is cut to spell the words "Ave Maria." Size: 21 cm x 31 cm Patty Rice Collection.

region near the Lithuanian and Latvian border around A.D.1000. Their jewelry forms followed geometric patterns with hanging attachments (see Fig. 3.35).

Gudynas et al. (1969), Lithuanian historians, described the folk cultures from various regions in Lithuania before the Second World War, with detailed accounts regarding the typical styles of

amber and other jewelry traditionally worn. For example, in the highland region of Aukstaitija, amber-, coral-, silver-, and colored glass-beaded necklaces were worn. Amber was popular for its healing powers. The amber necklaces often had pendants. The natives of each region wore their own style, just as they wore their own style of folk costume—amber's role being to communicate a distinct identity in terms of wealth, social status, and concern for dignity and aesthetics. The traditional folk costume, with its adornment, was important in annual family celebrations and religious festivals, such as weddings, church holy days, the harvest festival, and other special occasions celebrated with ceremony, in rural life. The form of adornment not only indicated status and wealth, but when it came to amber jewelry, it was also empowered with mystical properties.

Figure 3.34. Crocheted collar made of all varieties of natural Baltic amber. The threads go through the holes in the amuletic pendant-shaped pieces of amber. Made in Lithuania. Total wt. 238 g
Below: Figure 3.35. Typical Lithuanian folk jewelry. Amber chunk-style necklace with wire spacers and hanging chains. Made of opaque yellow amber. This style is similar to old Curonian-style jewelry. Patty Rice Collection.

In other provinces, such as the Vilnius region, tiny beaded amber necklaces were worn in two or three rows. Dzukija women wore several types of beads made of amber, coral, and bone. The very heaviest, dark amber necklaces were found among the peasants. Early artifacts from this region typically were flat, disklike amber beads with a large hole in the center. In the region of Zanavykija, many kinds and sizes of amber-, coral-, and glass-beaded necklaces were found. The small round beads were characteristic of the region and were

threaded on horsetail hairs or waxed threads. The women of Kapsai strung and wore flat graduated amber beads. Delicately beaded chokers were often worn in place of, or together with, a longer strand. Because it was valuable, amber was a highly prized heirloom and a sign of wealth. Most country women only owned one strand. In Lithuania Minor, though women wore amber, coral, glass, or silver necklaces, amber was most popular. The inhabitants of Klaipeda and Karaliaucuis, along the shores of the Baltic Sea, had for generations undertaken the polishing and fashioning of amber into necklaces. Those who could not afford to buy amber necklaces but lived near the coastline gathered unpolished pieces of amber and strung the seastone into strands. Villagers wore amber not only for its beauty but for its protective powers as well.

a

b

c

Figure 3.36. Women wearing amber jewelry with traditional Lithuanian costumes: a, Peasant woman from Semaitija Province, wearing long, large amber-beaded necklace;(Reprinted courtesy Luetuviu Luaudes menas, Drabuziai, Vilnius,1974.) b, Lithuanian-American wearing amber beads at Detroit Old World Market;1978 c, Lithuanian-American folk dancer, Diana Odenwalder, wearing Lithuanian style amber beads and earrings at San Diego festival, 1979.

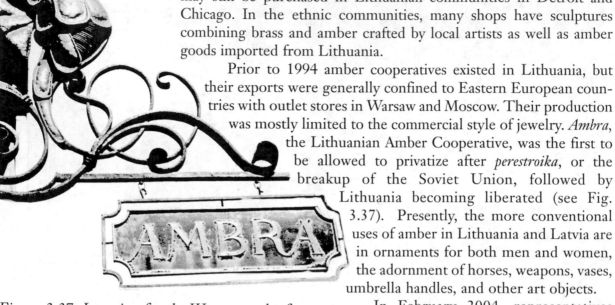

During international festivals in the United States, golden strands of amber and other amber jewelry are still traditionally worn by Lithuanian Americans and folk dancers with their national costumes. Many Lithuanian booths carry amber designed into various unique ornaments for sale (see Fig. 3.36). In the United States, objects such as sculptures, mosaic scenes, and jewelry constructed of amber may still be purchased in Lithuanian communities in Detroit and Chicago. In the ethnic communities, many shops have sculptures combining brass and amber crafted by local artists as well as amber goods imported from Lithuania.

Prior to 1994 amber cooperatives existed in Lithuania, but their exports were generally confined to Eastern European countries with outlet stores in Warsaw and Moscow. Their production was mostly limited to the commercial style of jewelry. *Ambra,* the Lithuanian Amber Cooperative, was the first to be allowed to privatize after *perestroika,* or the breakup of the Soviet Union, followed by Lithuania becoming liberated (see Fig. 3.37). Presently, the more conventional uses of amber in Lithuania and Latvia are in ornaments for both men and women, the adornment of horses, weapons, vases, umbrella handles, and other art objects.

Figure 3.37. Iron sign for the Warsaw outlet for AMBRA, the Lithuanian Amber Cooperative, Warsaw, Poland.
Below: Figure 3.38. a, Latvian necklace with silver rings encircling each golden amber ball; b, Latvian amber ring in sun-worship design. Festival, 1979.

In February 2004, representatives from Lithuanian amber factories exhibited in the Tucson Gem and Mineral Show with a wide variety of natural, processed, clarified, and heat-treated amber. The typical Lithuanian artistic styles and conventional mass-produced designs were both available.

Latvians living close to the Baltic shores also found amber an intricate part of their lives. The ancient inhabitants of this area were sun worshippers and even today jewelry designs reflect sun-worship symbols. The Latvian ring and necklace with the amber ball circled with concentric rings or with a ring of silver are representative of the sun-worship designs (see Fig. 3.38).

The Riga Museum of Ethnographic Studies published

Figure 3.39. Drawings from the Riga Ornamentation Society of Latvian jewelry: a, Drawing of amber jewelry styles; b, Drawing of amber jewelry styles; c, Traditional costume with amber; d, Amber earrings. (Reprinted courtesy Luetuviu Luaudes menas, Drabuziai, Vilnius 1974.)

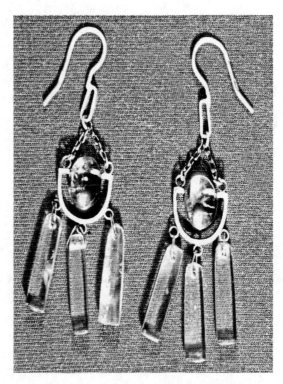

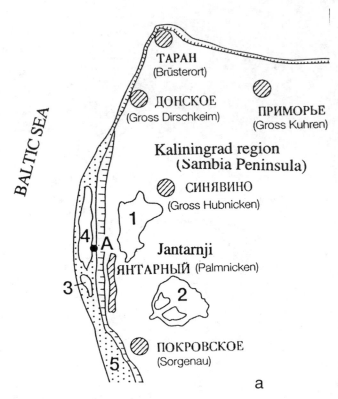

ТАРАН
(Brüsterort)

ДОНСКОЕ
(Gross Dirschkeim)

ПРИМОРЬЕ
(Gross Kuhren)

BALTIC SEA

Kaliningrad region
(Sambia Peninsula)

СИНЯВИНО
(Gross Hubnicken)

4 A
1

Jantarnji
ЯНТАРНЫЙ (Palmnicken)

3
2

5
ПОКРОВСКОЕ
(Sorgenau)

a

Figure 3.40. Map schematic of mining locations in 1997. A "Anna" mine, 1 Palmnicken (Yantarnyy) mine; 2 Primrose (Primorskoye) mine; 3 Plazhovaya or strand mine; 4 artifical lake and spill dam; 5 strand with pieces of amber. Reprinted with permission from Kosmowska-Ceranowicz 1995**

descriptions of the typical regional amber jewelry designs found in the collections of the Society of Latvian Culture at Riga (Latvju Raksti Omement Letton; Rauskas, 1944). The drawings illustrate the flat rings of amber used as a brooch and fine construction of dangling jewelry with amber beads linked to a flat amber spacer drilled with jump ring loops attaching smaller dangling pieces of amber. Often the flat pieces of amber were carved or incised with a circle and a dot in the center, reflecting the sun-worship symbols (see Fig. 3.39).

Amber Production in Kaliningrad from 1947 to 1997

During the period when the Kaliningrad Oblast was under the rule of the USSR, visitors were not allowed in the region as it was considered a militarized zone. Since decentralization, the area has been trying to organize itself for receiving international tourists and for international trade. Kostiashova (1997*a*, 44–5; 1997*b*), of the Kaliningrad region of Russia, reported on the amber production by the Russian Amber Company at Yantarnyy (the amber-bearing area known as Palmnicken prior to the war) at the *Baltic Amber and other Fossil Resins Symposium* held in Gdańsk in 1997.

In 1990 the Russian Amber Company employed about 2063 workers and extracted over 700 tons of amber. However, because of political changes, the company was about to become bankrupt. Production was halted. During 1993 and 1994, the cooperative was reorganized. Today, the company is a free joint-stock company operating under the name Russian Amber (*Russkij Jantar* in Russian).

In 1995 the Russian Amber Factory of Kaliningrad controlled the mining at Yantarnyy with open-pit operations at Primorskoye (Primrose), south of the previous German open-pit mine, and strand open-pit mining on the Baltic coast, south of the previous strand operation (see Fig. 3.40). At the Primorskoye open-pit mine the blue earth produces about 1 to 2 kilograms of amber per cubic meter.

Figure 3.41 shows the mining area in the Kaliningrad region of Russia and illustrates where the previous mines operated by the Königsberg Amber Works were located as well as where the current operating mines are located. Open-pit mining is conducted both at

Primorsky and on the strand location. The amber is considered to be in secondary deposits and occurs in several amber-bearing layers from Paleogene to Quaternary.

The amber production statistics of the extraction from the mines during the period from 1951 to 1994 are recorded in Table 3.1.

The Kaliningrad Amber Factory is currently the largest enterprise in the world manufacturing pressed amber as well as ornamental amber products. The factory has perfected the techniques for making pressed amber using small pieces of amber. They produce a fine-quality pressed amber by grinding clean amber into a fine powder that is placed in an autoclave where it is heated to about 350–370°C and made into pressed blocks and rods. This material can then be carved into art pieces, ornaments, or smoking products, such as pipe mouthpieces and cigarette holders.

Since 1947 when the factory was established, it has produced over 65 million jewelry items. When the Russian government took over in 1947, it did not value the amber that had been sold mostly in Palmnicken. At that time the workshop employed about 75 people working amber. The government experimented using amber in many experimental ways including chemical chemical treatment. However, there was little market because synthetic material was available. The enterprise was then transferred to the state. The production of pressed amber was the main output of the factory. The quantity of amber used for jewelry increased and ornamentation became the main product. Table 3.2 lists the quantity of amber measured by weight used in jewelry production.

During the1950s, production from the Kaliningrad factory included necklaces, rings, and pins, of both natural and pressed amber. In the late fifties, chess sets and animal souvenirs were made. In the 1960s the market declined, but it was revived as craftsmen began to emphasize the natural beauty and color of

Table 3.1 Amber Extraction by Kaliningrad Amber Company, 1951-1990

Year	Tons	Year	Tons	Year	Tons
1951	302.0	1964	428.8	1977	779.0
1952	271.0	1965	404.4	1978	741.0
1953	135.0	1966	411.6	1979	744.0
1954	229.0	1967	326.1	1980	761.4
1955	207.0	1968	541.0	1981	715.0
1956	275.6	1969	439.5	1982	723.0
1957	214.0	1970	453.6	1983	638.0
1958	171.7	1971	438.6	1984	656.2
1959	284.0	1972	452.0*	1985	732.0
1960	311.0	1973	506.3	1986	698.6
1961	347.6	1974	587.2	1987	581.0
1962	351.5	1975	682.0	1988	732.0
1963	361.5	1976	750.0	1990	760.0

Source: Data based on Z. Kostiashova, Russian Amber Company at Yantarnyy, Kaliningrad, Russia. Paper presented at symposium, Baltic Amber and Other Fossil Resins, September 1997, Gdańsk.

*Production only for the Plazhovaya mine (beach mine); data for other sections not available.

Table 3.2 Quantity of Amber Used in Jewelry Production

Year	Kilograms
1948	1166.0
1951	5171.0
1961	7803.0
1981	2075.7
1991	2223.0

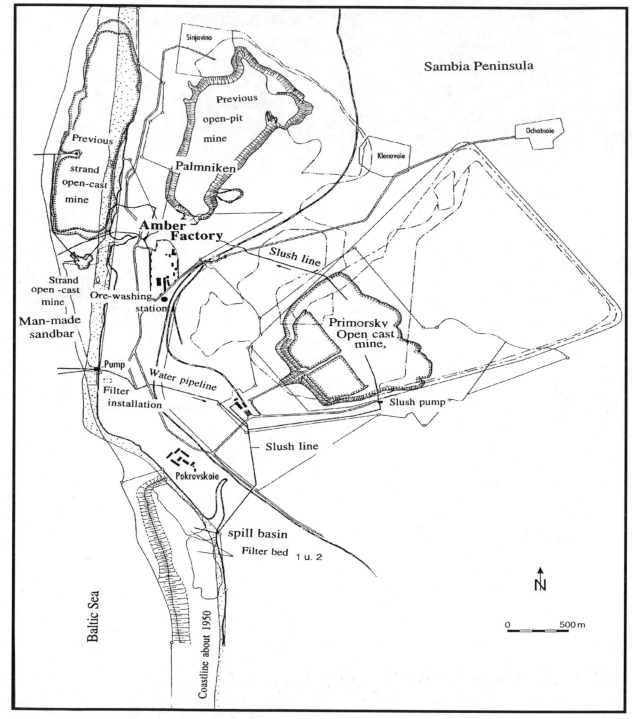

Figure 3.41. Schematic of amber mining company. Reprinted with permission from Slotta and Ganzolewski 1996, pg. 250.

amber. Experienced professionals, such as specialists from the Baltic republics and Leningrad/St. Petersburg, assisted in the more creative development of amber-processing. More individual works were encouraged rather than all factory-produced pieces. Amber and silver were used for novel ornaments. As an example, 1 million spiders of amber and silver were made at the factory. At this time 80 percent of production was made of pressed amber.

In 1962 production of natural amber jewelry and carved ornaments increased. Small natural amber nugget necklaces were made and amber was combined with wood. In 1968 the Russian factory produced pressed amber articles framed in gold—this contrasted with the work of the Polish craftsmen who preferred to frame their amber in silver. By 1982 production increased and over 100 different varieties of ornaments were included in the catalogue. With the change in government more emphasis was placed on the development of artistic creativity and amber-processing methods.

The Kaliningrad Amber Museum opened in 1979. Today it houses over 6000 amber artworks from carving to graphic displays portraying art of amber craftsmen. The museum exposition occupies 28 rooms of a reconstructed Teutonic Castle that covers an area of 1065 square meters. An important section of the exhibition is the *Amber in Art from 1700s to 1900s*. Some magnificent works of amber are now in the Kaliningrad Amber Museum. One such piece is the giant vase called *The Abundance*, which is almost 4 feet tall, made of mosaics of amber (Fig. 3.42). This exhibit includes an amber crucifix from the 1700s from St. Petersburg that was at one time in the Amber Room. Few items survive from the Amber Room. However, the museum featured in 1998 a display illustrating the current progress in restoration. Another fascinating portion of the collection is called *Amber in Modern Art* and focuses on the work of contemporary artists whose creations include amber.

In 1992 the Museum of the World Ocean in Kaliningrad, the main sea museum in Russia, established a company called Sea Venture Bureau, Ltd, whose purpose was to carry out scientific and business projects. One of its activities was to implement an Amber Program. Projects taken on by the Amber Program, as described by Andrey Krylov, Director, include:
- Creating a new data bank on amber geology in the Kaliningrad region, with the promotion of scientific studies of amber and methods of its exploration, extraction, and processing.
- Developing new technologies for extracting amber at sea as well as on shore; that is, drilling, hydroextraction.

Figure 3.42. "The Abundance" amber vase in the Amber Museum in Kaliningrad, Russia. Made of amber mosaic and chips. Sean Odenwalder, six foot tall, stands beside the amber vase for size comparison. The vase is about 4 feet or 1.22 meters tall Photo taken in 1998.

• Assisting the Museum of the World Occan to form a national scientific collection of biological inclusions in amber.

• Organizing joint business and commercial projects with the Kaliningrad Amber Plant. This state mining plant is the largest in the world, extracting about 700 tons of raw material yearly—more than 90 percent of world amber extraction.

• Producing amber jewelry, gift goods, souvenirs, carving caskets, and so forth and preparation of amber fossils.

• Engaging in commercial activities connected with fossils in amber and amber products (Krylov 1992)

By 2002 the plant employed 1,600 people, whose average wage was 700 rubles per month. Out of the six thousand inhabitants in Yantarnyy, Gediminas Pilaitis, reporter for Lietuvos Rytes, estimated at least one member from each family was in some way connected to the production of amber.

In 2004 the Lithuanian company, Masmedos prekyba, in cooperation with Lithuanian and foreign amber business and cultural institutions organized the first amber and jewelry exhibition, titled AMBER TRIP held in Vilnius, Lithuania. The AMBER TRIP was coordinated with the AMBERIF exhibition in Gdańsk so those interested in the amber world could participate in both seminars and conferences. The program's purpose was to develop cooperation, sharing of experiences and information among Lithuanian and foreign exhibitors and visitors as well as to develop new markets. The AMBER TRIP 2004 stated that traditionally Lithuania has manifested great interest in amber; via amber museums and galleries. Also, amber processing, souvenir, and ornament manufacturing companies are located in Lithuania. The brochure encouraged visitors to "take a close look at Lithuania, a land of unique traditions and distinctive culture, which has been described by the Roman historian, Cornelius Tacitus, as the motherland of amber as far back as the First Century AD."

Giedrius Guntorius, often known as the Lithuanian Amber Ambassador, was the organizer who encouraged the amber trade fair in Vilnius. He wanted to educate visitors about the artistic uses of amber and the process of refining raw amber. The Lithuanian decorative amber tradition involves letting the stone "speak more for itself." Guntorius stresses that the Lithuanian artists strive to work amber as little as possible and to keep its natural qualities intact. He maintains that it is important to perceive more than the simple monetary value of a piece of amber - the value is aesthetic -"It's about beauty." Lithuanian amber artists are often small craft retailers who produce some of the most treasured and exclusive amber articles, working from their homes as they have done for many generations. Guntorius stated that the amber trade fair was an excellent way to introduce Lithuanian amber culture to the world as Lithuania joins the European Union (see Fig. 3-43).   A visit to Vilnius and the amber museums and shops in 2006 proves amber is very highly valued as an art and materal for art objects.

Figure 3.43. Mariusz Gliviński and wife with author at their trade show booth "Amber Moda." Mariusz Gliwiriski is president of the International Amber Association 2004.

KALININGRAD AMBER FACTORY MINING OPERATION IN 1997

1. The Kaliningrad Amber Factory operates the mining enterprise, washing and sorting works, pressed-amber factory, jewelry factory, and an amber museum. This is the largest amber processing factory in the world and is located at Jantarnyy (Palmnicken), Kaliningrad Region, Russia. After closing in 1990, the factory reorganized, reopening in 1992. This region has only been open to visitors since 1990.

2a and b. The Primorskoye quarry is mined using an open-pit operation. At the new Jantarnyy (Palmnicken) mine, the blue earth amber-bearing layer is found at a depth of about 45 to 50 meters. These amber-bearing deposits from Upper Eocene and Lower Oligocene layers were accumulated in cycled bundles that were given local names, such as "wild earth," "blue earth," "white wall," and "green wall," by the German geologist, Zaddach, in the mid-1800s. The blue earth, where industrial quantities of amber are found, ranges from 0 to 17.5 meters. Often described as glauconite clay, it is interesting that the blue earth contains 3 percent glauconite, while the wild earth about 50 percent, and the green wall about 38 percent (Kharin 1995, 50).

3. *The Primorsky quarry water pipeline and slush basin. Russian geologists conducted geological surveys using borings (444 cores) at various locations at Primorsky quarry to determine the lithologico- stratigraphic profile of the deposits. They found that the blue earth glauconite contained traces of oxidation, indicating that it was brought from eroded old sediments. They describe the blue earth as being formed in three layers. The upper layer of the amber-bearing deposit, or layer I, of the blue earth is about 110 centimeters thick, with medium- and fine-grained sands, characterized by quartz carbonaceous-aleuritic aggregates, glauconite, feldspar, and mica. Heavy minerals such as muscovite, green biotite, and epidote-zoisite form a subfraction of the layer. A small quantity of amber is deposited in this layer.Layer II is about 280 centimeters thick and is represented by fine-grain aleuritic ooze containing increasing amounts of glauconite with black marine incrustations and fossils. Layer III is about 110 centimeters thick consisting of thin interbeds of large-grained sediments, fine grain sand, and aleurite fractions. This is the layer containing industrial quantities of amber.*

4. *Dr. Jan Burdukiewicz, Polish paleontologist and archeologist, and Dr. Siegfried Ritzkowski, German geologist, sample the blue earth in the bottom of the Primorsky quarry.*

5. *Primorsky quarry blue earth containing fragments of amber in situ being photographed by Dr. Anna Zobel, chemist from Trent University in Peterborough, Ontario, Canada. The amber lumps appear coated with a black grainy crust. Fragments of amber can be found enclosed in the center of sideritic concretions and when exposed to oxidation the amber is covered by a crust of iron-hydroxides.*

(Kharin 1995, 50)

6. *The Baltic Sea coastline and strand on the north-western shore near Yantarnyy (formerly Palmnicken), Russia. After storms, the sea tosses amber ashore, depositing it with the flotsam on the strand, as the sea washes against the exposed amber-bearing layer on the Baltic shelf.*

7. *Russian geologist from the Yantarnyy Amber Factory explaining the geologic layers of the cliffs along the strand. Amber is sometimes found in Quartenary deposits having been redeposited by the sea and also the blue earth layer is near the surface at the foot of the cliffs along the shore.*

8. *Geologists examining the geologic strata during the amber symposium tour to the Kaliningrad amber region. Dr. Barbara Kosmowska-Ceranowicz, geologist from Muzeum Ziemi, Warsaw, Poland, and Dr. Anna Zobel, chemist from Ontario, Canada, and Sean Odenwalder, student from San Diego, California, take a closer look at the exposed earth layers of the cliffs eroded by the Baltic Sea along the coast near Yantarnyy, Russia.*

9. *Exposed sand cliff jutting up from the Baltic Sea along the amber mining region near Yantarnyy with vegetation growing on the top. The region is covered with trees and shrubs in the unpopulated areas. Dr. Siegfried Ritzkowski, of Goetingen and his wife (bottom left), Dr. Barbara Kosmowska-Ceranowicz, of Warsaw {upper left), Sean Odenwalder, student from San Diego, California (upper right), and other participants of the symposium (foreground) collecting samples of the amber region shore exposures.*

10. *Open cast mining along the Baltic strand. This region of the strand was being eroded by the sea, so the environmental engineers decided to use power shovels and earthmovers to build a sandbar to stop the erosion. As the earth was removed to make the sandbar, the blue earth layer was exposed and amber-extracting became a secondary product. The mine is operated by the Yantarnyy Amber Factory.*

11. *Baltic Amber Symposium geologists and researchers touring the most recent amber extraction facilities in the Kaliningrad region. Slightly southwest of the Amber Factory operations, amber was found in the countryside by the peasants. The state determined that this area was not appropriate for large-scale mining because it would ruin the environment. The local residents found they could dig pits and reach the amber in the fields.*

Primitive holes in Muromskoye (wild exploration)

12. *A local amber miner digging through the earth layers to the blue earth to locate amber. Several pits, about 4 to 6 meters deep, were dug in the field. When one pit became exhausted of amber, the residents would dig another nearby.*

AMBER CRAFT DEVELOPMENT IN POLAND

During the fifteenth through eighteenth centuries, as previously described, masterpieces of sculptural art in amber were produced at craft centers in Gdańsk, Słupsk, and Elbląg, all located in what had been East Prussia before World War II. After 1945 the Free City of Danzig and portions of East Prussia were added to Poland and city names were changed to their Polish counterpart. In addition to the famous workshops previously mentioned, there were smaller amber workshops in Poland's coastal regions, often with equally talented artists practicing crafts of amber-working and wood-carving. These masters passed on their skills to understudies, yet the handwork gradually declined in the nineteenth and early twentieth centuries.

Amber sculptures and altar pieces in Polish churches were mostly destroyed during World War II, but today several museums in Poland contain historical exhibits of amber. The archaeological museums of Lódz and Gdańsk display collections of medieval amber products and amber artifacts from archaeological digs, with detailed descriptions for each period. A collection of artistically sculptured amber pieces from the sixteenth and seventeenth centuries, as well as modern carvings, is on exhibition in the Malbork Castle Museum. The largest scientific collection, however, is in the Muzeum Ziemi (Museum of the Earth) in Warsaw. The exhibit not only explains amber's formation, but also the history of its use from antiquity to the present. The region of Kurpie, Poland, where a cottage industry in amber has existed since medieval times, is represented by a collection of handmade articles in Lomża.

It was the folk artists who kept the traditions of the art of amber-working alive. Folk artists from Kurpie, Warmia, Masuria, and Kashubia, living in the region near the Narew River, remained familiar with quarrying and working amber. They preserved techniques of old time amber workers and local customs of using amber. According to Chętnik (1981, 31–8), the Polish ethnographer who researched the customs, life, and conditions of the Kurpian culture, from 1860 to 1880 villagers near Myszyniec used to spend more time working amber than farming. There were well-established workshops in Ostrołęka, Myszyniec, Nowogród, and in the area of Przasnysz and Lomża. Grabowska (1983, 29) states "One can detect in the products themselves and in the way they are used, [the artists'] admiration for the natural beauty of the material and faith in its mysterious power [to] 'bring relief in the various sufferings of man'."

Interestingly, the Kurpian amber workers classified amber according to where it was found. For example, ground amber was hard and occurred in large lumps in the ground; water amber was fished from rivers and lakes, and was often clear and well polished; mud amber was hard, with botanical debris embedded, and was often used for a component of incense; and hill amber was recovered from sandy knolls. The Kurpians highly prized reddish amber, whereas yellowish-white cabbage amber was not quite as fashionable in the Kurpie region (see Fig. 3.44).

Figure 3.44. Amber fishermen on the Narew River in Kurpie region, Poland. Courtesy Muzeum Ziemi, Warsaw, Poland.

Amber was found along the Narew River and its tributaries as the peasants farmed, often while ploughing or digging trenches. When searching actively for amber, the farming people knew how to recognize signs indicating where amber might be found. They would take a sample, usually by digging a 1 meter deep trench and if they found blue earth, they would begin work in that location. Recognizing the blue earth as the layer of green sand of marine origin where amber was most abundant, they dug down to find amber nests or kettles or sometimes amber stripes with concentrations of amber. The amber found was divided among the members of the group, which sometimes ranged between 10 and 50 people. During the nineteenth century, groups of diggers were hired by landowners or entrepreneurs who made large profits from the sale of amber. This exploitation of the land was completely without planning and arable land and forests were destroyed by digging pits as amber mines. Similar activities took place in Masuria and on the coast. While amber prospecting had been limited by a monopoly of the Prussian government, in 1836 the decree was withdrawn and the coast was thrown open to all. As early as 1820, Masurian mines had been leased to Sambian miners. However, in 1867 the land was reforested, and the pit mines were closed down. Amber was still recovered on private lands and this appears to have completely met the demand for the period.

Even with the rigorous prohibitions and regulations related to the use of amber, holiday costumes were decorated with golden beads and several lumps of amber were often presented to the church for incense. Juniper berries and pine needles were often added to small chips of amber to create a fragrant scent while burning amber incense. On Sundays the villagers would place a piece of amber or even a single amber bead in the collection plate. This offering was eagerly accepted by the priests, because they could sell the amber in a larger town, making a sizable profit. On holidays Kurpie peasants would bring their baskets of food—bread, eggs, and geese—including amber fragments and objects made from them to be blessed by the priests.

Kurpian peasants would build wayside chapels to commemorate an event that needed special blessings. For example, if a loved one was killed at a specific location, a wayside chapel would be built along the road in that exact location, or a chapel covered with a thatched roof would be built in front of the home to bring special blessings. The chapel for blessing the home would be smaller, as is the one built from wood with an amber figure of a peasant kneeling and the Holy Mother with her hand raised for blessing after the famous icon, *Matka Milosierdzia* or Holy Mother of Mercy. Figures of the Mother of Mercy similar to this are often found in wayside chapels to bring mercy or blessings to the person kneeling at the chapel (see Fig. 3.45). Obviously, to the Kurpian people, amber was more than just an attraction to the material itself, it also possessed powers to heal and protect.

In 1918 the government again attempted to nationalize the gathering of amber in the Kurpian region. The following decree was issued:

Sequestration of amber.

Natural amber and waste amber from processing must be registered with the Chief Administrative Officer of the county and the President of Police. Deposits of amber must also be registered, and this is compulsory not only for land owners, but also for all those who know of such deposits. All trade in amber is forbidden, especially "chain" trade. Raw amber will be purchased by the Chief Administrative Officer of the county or the President of Police at pre-war prices, payable by the state works in Königsberg. Breaking these regulations is punishable by a fine of up to 20,000 marks.

However, Grabowska (1983, 30–1) assures us that the "local people took good care of their treasure." She reports that during this period, inhabitants of Lipniki were rounded up to dig for amber on the land of the *soltys*, or village headman. In spite of information that there was amber on this land, several dozen Kurpian peasants worked all day without finding a single lump! She quotes them as saying, "If we dig up amber, then the Germans will wear us out with digging all the time, and take the amber."

Amber was so revered in the Kurpian region that it was presented to daughters as part of their wedding dowry. The wedding necklaces were supposed to be handcrafted by local amber workers, not simply bought from strangers nor "merchant amber," mass-produced in factories. Since the wedding necklaces were often kept for several generations, they became mellowed over time from body oils and oxidation. Gradually, the amber developed a glowing patina of darker orange to reddish color.

Kurpie girls and women of all ages wore a typical style of amber necklace, usually constructed with three strands of amber beads. The first strand was composed of small round slightly flattened or oval beads. The next two strands were made of larger beads, often cut in the faceted style. The beads were usually graduated with the very largest bead in the center. This bead, sometimes up to 3 centimeters in diameter, was ornamental and of clear amber (see Fig. 3.46).

Single-strand necklaces with special medallions called *pasyjki*, or passion pieces, in the center were very popular in Kurpie. The amber *pasyjki* were shaped in the form of a leaf with a carving of Christ on the cross and delicately etched with ornamentation of rosettes (see Fig. 3.47). These amber beads were worn by Kurpie women as their holiday costume accessories. Women from nearby Mazovia wore similar style amber beads. Not to be outdone, the young men of Masuria would decorate their holiday costumes with white amber on their shirts.

Grabowska (1983, 32) also reports that as late as "the 1960's Jozef Drewa from near Kartuzy was turning round, smoothly-polished beads, exactly the same as those once worn by fisherwomen from Hel Peninsula and the Vistula Sand Bar and which were threaded on thick thread to make amber rosaries. These were made from two types of amber—transparent and opaque—and in this way, the sets of ten prayer beads could be divided."

The family amber necklaces were so highly regarded that when there was no heiress to receive them, they were placed in the last owner's coffin. The women who died were buried in clothing decorated with amber beads. This has been the custom for many years, as

Figure 3.45. Wayside chapel (wooden) with amber figure of a peasant and one of Holy Mother with her hand raised praying for peace in Poland. The peasant carving is from semitransparent fatty amber. The Holy Mother figure is from opaque bone yellow amber. The wooden chapel and two carvings have the artist's mark and are dated 1974. Purchased in Toruń, Poland. Size: Mass of amber carvings: 45 g, wooden chapel 15 in. x 3.5 in. x 3.5 in., Amber carvings: Holy Mother figure: 5 cm x 3.4 cm x 3.4 cm, peasant figure: 7.7 cm x 3.4 cm x 2.2 cm. Patty Rice Collection.

indicated by finds of grave amber when digging in the cemetery at the Gothic church in Nowogród. These old amber beads also show the traditional shape and methods of processing amber in this region. Advice from Kurpie was that amber "should be kept in a cool and dark place, wrapped in a soft cloth, if it was to last for a long time" (Grabowska 1983, 32).

The peasants were not the only ones who desired Kurpian amber-workers' goods. Local gentry would order amber buttons for their Polish national dress. These workers also made special-order pipe stems and later cigarette holders. Drilling the long, thin hole in the items was very difficult. Often the amber would split and the work would have to begin again. The amber would be gently heated to create the slight bend needed for the pipe stem. Most amazing was the production of watch chains, carved from one large lump of clear amber. This required great skill as the individual links were carved interlaced with no joints. At this time, the technique for gluing amber was unknown. Beakers were made from single lumps of clear amber, while the whitish-yellow cabbage amber was most often used for snuffboxes.

Amber craftsmen from the village of Ostrołęka proudly presented the Polish author, Jósef Ingacy Kraszewski of Krakow, a set of writing tools made of amber on the jubilee of his career as a writer. The creation of these amber tools was remarkable since it was all done by hand with primitive tools. The processing first called for tumbling the amber in water containing sand to remove the external crust. The piece was then shaped using various types of knives as scrapers or as a saw. Next a hole was drilled using a "drill made up of a vertical axis with a wooden circle affixed, linked at the top of the axis with strong string. The instrument was worked with a crosswise spar moved backwards and forwards by hand, making it possible to drill a hole using a steel gimlet fastened to the end of the wooden rod" (Grabowska 1983, 32).

Figure 3.46. Kurpie woman wearing a single strand of large amber beads in Torun, Poland, 1982.

Various types of lathes for turning the amber were described as "hand-worked string lathes," "rod drills" (used until the 1880s), and even the wooden spinning wheel, primarily used for spinning wool. The amber would be attached to the wheel so that it would spin as the wheel was turned. The craftsman would hold a piece of lamp glass in his hand to shape the amber while it was turning. When the amber was the correct shape, it was smoothed again with glass, then polished with a soft sheepskin cloth. Chalk or amber dust produced during turning was used as a polish. Last of all the piece was rubbed with a soft cloth (Grabowska 1983, 32).

In the Kurpie region, amber was used for barter and was considered a reliable means of payment. When the peasants dug amber from their own land, it was put aside for a rainy day, because when money lost its value amber could be used to secure goods for the family. When going to town, a person from Kurpie would always have a few lumps of amber to trade with the merchants. However, in the government stores money was needed to pay for items purchased.

Kurpie is the region in Poland where the magical and healing powers of amber continued to be believed. On St. John's Eve special herbs were gathered to mix with amber dust and small fragments that were burned to fumigate the room of the sick. It was believed all types of illness would improve by inhaling the resinous aroma of burning amber.

Believing in amber's power to protect, mothers would suspend a large amber ring over their babies' cribs or play areas, to protect them from spells or convulsions. Also the glistening amber ring was a useful toy to keep the baby amused or to bite on when teething, thus babies were more calm. When amber was given at a wedding, it assured fertility of the bride and groom. Another Kurpian folk custom was to wear amber earrings to cure headaches or even eye complaints. A traditional saying, often heard when an old Kurp died, was "he had gone 'to the other world to dig for amber.' Thus, the treasure of his land accompanied him from cradle to grave" (Grabowska 1983, 33; see Fig. 3.48).

Supplies of natural amber in Poland are not as plentiful as in the Sambian peninsula, however, research has revealed that the amber-bearing Sambian Delta extends into Polish territory. Kosmowska-Ceranowicz, (1996a, 4) reports extraction of Sambian deposits began taking place on a large scale in the mid-nineteenth century beginning

Figure 3.47. Kurpie amber necklace with extended beads and medallion pendant. Photographed in Muzeum Ziemi, Warsaw, Poland, by author.

*Figure 3.48.
Beachcombers in the past
recovering amber from the
Baltic shores in Poland
about 1980. Courtesy
Muzeum Ziemi, Warsaw,
Poland.*

in the form of dredging in the Courish Lagoon as described previously. The map of Poland shows regions where amber is generally found (see Fig. 3.49).. Kosmowska-Ceranowicz reports core samples taken near Chłapowo, northeast of Gdańsk, were assessed to estimate the Polish deposits at around 643,820 tons, with individual layers containing from 132 to 5976.5 grams per cubic meter. The greatest difficulty is that these Polish deposits lie at a depth of 120 meters, making extraction at the present time impossible. Yet there are still alluvial deposits and glacial nests that continue to be exploited. The industry is regulated by the state with collection of raw amber controlled by a committee of the National Council of Gdańsk, which grants licenses to individuals to mine amber near Gdańsk. In the 1980s amber was washed out of the earth using large, high-pressure water hoses to inject water into amber-bearing layers, forcing the amber lumps to the surface. Unfortunately, this method disrupted the ground, causing much controversy between environmentalists and amber authorities. (See Photo Documentation of Hydraulic Extraction of Amber at Wislinka Outside Gdańsk in 1997 where the only approved extraction is in an experimental stage.)

In the vicinity of Gdańsk, some small firms and cooperatives wash layers of sand to recover amber from the coastal area. One small firm outside Gdańsk turns over about 8 tons of raw amber per year by using a work-force of 15 miners and processors. Such a firm may supplement its own pro-duction by importing raw amber from Kaliningrad on the Russian side of the bor-der.

In addition to the private firms produc-ing amber goods, while under Soviet control, Gdańsk was also the loca-tion of the State Amber Works or *Państwawa Wytwórnia Wyrobów Bursztyowych*. This firm produced natural amber jewelry, including numer-ous items for export.

Figure 3.49. Map of Poland showing regions where amber is generally found. Map locates some of the finds of amber and former amber mines. Sites where the presence of amber has been recorded are marked as follows: a, A1-Amber found in Paleogene sediments; b, B1-Amber found in ice flows containing Tertiary sed-iments laid down during the Pleistocene. The hatched area along the Baltic coast indicates areas where the greatest concen-trations of amber from Holocene deposits occur. Reprinted from Kosmowska-Ceranowicz and Pietrzak 1982, Kosmowska-Ceranowicz 1986***, etc.

Muslim praycr beads were exported to countries in the Near East. Pressed amber necklaces, bracelets, and pins, as well as large statuettes of artificial amber were also produced for export to countries throughout the world. Many of these souvenirs are purchased by tourists who do not want to spend large sums for natural amber, but want a memento of Amberland. Many facto-ries also manufacture artificial amber articles, using synthetic resin to embed small natural amber chips in a product called "polybern." Polybern beads, bracelets, brooches, vases, and sculptures are not only sold to tourists in Poland but are also exported. By the time the items are sold in the United States, dealers often fail to explain that they are synthetic, but simply tell uninformed buyers that the beads are amber. Numerous types of pressed amber products are also made and exported. American Indians often buy pressed amber rods to carve fetishes.

A new mechanized workshop for processing amber has been in operation since 1939 at Słupsk, in the location of the old Stolp Guild, which closed its doors in 1883. In Sopot, the government-controlled wholesale export amber firm, the Art-Region Cooperative, employed artists and craftsmen in a factory and also farmed out work to many home pieceworkers. This firm also had a factory at Wrzeszcz, a suburb of Gdańsk (Szejnert 1977).

While under Communist control, all direct export of amber goods from Poland was done by Coopexim-Cepelia. Craftsmen in amber-working cooperatives near the Baltic all produced essentially similar products for export, channeling their articles through Cepelia. Tourists were requested to buy amber only from these government shops where an 80 percent duty was added to the price. No raw amber could be taken from Poland without official certification by a museum that the material was for scientific purposes. The small shops and independent artists often produced much more creative and artistic amber articles than did the mass-production factories. These artistic pieces could be purchased by foreign tourists, then appraised by government officials, who levied duties to be paid upon leaving the country.

Political restrictions were imposed by nationalizing the amber industry between 1949 and 1956. Some of the amber-working equipment and workers from the liquidated amber plants were taken over by the state-owned amber-working factory, which employed over a hundred people and imported a regular supply of 3000 kilograms of amber from Sambia each year. After 1956 political restrictions were slightly lifted and a number of private factories began producing amber goods (Gierłowski 1997, 47–9).

By 1970 there were over 400 workshops in the Gdańsk region and over 50 in the rest of Poland. However, the authorities did not allow larger private companies to develop. Individual artists and craftsmen had to have designs approved by state-appointed commissions. Thus employment in the amber industry totaled only about 600 workers. In 1968 development operations at the port of Gdańsk began extracting about 20 to 40 tons of amber annually from the Vistula estuary, thus, supplies of raw material from Sambia were not needed.

From 1971 to 1989, Polish authorities permitted amber-working factories to use hired labor, thus small companies could expand and provide amber goods for export. However, up until the fall of socialism, goods being exported were controlled by the state-owned firm. Gradually private producers of artistic handicrafts were allowed to increase their share of foreign currencies earned by exporting their goods. Between 1990 and 1996 radical changes took place in Poland's economy, which allowed the workshops to become involved in trade on the world amber market. In fact, Polish amber companies became the leading suppliers of ready-made amber goods. In 1997

Figure 3.50. Charts illustrating amber workers growth in Gdańsk and sources of raw material and export: a, World exports; b, Sources of raw amber for Gdańsk artists; c, Amber workshops in Gdańsk. Courtesy Gierlowski 1997. Later published in Gierlowski 1999.

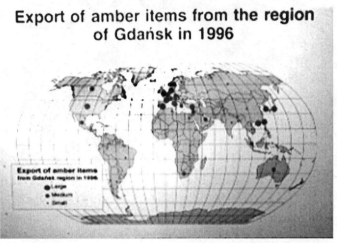

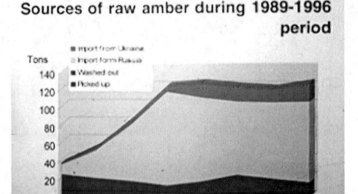

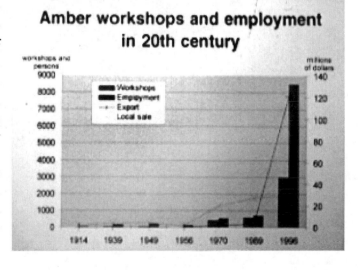

Poland exported about 70 percent of the finished natural amber jewelry on the world market, valued at about $300,000,000. About 80 percent of the companies exporting amber are located in the Gdańsk region. In the city of Gdańsk there are over 300 shops selling amber goods (see Table 3.3).

Wiesław Gierlowski, of the Polish Amber Workers union, Stowarzyszenie Bursztynnikow w Polsce, reported on the growth in amber-working shops, exports of amber, and resources for raw materials at the 1997 Baltic Amber Symposium indicating that Poland uses raw amber both from Poland resources extracted by picking up from the shores and by washing out using hydraulic methods (see Fig. 3.50). However beginning in 1990 or 1991, Polish amber workers became more dependent on importing amber supplies from the Russian Amber Factory. After 1992 almost four-fifths of the raw material was imported. As the number of amber workers grew, the amount of raw material needed increased. This increased demand was satisfied by importing amber, not only from Russia, but also from the Ukraine.

One fine collection of unique creative amber sculptures, purchased from a variety of individual Polish artists during the years from 1955 to 1986, is the Burgess-Jastak Collection,

Figure 3.51. Baltic amber carved bust of a Polish gentry or szlachta. Made in Gdańsk, Poland. Size: Total ht. 8 cm, wt. 140 g. Courtesy Burgess-Jastak Collection.

now belonging to the University of Delaware Gallery. The amber sculptures in the collection include twentieth century busts of characters from the medieval ages, such as merchants in the Hanseatic League, scholars of the past, figures from folklore, religious scenes, and sculptures depicting mythology figures. All are unique examples of twentieth-century amber carvings and illustrate that amber-working did not die out in Poland but simply changed in form.

A typical Baltic amber carving in the collection depicts a bust of an old man wearing a traditional headpiece representative of Polish nobility or *szlachta*. The bust is carved with detailed facial features, hair and beard, and flowing folds in the hat. The workmanship on the amber illustrates the craftsman's expert knowledge of techniques for bringing out the highest polish and still keeping the characteristic natural amber markings, that is, indentations and patches of crust on the back and base of the sculpture. The head is carved of one piece of yellow translucent amber with mottling of white opaque amber, with the base being a separate piece of amber joined to support the carving. The mixture of whitish opaque amber to translucent amber is called cabbage amber (see Fig. 3.51).

Another twentieth-century bust is an example of an Aesop Fable character, depicting a blustery old man with a detailed curly beard, wind-blown hair, and delicately carved facial features. The figure clearly portrays the force of the wind in the fable of the *Sun and the Wind*. The head of the figure is carved from yellow opaque amber with the artist effectively leaving orange compacted crust on the back of the piece. The artist has interestingly attached the head to a base of translucent yellow-orange amber leaving natural crust indentations on polished surfaces (see Fig. 3.52).

Poland's history is a recurring theme in the Polish artists' work. This is illustrated in the Burgess-Jastak Collection by one of the busts that portrays Queen Jadwiga (1373–1399), who brought religion and learning within reach of the people. For five centuries, Lithuanians and Poles have regarded Jadwiga as their greatest queen. Even though she was very young, she was skilled at negotiating and peacemaking. She married Jagiełło, Grand Duke of Lithuania, and united the two countries when he was elected king of Poland. The 5 inch tall amber relief carving is exquisitely done, complete with a crown, long flowing hair, facial features, and folding details on the bust. The hue of the amber is pale yellow to mottled yellow ranging from translucent to opaque. This relief carving was created from a rather flat slab of amber since the finished carving is about 5 inches (12 cm) tall, 2.5 inches (6.5 cm) wide, but only 1 inch (2 cm) thick. The artist, in a naturalistic manner, left bark or crust markings on the back of the amber piece (see Fig. 3.53).

Nestled among the many amber jewelry shops lining St. Mary's Street in Gdańsk, Poland, today is the workshop of Zdzislaw Kycler, who specializes in amber jewelry and carvings. Amber pendants are set in fused silver with hand-crafted chains. Carvings are done in typical Polish style with natural features of the crust included in the composition of the sculpture. His love for the age-old gem radiates through the fine work of his sculptures. The Burgess-Jastak Collection contains many fine sculptures from the Zdzislaw Kycler workshop. One such amber sculpture, purchased in 1989, is *Atlas, Holding the World on His Shoulders*. This figure, 16.4 centimeters tall, is constructed of three amber pieces varying from white opaque to translucent yellow amber. The natural black lignite markings add to the earthy appearance of the base. The carved Atlas is three-dimensional, translucent yellow, butterscotch amber bending with the burden of the world on his shoulders—the world being a sphere of transparent yellow amber with brown markings engraved in the shape of the continents. The sculpture weighs 63.5 grams (see Fig. 3.54).

An illustration of the religious themes depicted in amber by contemporary Polish

Figure 3.52. "Sun and the Wind" figure. The head of the figure is yellow opaque amber with orange compacted crust on the back. Made in Gdańsk, Poland. Size: Ht. 10 cm, wt. 474.5 g. Courtesy Burgess-Jastak Collection.

Table 3.3 Production Statistics for the Gdańsk Region

Year of Production	Number of Employees	Raw Material(tons)
1914	20	60
1939	40	12
1949	60	10
1956	7	6
1970	400	25
1989	600	40
1996	3000	193

Source: Data from W. Gierlowski, Gdańsk's Craftsmen, Workshops and the Management of Amber Resources in the Twentieth Century. In Baltic Amber and Other Fossil Resins, Barbara Kosmowska-Ceranowicz, ed. (Warsaw: Muzeum Ziemi), 1997: 47-9.

Figure 3.53. Baltic amber relief bust of "Queen Jadwiga". Made in Gdańsk, Poland. Size: Ht. 12.3 cm., wt. 110 g. Courtesy Burgess-Jastak Collection.

artists is the work of work of Stanisław Podwysocki, of Gdańsk. His amber sculpture of the *Pieta*, replicating Michelangelo's *Pieta* in Rome, was purchased in 1992. The sculptor created the *Pieta* with detailed workmanship, producing a high polish and smooth flowing lines while carving a single lump of translucent creamy amber. The orange crust markings were carefully left on the back of the piece to preserve the naturalness of the material. This exquisitely detailed sculpture is very rare for current work being carved of one piece of amber, since the finished piece weighs over 525 grams and stands 5.5 inches tall (see Fig. 3.55).

Another fascinating piece in the collection depicts *Adam and Eve in the Garden of Eden*, carved in deep relief with piercing making the figures three-dimensional. The artist carved elaborate details, defining leaves, branches, and apples on the Tree of Life with a serpent winding around the tree. Eve stands picking an apple, while Adam sits on a low stool holding an apple. The entire sculpture is from one lump of opaque light yellow amber, highly polished, but retaining the compact crust with a slight polish on the back of the carving. The amber sculpture is set on a fused sterling silver base. For all the intricate pierced carving and three-dimensional effect, this piece is rather small, the amber figure being only about 4 inches tall, 3.25 inches wide, and 1.25 inches thick. The figure is mounted on a fused sterling silver base. This amber treasure was purchased in 1960 in Gdańsk, Poland (see Fig. 3.56).

Artists continue to produce amber art in the classical style, using Greek and Roman gods and goddesses and myths as themes for their art. During the Neoclassical period of art, artists were encouraged to return to classical themes. Amber artists in the mid-1700s were also including classical themes in their sculptures and decorations of the caskets, and compotes. Gdańsk artists of today continue this theme in elaborate sculptures as seen in the sculpture of the *Three Graces*. This piece shows that amber artists are not only experts in the working of amber but are skilled silversmiths as well. The figures are carved of cabbage, mottled yellow amber and given a high polish. The amber sculpture was then designed into a sterling silver frame and pedestal. The amber figures were carved from a large cobblestone of amber resulting in a finished piece that stands 5.75 inches tall, 5.25 inches across, and only 1.5 inches thick (see Fig.

Figure 3.55. Baltic amber carving of "The Pieta:" a, Front view; b, Side view. This exquisitely detailed carving is constructed using a translucent cream-colored single lump of amber. Carved by Stanislaw Podwysocki of Gdańsk, Poland. Size:13.8 cm x 6.5 cm x 14.9 cm, b. wt.5 25 g. Courtesy Burgess-Jastak Collection.

3.57).

Figure 3.54. Baltic amber figure of "Atlas, Holding the World on His Shoulders" constructed of three amber pieces, varying from white opaque to translucent yellow amber. The base has natural black lignite inclusions. Atlas is three-dimensional, translucent yellow, butterscotch amber, the sphere is of transparent yellow amber with brown markings engraved in the shape of the continents. Made in Gdańsk, Poland. Size: Ht. 16.4 cm, wt. 63.5 g Courtesy Burgess-Jastak Collection.

Figure 3.57. Sculpture of the "Three Graces" constructed of a silver base with deep relief amber figures. Purchased in Gdańsk, Poland, 1987. Size: 14.4 cm x 13.2 cm x 3.6 cm, wt. 175 g. Courtesy Burgess-Jastak Collection.

Figure 3.56. "Tree of Life" carving of Baltic amber, carved in deep relief with piercing that makes the figures appear three-dimensional. Purchased in Gdańsk, Poland. Size: 7.7 cm x 8.75 cm x 3 cm, wt. 122 g. Courtesy Burgess-Jastak Collection.

SPECIAL PHOTO DOCUMENTATION

HYDRAULIC EXTRACTION OF AMBER AT WISLINKA OUTSIDE GDAŃSK IN 1997

1. One firm in Gdańsk is experimenting with the controlled extraction of amber in the region of Wiślinka in the countryside near Gdańsk. The area in a field near a stream was bulldozed to remove topsoil and form a catch basin for the flooding of water during the extraction process. Amber is washed from the earth using high-pressure water hoses to inject water into amber-bearing layers, thereby forcing amber lumps to the surface. Participants from the Baltic Amber Symposium held at the Gdańsk Archaeological Museum in 1997 observe preparations for the extraction of amber.

2. A high-pressure pump is used to pump water from a nearby stream.

3. A large hose is attached to the high-pressure pump to bring a pressurized force of water to the mining extraction tool, which consists of a pole or rod with a head that is also attached to the other end of the 3 inch diameter hose.

4. The worker controls the pressurized hose by means of the 18 foot long metal pole. The area where the extraction is going to take place is framed with an aluminum guard forming a circle with an opening on one side. The circle allows the water with the debris to flow out at the opening. The opening is covered with a net to catch the large chunks of material raised by the force of the water.

144

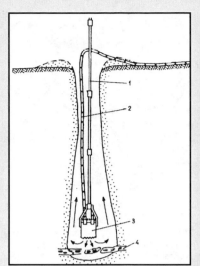

5. *The long pole is raised forcing the water pressure straight into the ground. The worker pushes the pole down deeply into the sediments and mud. The water force helps to dig the pole and hose into the ground. (See diagram at right, based on a drawing from Kosmowska-Ceranowicz: 1, aluminum shaft about 18 feet long; 2, water hose; 3, head piece; 4, sediment layer with amber.)*

6. *The water pressure forces debris and amber lumps to the surface and water begins to flow through the opening in the circular fence. Workers keep the net area aligned so it catches any large piece of material. A water hose is used to spray the net to rid it of any excess mud. When the net is full, it is dumped on the shore and workers search through the debris to find the amber lumps.*

7. *At Wislinka, two areas can be worked at one time because the pump is powerful enough to supply two hoses.*

8. *Workers have to wade into muddy water 2 to 2.5 feet deep so they wear tall rubber boots.*

9. *Baltic Amber Symposium participants searching through the "net catch" to find amber nuggets.*

THE AMBER BELT OF DENMARK, SWEDEN, AND THE FRISIAN ISLANDS

Along the west coast of Jutland near Oksby, after stormy nights, just as the early morning tide ebbs from the dunes, villagers still comb the beaches looking for amber lumps. This is the present-day amber belt of Denmark, traversing the coastal region from the German border to the northern tip of Jutland, with the most abundant strand deposits from Fanö in the south to Nissum Bedning in the north. Amber collectors of all ages, carrying lanterns to light their way, may be seen in the predawn darkness searching the strand. On a good morning, that is, after a severe storm, there may be as many as a hundred collectors assembled from the nearby towns. As the generally small pebbles of amber are found, they are tucked safely away into leather pouches tied to the collectors' waists. Occasionally, a lucky searcher may discover a fist-size piece. However, even after a storm, the total collection combined from several searchers in the Oksby region might only amount to 22 pounds (10 kg). Most of these amber-collectors are amateurs, collecting amber for their own small-scale uses. Yet a few native amber craftsmen make a living from collecting their own amber and cutting and selling amber jewelry. So little native amber is available that exports from Denmark to

Figure 3.58a. Collectors searching the strand for amber on the Baltic coast.. Photo by Leif Brost, Swedish Amber Museum.

Figure 3.58b. Leif Brost with barrel-shaped lump of colophony fished from the sea by a trawler. Photo provided by Leif Brost. Courtesy Geoscience Press.

the United States consist of about 80 percent raw material imported from Poland, Kaliningrad, Lithuania, and Germany. However, recent reports by Leif Brost, owner of the Swedish Amber Museum in Höllviken, Vellinge District, the sea annually washes up approximately 30 kilograms of amber along the coastline of Skane, the southwest region of Sweden, and further inland along the Alnarps River (Dahlström and Brost 1996, 48; see Fig. 3.58b).

Interestingly, beachcombing along the Baltic also produces recent resins or even pieces of subfossil resin. These lumps are like amber in appearance and are lightweight, yellow, and feel different than other stones on the beach. Though some call them "young amber," they also are called "colophony," or even "copal." The largest piece of resin, surprisingly in the shape of a barrel, was fished out of the Baltic in 1988 by a trawler. To their amazement the barrel-shaped resin lump weighed 238 kilograms. In order to determine if this large lump was "copal," a resin often used in manufacture of varnish, a small fragment of the nodule was submitted for detailed analysis and results were reported by Stout, Beck, and Kosmowska-Ceranowicz in 1993 (130–148) as resin from perhaps a varnish industry from as early as the Middle Ages or even up to the early twentieth century. The boulder was purchased by Leif Brost. It was split into four parts in transporting and is now located in the Swedish Amber Museum in Höllviken, Kampinge, Sweden (see Fig. 3.58). It was also examined in the Quaternary Geology Laboratory at Lund University using carbon-14 testing, which indicated the resin was "from anywhere between the seventeenth century to 1950." (Dahlström and Brost 1996, 108).

Historically, the west coast of Denmark, as well as adjacent

German lands, comprise the region where the earliest Bronze Age amber trade route to the Mediterranean originated. The earliest mention of this region seems to be Pliny's statement: "the Gutones, a people in Germany, inhabit the shores of an estuary . . . one day's sail from this territory is the isle of Abalus upon the shores of which amber is thrown by the waves in the spring . . . the inhabitants use it for fuel and sell it to their neighbors, the Teutons." Some historians believe that Helgoland, the only island in the North Sea a day's sail from the German coast, is the classical Isle of Abalus. Furthermore the Eridanus River, referred to by classical writers, is thought to be the Elbe, which empties into the North Sea in this region. The Elektrides, or amber islands mentioned by the early Roman writers, were perhaps the Frisian Islands (Ley 1951, 11).

During the 1800s amber was more abundant in this region, for example, approximately 3000 pounds (1360 kg) of amber were collected per year along the North Sea coast of Denmark. The Islands of Römö, Sylt, and Fanö produced amber, with some lumps weighing several pounds each. From 1822 to 1825, which included many stormy, productive years, a Danish merchant reported collecting 686 pounds (311 kg) of amber along the shore near Ringkjobing, mid-west of Jutland adjacent to the North Sea (Ley 1951, 11). Just across the narrow opening to the Baltic Sea, the southwest promontory of Scandinavia—including the stretch from Copenhagen to Malmo, Sweden—provides minor amounts of amber that can be picked up along the coast and in the waters of the Baltic after storms. Small nodules are most often found, but these are of good quality and are put to use by local artists. An account of Scandinavian Baltic amber by Dahlström and Brost described Sweden's reserves located on its southeastern coast near Skane with the primary sites for amber finds located in Kampinge cove and the Falsterbo peninsula, in southern Sweden. The old amber beaches between Kaseberga and Trelleborg in the south, Mille and Landskrana in the west, and the east coast receive only small pieces tossed upon the strand from the sea (Dahlström and Brost 1996, 51). In 2005 communication with Brost, he states that most amber in Denmark occurs in the western part around the tip named Blavandshuk, just north of Fanö, but a small amount can be found almost anywhere in Denmark. The Limfjorden in north Jutland is also a good place to find amber. He states among the sand removed during construction in Malmoe harbor on the Danish side of Oresund across from Falsterbo, Sweden, quite a lot of amber was found.

The Scandin-avian scientist who has done the most in-depth research on Denmark's amber is Sven Larsson, whose 1978 work

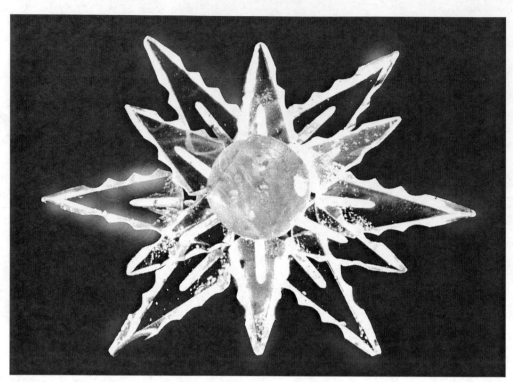

Figure 3.59. Transparent yellow Baltic amber starburst pin made by German lapidaries. Size: 2.5 in x 2.75 in or 6.3 cm x 7 cm. Patty Rice Collection.

reports scientific studies done on amber found in various locations in Jutland. Interestingly, he has found that amber recovered from the area has not all been produced by the same source tree. For example, the oldest Tertiary amber known in the Danish region was found deposited in *cementstone* in the North Jutland mo-clay, a marine deposit consisting of diatomaceous shells. He gives a complete description of its formation taking place during the time most of Denmark was part of the North Sea. Sending his amber samples for testing by infrared spectroscope by Beck in 1975, it was found that the resin of the mo-clay was not succinite, or so-called Baltic amber, but belonged in the retinite group of resins and was not exuded by the same trees that produced Baltic amber (see chapter 6). Larsson (1978, 32) states: "This investigation indicates that the problems of the botanical taxonomy of the amber forest are far from finally solved, at least as far as the western Baltic areas are concerned."

The Zoological Museum of Copenhagen contains a large collection of insect-bearing amber specimens, many of which were collected from the shores of Denmark. This comprises one of the largest collections outside of Poland and Lithuania. Larsson's 1978 book, *Baltic Amber: A Palaeobiological Study*, is based on the study of this extensive collection of amber fossils and summarizes the scientific work in amber for the past 150 years, giving a complete chemical and geologic description. Today the museum collection comprises more than 8000 specimens, including over 500 spiders, and is reported to have originated almost exclusively from the Danish coasts. However, about 700 pieces are Baltic amber from the Prussian era. Many of

these are very old and have aged from being exposed to air in various museums over generations.

Amber jewelry, worn by the earliest inhabitants of Denmark, is found in peat bogs buried with the dead as a sacrifice. Archeological pieces are among the treasures in the Swedish Amber Museum in Kämpinge, and they are often formed to represent symbolic figures significant to the ancients' beliefs.

Even though Denmark has had such a long history of amber influencing the lives of its inhabitants, today only a few local artists produce amber on a scale large enough for export. Scandinavian artists set amber in either gold or silver mountings in typical Scandinavian styles.

GERMANY AMBER LAPIDARIES

Beginning in the Middle Ages, skilled amber craftsmen established themselves in Germany away from the main source of raw amber, and this ancient industry still flourishes. Today, German lapidaries manufacture fashionable amber beads and other goods, which are characterized by fine quality and exquisitely fine finish (see Fig. 3.59). Many import-export firms deal exclusively in natural Baltic amber, and large quantities of amber are imported from Poland, Russia, and Lithuania to provide raw material for the several firms scattered throughout the country and to the long-famous amber lapidaries located in Idar-Oberstein, the well-known center of the German gem industry.

Although many amber necklaces are made from tumbled amber pieces, German factories still produce hand-finished beads in round and olive shapes or faceted pendants and faceted beads. The latter, however, are becoming much more difficult to obtain than in the past because the skilled handwork necessary in their production sharply increases costs. Amber specialists, such as Otto May and Gunther Herrling (Westfalica Amber Jewellery) mount amber in gold settings as well as in the typical silver mountings often used in Poland, Lithuania, and Russia.

German craftsmen continue old traditions of carving amber, but on a lesser scale than in the past. Currently, the market is open to more competition from Poland, Russia, and Lithuania manufacturers. Although competition is greater, customers are becoming more aware of amber jewelry, and this new awareness will increase the demand for the fine quality, elegant amber jewelry produced by German industries. Miniature ornaments in the form of tiny animals and flowers carved in fine detail that require painstaking work on the part of the artist is one German specialty. Also large lumps are carved in a manner that uses each piece to its best advantage, as large

a

b

Figure 3.60. a, Compote style amber mosaic vase made by Russian restoration artist at Catherine's Palace Amber room. b, Jewelry chest made with thin plates of amber using the "incrustation" technique and constructed by following drawings of 18th century objects.

pieces of raw amber are now scarce.

Berlin's Natural History Museum houses one of the largest single lumps of amber ever unearthed in East Prussia. This fine-colored specimen weighs over 21 pounds (9.5 kg) and was found in 1860. In 1958 it was valued at over $4200, and it is much more valuable today.

Amber carvings and amber-veneered or encrustation objects made during the fifteenth, sixteenth, and seventeenth centuries can be found in museums and churches throughout Germany. Museums feature cabinets encrusted with amber (see Fig. 3.60).

Stuttgart's Staatliches Museum für Naturkunde houses the largest piece of amber in the world that was actually found in Malaysia. On December 3, 1991, a piece of amber weighing more than 150 pounds was discovered from the Miocene Nyaluau Formation of Sarawak, Malaysia. The piece actually had to be sawed into several sections to transport it to the Museum für Naturkunde where it is on display (see section on Asian amber in Chapter 7). The Stuttgart museum contains the most comprehensive collection of amber from around the world, including a large collection of fossiliferous pieces from the Baltic as well as from the Dominican Republic and Lebanon. Dieter Schlee, the scientist who curated the collection in Stuttgart, has published numerous reports on amber in nature and has conducted in-depth scientific studies on fossil insects and the composition of Dominican amber and Lebanese amber, as well as studies of Baltic amber (Grimaldi 1996a, 40).

Institut für Palaontologie, Göttingen, Germany. now houses a portion of the famous Königsberg collection of fossil insects in amber collected by Klebs in the last half of the nineteenth century. In 1890, Conwentz produced botanical monographs of this collection, which illustrate the flora and fauna with detailed drawings. Much of this collection was destroyed by bombing raids during World War II; however,

the collection was divided up and put in various locations for safe-keeping. Now that Germany is reunited, this collection is again available for today's scientists to study.

ENGLISH AMBER

Amber occurs sparingly at many localities along the east coast of England, but mainly along the north seashores of Norfolk and Suffolk. Seaside resorts in Aldeburgh, Cromer, Felixstowe, Great Yarmouth, Lowestoft, and Sathwold specialize in amber products since they are located along the area where amber is occasionally found. Local shops carry fine amber jewelry imported from Germany, Denmark, Poland, and Russia, as well as handcrafted pieces of local origin.

English amber is usually golden or cloudy yellow and generally occurs in small lumps, although large pieces of up to 2 to 3 ounces have been reported. English amber is classed as succinite similar to Baltic amber, but it commonly contains shells and reveals different types of flora and fauna than those of the amber from Sambia. Its source is not known, but presumably it has washed from a Tertiary bed now submerged under the North Sea.

Amber artifacts found in prehistoric graves in England evoked disagreement among scholars as to whether the amber originated in England before it was worked or was imported from the continent. Since English and Baltic amber both contain succinic acid, one can only speculate whether Stone Age beads and ornaments came from English or Baltic shores. Most Stone Age sites in England are in Wessex, where the Wessex Culture left evidences of its existence in the remains of their now-famous monument Stonehenge. The barrows, or burial mounds, of this period often contain gold, ivory, and amber. Being sun worshippers, these ancient people believed amber was related to the sun, and for this reason, amber pendants set in a gold circle were worn as solar amulets (Hunger 1977, 22).

Some of the oldest British amber artifacts which date to as early as 10,000 B.C. were found at Gough's cave in Cheddar and other from about 8,000 B.C. in Star Carr, Yorkshire. Archeologists believe the raw amber from which these artifacts were made originated from the English coast (Ross 1998b, 19).

One of the most archeologically significant amber artifacts found in a Bronze Age tomb in Dorchester is an amber cup of a reddish golden color (see Fig. 3.61). The cup was formed from one piece of amber. The surface is smooth and seems to have been turned on a lathe, but its solid handle was carved from the original piece of amber. Grimaldi (1996, 147) indicates the cup dating from 1285-1193 B.C. The cup was discovered in the mid 1800's during construction of a railway station, when part of a tumulus mound at Hove, near Brighton, was being removed. The workmen uncovered an old grave and found a crude oak coffin, hollowed out of a tree trunk. The coffin wood crumbled with decay and the bones were decayed. The wood of the coffin was subjected to Carbon-14 dating and placed the burial around 1500 B.C. (Fraquet 1987, 12). Surprisingly, in the coffin was the amber cup, positioned as if it were placed on the breast of the body. Also in the coffin was an axehead of obsidian shaped in a Scandinavian style, a whetstone amulet, and a bronze dagger. The amber cup may have been used in ceremonies since the Wessex Culture highly revered amber and graves reflected the prestige of the one buried. Fraquet (1987, 11) claims the survival of the cup is remarkable because in Medieval times it was traditional for the young to dance on the tumulus on Palm Sunday and Good Friday, but indicates this dance had its origin with the Saxon goddess *Eostre*.

Figure 3.61. Amber cup from a grave at Hove, England.
Size: Ht. 6.4 cm or 2.5 in
Width. 9 cm. or 3.5 in.
Weight. 100 g.
Brighton Museum, England.

In a later English period, Anglo-Saxon craftsmen discovered that amber could be easily carved and polished. It had abundant uses and was believed to possess the power to protect its wearers against witchcraft and evil spirits. It was unnecessary to wear an entire necklace of amber--a single piece sufficed.

While studying artifacts from sixth-century graves in England, archaeologist, T. C. Lethbridge (Hunger 1977, 28) observed that amber was more common in graves dating from the second half of that century. He speculated that the increase was a result of improved trade, possibly with inhabitants of the North Sea coast. Abington Cemetery, which lies along one of the major English trade routes, yielded graves dated from later than the sixth century that also contained amber in more abundance than in earlier graves, possibly indicating not only an increase in availability, but an increasing esteem based on amber's magical properties.

As early as 1450 English literature, in the *Book of Courtesy*, indicated the importance of amber beads for ladies' wear. During the sixteenth century, the celebrated English poet and writer, Francis Bacon (1561–1626) described amber inclusions in his *Histories of Life and Death*:

> The Spider, Flye and Ant, being tender dissipable substances falling into amber are therein buryed, finding therein both a Death and Tombe, preserving them better from Corruption than Regal Monument.

Sir Thomas Moore (1779–1825), the poet of Ireland, was also intrigued by amber, and, referring to Sophocles' belief that amber was produced by the mourning of mythical birds, wrote *The Lament of the Peri for Hinda*, in *The Fire Worshippers*, verse 8:

> Around thee shall glisten the loveliest amber
> That ever the sorrowing sea-bird has wept;
> With many a shell, in whose hallow-weathered chamber
> We, Peris of Ocean, by moonlight have slept.

Other English poets also were fascinated by amber. References to the gemstone were included in the poetry of Christopher Marlow, Sir Walter Ralcigh, Pope, and Tennyson. In fact, John Milton was so impressed with amber when writing *Paradise Lost, Book VI*, that he portrayed the Chariot of the Paternal Deity as "inlayed with pure amber."

During the period shortly after World War I, amber enjoyed a renewed popularity in England, and beaded necklaces were very much the fashion. Being lightweight and warm to the touch, long necklaces of faceted amber beads were worn with great comfort. The more the beads were worn, the more beautiful they became as they mellowed with age. Amber gradually darkens as it is exposed to oxygen and body oils.

London's Albert and Victoria Museum contains several carved altar pieces and jewelry caskets from the sixteenth and seventeenth centuries. Because amber has long been a favorite in English jewelry, antique shops often have amber articles for sale.

For 51 years, London's Natural History Museum was distinguished for housing the largest known piece of amber ever found. The piece weighs 33 pounds, 10 ounces and is thought to be Burmese amber. It was acquired by John C. Bowing in 1860 for £300 in Canton, China, and was placed in the museum in 1940. However, as mentioned earlier, on December 3, 1991, a much larger piece, weighing more than 150 pounds, was discovered in Malaysia (see Chapter 7).

Figure 3.62.a. Amber and silver bee created using amber pieces shaped to form the body and wings using silver bezels and silver legs and head; b (next page)Amber bust made by Russian Restorer artists at Tsarskoye Selo, Pushkin, Russia. Photographed in Tucson Exhibition in 2004.

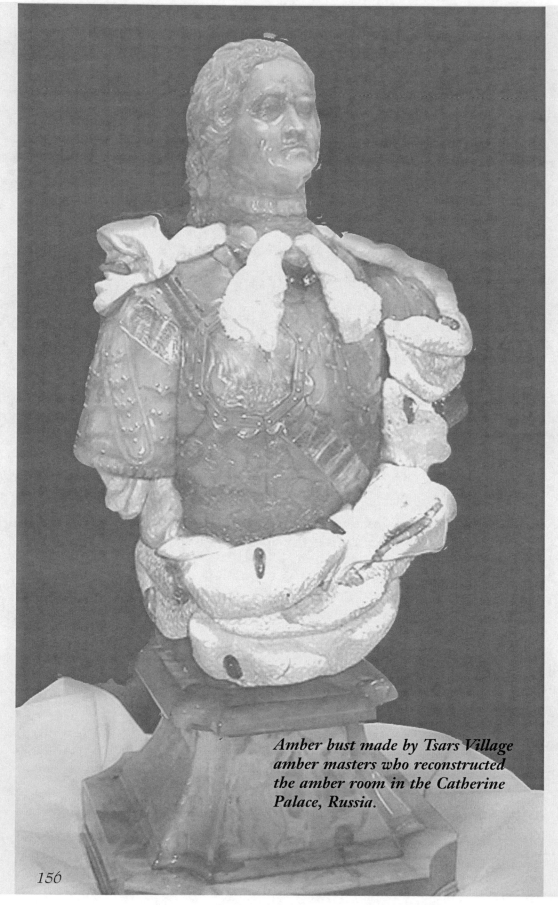

Amber bust made by Tsars Village amber masters who reconstructed the amber room in the Catherine Palace, Russia.

THE CURIOUS LORE OF AMBER

Oh, listen in the evenings
When the sea is restless
And sprays the shore with amber
The depths unseen palm . . . .
—*Maironis*

ANCIENT MYTHS

For many centuries, cultures acquainted with amber wondered what this beautiful gift of nature was and how it originated. True to the customs of the times, many myths and legends relating to amber resulted from attempts to explain the origin of this mysterious substance. Most of the folklore was simply carried from generation to generation by word of mouth, as older members of the culture shared their wisdom with younger members. Some legends found their way into the literature of antiquity and are preserved for us today in their poetic versions. Folklore relating to amber also is recorded in the mosaic folk art of the Lithuanians, the incrustation and bas-relief carvings of the Medieval Amber Guildsmen, and sculptures and carvings of the amber artists today (see Fig. 4.1).

Tears of the Heliades and Other Greek Myths

It was the Roman poet Ovid who recorded the prevalent Greek myth, the Tears of Heliades, attributing divine origin to amber. The Greek legend recounts the adventures of Phaëthon, who grew to young manhood without knowing that one of his parents was immortal. Prevailing upon his mother, Clymene, to inform him who his father was, Phaëthon was astonished to discover he was the child of the sun god Helios. Doubting his mother's word, he sought out the sun god to seek proof of his parentage.

Phaëthon journeyed to India, where Helios' palace stood splendidly glimmering with precious stones and gleaming gold. Upon entering the hall, the boy was blinded by light from the sun god and found it impossible to see. But upon seeing Phaëthon, Helios dimmed his radiance and commanded Phaëthon to approach.

Phaëthon demanded: "Helios, if you truly are my father, what proof will

Figure 4.1. Baltic amber sculpture from pale yellow to white opaque amber. Made in Poland.

you give so I may be known as your son?"

"You deserve not to be disowned, my son. Whatever you ask, I will grant you," promised Helios.

Thereupon, the boy asked to be permitted for one day to drive the flaming chariot of the sun across the arch of the heavens.

Knowing the great dangers involved, Helios offered Phaëthon other gifts, attempting to persuade the would-be charioteer of the skill necessary and the dangers to be encountered. But possessing the willfulness of youth, Phaëthon could not be deterred. With great reluctance, Helios gave his consent.

On the appointed day, the daughters of Helios, the Heliades, helped Phaëthon yoke his father's steeds to the chariot. As dawn opened the doors of the east with the stars and moon retiring, Phaëthon leaped into the golden chariot. Delightedly grasping the reins of the fiery horses, he sped off to the west. Soon the palace was far behind him as the impatient horses sprang swiftly forward, outrunning the morning breezes.

Feeling a lighter load than usual, the horses soon realized inexperienced hands held the reins, and they ran wild, straying far from the traveled path. Seized with fear, poor Phaëthon forgot his horses' names and knew not how to guide them. He dropped the reins in terror of the height and became overwhelmed with dizziness. The uncontrolled steeds dashed on without restraint, rushing hither and thither, wherever they chose. As they approached too near the Earth, clouds began to smoke, the harvest blazed, fields were parched and the rivers dried up. Even the sea shrank because of the awful heat (see Fig. 4.2).

Fearing the world would be reduced to ashes, the goddess of earth begged Jupiter to take pity and give her relief. In an effort to save the Earth from being consumed, Jupiter launched a thunderbolt at Phaëthon, causing him to fall headlong and flaming into the Eridanus River. Finding him dead, the river nymphs reared a tomb to Phaëthon's memory along the shore.

Lamenting over Phaëthon's untimely end, his sisters, the Heliades, accompanied by their mother, a daughter of Oceanus, all wept bitterly. As punishment for encouraging Phaëthon's reckless ride and for assisting him in yoking the steeds to the chariot, the Heliades became rooted to the spot and changed into poplar trees, from which tears continually fell. These tears became hardened by the sun and turned into amber (Haddow 1891).

This early Greek myth poetically illustrates that the relationship between trees and amber was observed even in antiquity. In an attempt to explain the naturally occurring annual phenomenon of the scorching hot summer sun, when the Earth seems about to be burned, the ancients blamed the unskilled driver of the sun chariot. As fall approaches, the hot spell ends; thunderbolts cross the sky in a siege of lightning and rain, indicating that Mother Earth's appeal has been heard. Phaëthon's sisters taking root as poplars weeping tears of amber shows that though some ancients believed amber to be a product of trees from a river area, they confused the source, relating amber to the black poplar tree rather than to a coniferous tree (Ley 1951, 11).

The relationship of trees and the origin of amber appeared many times throughout history. For example, an old herbal, the *Hortus Sanitatus*, published just before Columbus discovered America, contained a quaint old woodcut of "the tree that exudes amber" (see Fig. 4.3). The tree was depicted with enormous drops of sap generously flowing from the trunk of a gnarled tree of some fabulous species (Beard and Rogers 1940, 272–5).

Other ancient philosophers and poets were not as observant as the creator of this myth and held different views regarding the origin of amber. Pliny, the Elder's work *Naturalis histori* (Book xxxvii), in his treaty of amber (sucina), records several other bits of lore about amber-- e.g., the fact that Demonstratus, an early 1st century historian and Roman senator, called amber *lyncurius* and considered it to be solidified lynx urine, with the brighter, darker colored amber produced by the males and the lighter colored amber by the females. Nicias of Greece, for example, regarded amber as a liquid produced by rays of the sun striking the surface of the soil with tremendous force, producing an "unctuous sweat" that was carried by waves to the sea. According to Nicias, it then solidified in the water and was thrown back to the shores of Germania (Williamson 1932, 35-36) Another fanciful tale declared amber to be honey melted by the sun and dropped into the sea from the mountains of Ajan, whereupon it became congealed by the water (Barrera 1860). The Greeks were not the only people who identified amber with burned honey. Far to the east in ancient China, hundreds of years before the

Figure 4.2 Phaëthon driving the Chariot of the Sun. Redrawn from Haddow 1891.

Figure 4.3. The tree that exudes amber, a woodcut from Hortus Sanitatus 1491. Reprinted from Cuba 1491.

beginning of the Christian Era, a Chinese scholar, who may have studied the translucence of amber in sunlight and, perhaps, even obtained a piece with a bee preserved within, recorded the following myth: "Somewhere are cliffs, the cliffs of Ning Chou, in which dwell thousands of bees. When the cliffs crumble, the bees come out. People burn them and make them into amber" (McDonald 1940, 156–8).

Another ancient myth involved amber being gathered in the "Gardens of the Hesperides," where golden apples of immortality were guarded. This belief was influential enough so amber itself was bestowed with powers of immortality. Sophocles, possibly having received one of the few pieces of amber with feather inclusions, believed amber was produced in the countries beyond India from the tears shed for the hero, Meleager, by his sisters. According to the legend, Meleager's sisters were transformed into birds, called meleagrides, who left Greece once a year, flying beyond India and weeping tears that turned into amber (Spekke 1957, 36).

Early classical cultures were not the only ones to produce philosophers who attempted to provide an explanation for the origin of amber. The Aisti, or Old Balts who were the ancestors of the present-day Lithuanians and Estonians, having immediate experience with the amber nodules thrown onto their sandy shores by the Baltic Sea, also created magical legends about amber's origin. Amber-related myths, born in Lithuania and inspired by the restless Baltic, are in their own way as poetic as those originating in the classical Mediterranean cultures.

The Tale of Juraté and Kastytis

In one Lithuanian legend, amber is portrayed as the fruit of a passionate and tragic love. According to legend, during ancient times, Perkunas, god of thunder, was the father god. The fairest of all goddesses was Juraté, a mermaid goddess of the sea, who lived in an amber palace at the bottom of the Baltic. Kastytis, a courageous fisherman living along the Baltic coast near the mouth of the Sventoji River, would cast his nets to catch fish from Juraté's kingdom. Displeased by this intrusion, Juraté sent her mermaids to warn Kastytis to leave her fish alone and disturb the sea no more. Paying no heed to her warnings and impervious to the charms of her mermaids, Kastytis continued to cast his nets and bring in fish. Watching the fisherman haul his catch into his boat, Juraté saw how handsome Kastytis was and admired his great courage. Possessing human failings, Juraté fell in love with the mere mortal, Kastytis. In spite of great differences between them, Juraté took the fisherman to her amber palace.

Perkunas, knowing Juraté was destined to be the consort of

Patrimpas, god of water, became greatly angered upon discovering the immortal goddess in love with a mortal. In his fury, the god of thunder sent a shaft of lightning from the skies, striking Juraté's palace, demolishing it into thousands of fragments and killing her lover, Kastytis. Juraté, crying tears of amber for Kastytis and their tragic love, was punished by being chained to the ruins of her castle. Thus, it is said that when storms churn the cruel Baltic, Juraté is being tossed to and fro by the waves—and even to this day, while storms are raging at sea, the sound of Juraté wailing in the depths may still be heard as she mourns for Kastytis, a son of the earth. As her weeping becomes emotional, the peaceful depths of the sea grow restless and stormy and lumps of amber from her demolished palace are spewed up from the sea onto the Baltic shores.

To the Lithuanians, the small, tear-shaped pieces of amber are the tears of Juraté, as clear and pure as her tragic love. The legend lives today through a variety of beautiful sculptures, mosaics in amber and incrustations of amber designed by Lithuanian artists (see Fig. 4.4).

Though passed along by Balts for centuries, the Juraté and Kastytis tale was first recorded by Lindvikas Adomas Jucevicius, a historian of Lithuanian literature, in 1842. Many artistic works of Lithuanian poets, artists, and composers have been inspired by the legend. Upon entering the Palanga Amber Museum, the visitors are transported into the immortal poetic world of Lithuanian folklore by a symphony based on the legend of Juraté and Kastytis. Music, poetry, and visual art in the form of a bronze sculpture and a colorful stained glass picture introduce the visitor step by step to the origin of amber—the "Gold of the North" (Gudynas and Pinkas 1967, 4).

Figure 4.4. Amber and bronze sculpture depicting the legend of Juraté and Kastytis.Size: Ht. 45.7 cm or 18 in. Courtesy M. Kizis Collection.

The Tale of Freya and Odur

The Norsemen, another ancient race familiar with amber as a product of nature, relate their own version of the divine origin of amber, in the myth of Freya and Odur (Green 1962, 75–9). Freya, the beautiful, blue-eyed, blonde goddess of love, beauty and fertility, had a great fondness for jewels. She was wed to handsome Odur, the sunshine, whom she loved dearly, and they dwelt happily in the land Asgard in Freya's palace, Folkvanger, with their two lovely daughters. Little did Freya know her love for jewels was soon to bring her great sadness.

One day, as Freya walked along the border of her kingdom where the Black Dwarfs dwelt, she spied them making the most wonderful necklace, glistening as bright as the sun. It was called Brisingamen, or the Brising necklace. Freya caught her breath on seeing the beauty of the golden necklace.

"Oh, please sell me this necklace for a treasure of silver," she begged. "I cannot live without it, for I have never seen one as beautiful."

The dwarfs replied, "All the silver in the world cannot buy the Brisingamen from us."

Believing her life could not be endured without the beautiful Brising necklace, Freya asked, "Is there any treasure in the world for which you would sell me the necklace?"

"Yes, you must buy it from each of us," answered the dwarfs, "for the treasure of your love. If you are wed to each of us for a day and a night, Brisingamen shall be yours."

Awed at the sparkle of the world's loveliest adornment, Freya was overcome with madness. She forgot Odur, her beloved husband. She forgot her two fair daughters. Indeed, she even forgot she was queen of the Aesir! She agreed to the pact. No one in Aesir knew about the dwarf weddings except the mischief-maker, Loki, who always seemed to know where evil was brewing.

After four days and nights of these unholy unions, Freya returned to Asgard to dwell again in her palace. Ashamed of what she had done, she hid the shining golden necklace as she went to her chambers. But, Loki sought out Odur to inform him of the happenings in the country of the dwarfs, whereupon Odur demanded that Loki produce proof of his tale. To do so, Loki set out to steal the Brising necklace. Turning himself into a flea, he flew into Freya's chambers, and, finding her asleep, he bit her on the cheek, causing her to turn so he was able to unfasten the necklace. Carefully slipping the Brisingamen away from Freya, Loki went straight to Odur to show him the necklace.

Odur, upon seeing the necklace, tossed it aside and wandered out of Asgard into the world, far into distant lands.

In the morning, Freya woke to discover Brisingamen gone. Intending to tell Odur, she sent for him, only to find he was gone as well! Weeping bitterly, she went to Valhalla to confess to Odin, the father god, whose palace was near the amber valley paradise of Glaesisvellir. At the gates of Valhalla was an amber grove called Glaeser, with trees that dripped glistening amber.

Odin forgave Freya for the evil she had done. Taking the

necklace from Loki, Odin decreed that she must wear Brisingamen forever to remind her of the past. Fastening the Brising necklace around her neck, Freya went forth into the world in search of Odur. As she wandered, she continued weeping. The teardrops falling on the land turned to gold in the rocks, while those falling into the sea turned to drops of amber.

Again we see the relationship of the old tales to the natural features related to amber observed by the ancients. Freya's tears turn into the teardrop shapes of amber as they fall into the ocean. The words, Glaeser and Glaesisvellir, were derived from the old German word for amber, *gles (gleas)*, which later, when glass was introduced into northern Europe, became the word for glass. In many Baltic countries, the folk belief was that an amber necklace would choke the wearer who told an untruth, thereby reminding one not to do evil. It is perhaps from this source that, in the language of gems, amber has become known to mean "disdain" according the *Dictionary of Mythology, Folklore, and Symbols* (Jobes 1961, 81–2). Most other sources emphasize amber's protective and healing powers.

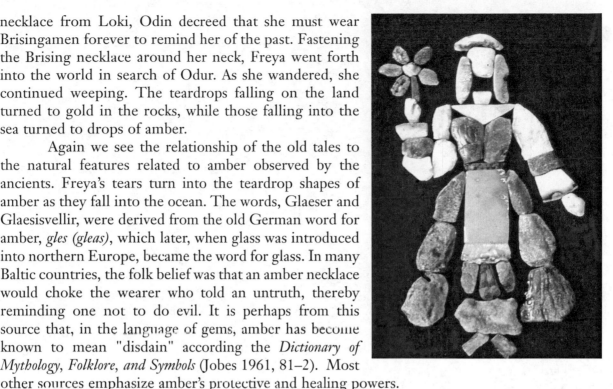

Figure 4.5. Amber mosaic of "Amberita." Size: Ht. 53.3 cm or 21 in W. 45.7 cm or 12 in. Courtesy V. Tomaslunas Collection.

Other Folklore and Myths Depicted in Amber

Lithuanian folklore contains another imaginative tale, *Amberita*, written by J. Narune, that explains why amber is tossed on Lithuanian shores as a gift from the generous sea. The story repeats the theme of a mortal taken by a god to a beautiful amber palace beneath the sea and, then, returning the captured damsel to the surface once a year, allowing her to throw amber chunks upon the shore as a gift by which to remember her (see Fig. 4.5).

From the Medieval Ages, particularly during the Neoclassical period of art, until today, amber craftsmen have followed the theme of classical mythology in many designs created in amber art. Greek gods and scenes of mythological unions between gods and mortals were carved as sculptures and in bas-relief as well as etching on transparent amber. One example, illustrated in Figure 4.6, is an amber sculpture of *Leda and the Swan* created by a Gdańsk artist of the mid-twentieth century. According to the Greek myth, Leda was the most beautiful maiden in all of Laconia. It is said her golden hair, her milk-white skin, and her tall, stately stature placed her as far above other maidens in beauty as "the moon is brighter than the stars." Leda, married to the king of Laconia, was a gentle and gracious queen. One day as Leda sat by the riverside, she protected a white swan from the

Figure 4.6. Baltic amber relief carving of "Leda and the Swan." The natural features of the amber are worked into the sculpture with crust markings as part of the design on the swan's wings. The amber hue is light yellow mottled from opaque to translucent with dark yellow crust. The artist used the technique of piercing and created finely detailed full figures, although the back of the piece is only polished. Size: 7.7 cm x 8.9 cm x 3 cm
Courtesy Burgess-Jastak Collection.

claws of an eagle. The swan began to talk to her saying. "Lady, I have journeyed farther than any swan has ever flown to look on one who is whiter than any swan."

Immediately Leda knew she was talking to one of the gods and bowed herself humbly before the great bird. The swan bent his arched neck and laid his head gently on her arm saying, "Because you saved me from the eagle, ask what you will and I will do it."

Leda spoke saying, "Gracious Zeus, I have everything a queen can wish for except children."

Zeus promised that when her children from King Tyndareus came the swan's gift would come with them. Weeks went by and soon Leda had a little son and a little daughter. As the children lay in their cradles, she heard the sound of flapping wings. As Leda turned to look, she saw a swan's egg along the pillows in each cradle. Instantly the shells cracked and inside were little swan children who looked exactly like her own children. Leda resolved to raise them as her own and gave thanks to Zeus. The princesses were named Helen and Clytemnestra, while the twin boys were named Castor and Polydeuces. Not even Leda herself knew which boy was the swan child.

The twin boys became the Heavenly Twins, the saviors of men, who would lead ships in peril at sea to safety or warriors to triumph (Jobes 1961, 81–2). What better material to use to create a sculpture that tells the story of protection and its lasting effect? Amber has always been bestowed with magical powers to protect the bearer.

In the Kurpie region of Poland a legend about the origin of amber,   similar to the theme of the Greek myth about Phaëthon's sisters shedding amber tears, has survived. The Kurpian myth relates that when god sent the flood to destroy sinful mankind, the people wept in despair as the rain continued for forty days and nights. As their tears fell upon the rising waters, they turned into amber. Of course, the amber that was clear as crystal was made of innocent people's tears and could be made into beautiful jewelry and other items for adornment. The darkened and opaque amber made of sinners' tears was useful mostly for pipes and smokers supplies. While the impure, useless amber most often sent to the smelters for making varnish was formed from the tears of evil men and drunkards (Kosmowska-Ceranowicz and Konart 1989, 200).

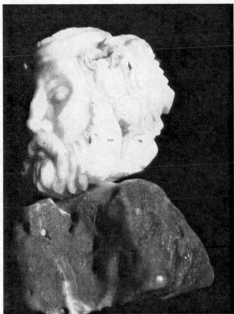

Figure 4.7. Baltic amber two-faced figure: a, Bearded old man with detailed mustache and facial features on the front of the figure; b, An upside down youth face with flowing hair on the back. The base and head are made of two separate pieces. The base has features of the compact reddish crust intact. Purchased in Poland. Size: 11.2 cm x 8.5 cm x 7.3 cm, wt. 279.4 g Courtesy Burgess-Jastak Collection.

a   Another interesting tale from Poland, told to me by Marion Oczarski, artist and folklorist, at Orchard Lake Schools, Michigan, is the story of *The Two Faced Man* often depicted in two-faced figures. The legend is:

> Once upon a time, in the seventeenth century, a poor nobleman had a beautiful daughter who was in love with a handsome, young peasant. Unfortunately, her father would not let her marry him because he was only a peasant without money. A rich, old gentleman offered gold and silver if he could marry this beautiful young lady. Thereupon, her father arranged for the old man to meet her in the garden and, if he liked her, she would be his.
>
> While the girl waited in the garden, she prayed to St. Władysław z Gielniowa, the patron saint of Warsaw, promising to build a chapel in his honor if the old suitor did not accept her.
>
> Well, it happened that the old gentleman took a wrong turn, passing a garden where a wicked, old witch was pulling weeds. Seeing the witch, he mistakenly thought she was the daughter and hastily turned his horses. Upon returning to his castle, he refused to marry the nobleman's daughter.
>
> The beautiful girl was then married to her young love. They prospered in love and gained wealth. Since she had married her true love, she built a statue for the chapel to St. Władysław that was carved with a head of the youth and the old,

wrinkled witch!

The figure stands in a chapel in St. Anne's Church in Warsaw. The amber carving in Figure 4.7 is a bust of the two-faced figure, a bearded old gentleman with mustache and an upside down face of a youth with flowing hair, constructed of yellow opaque amber with orange-red compacted crust. The two-, three-, and four-faced figures have their origin in pagan Slavic religion. In folk art, these figures often represent the phases of a person's life.

POWERS AND SYMBOLISM OF AMBER

Figure 4.8a. "Kierce" made from amber beads. b. Amber spider from Kurpian household made with amber beads and photographed by Gabriela Gierbowska, Gdańsk, Poland, 1995. Courtesy Barbara Kosmowska-Ceranowicz.

Fernie (1907), in *Precious Stones: for Curative Wear, and Other Remedial Uses*, wrote:

> *Of Mickle is the powerful grace that lies*
> *In herbs, plants, stones, and their true qualities.*

Since Neolithic times, amber has been revered for its mystical powers. Amber not only was thought to protect the living, but it was believed to speed the dead on their journey into the shadows. In ancient Scandinavia, for example, an amber axe was placed in tombs along with other treasured possessions to protect the soul during its journey and to confer immortality (see Figs. 2.2 and 2.10). The word designating the fabled drink of immortality, *ambrosia*, is cognate to

a   b

amber, as is the Greek word *ambrotos*, meaning immortal.

The association of amber with immortality continued in the older folklore of Lithuania, Latvia, Estonia, Poland, the British Isles, and Scandinavia, with "amber mountains" representing the land where the dead dwelt at the end of the world. These mountains, as well as "amber islands," appear to be the forerunners of the glass mountains and islands so commonly found in fairy tales. Because amber in early times was called *gles* in northern regions of Europe, it became confused with glass in stories when this same word began to be used to designate the latter. Therefore, as time passed, amber mountains and islands were forgotten, and instead they became glass mountains and islands (Fernie 1907, iii).

As tales were passed from generation to generation, the climbing of the amber mountain by characters in the story often developed into a test for the hero to pass to win his princess. The land of the dead of early tales later became transformed into a magical place of protection or safety where the princess was safeguarded for immortality or until the hero was able to win her. Since in the fairy tales, witches were often characterized as living on such mountains, amber amulets were used as a protection against evil witchcraft. This practice spread as far south in Europe as Italy (Leach 1949, 456).

An amber pendant enshrining a small insect was especially prized for its imagined powers of protection against witchcraft. Thus, such a talisman brought an enormous price Beard 1940, 272-275). This belief was so widespread that during Pliny's time, for instance, even young children wore amber around their necks to protect them from evil influence. Mothers in the Kurpie region hung amber over their children's cribs to protect them from harm. The Moslems of the Middle East made amber bracelets for their babies to wear to keep away evil mischief. On the other hand, ancient legends relating accounts of witchcraft also attribute to amber "the power to warn its masters, the favored of the witches, of their danger and when handled by its master to turn pale, to gleam with rippling light, and to emit a perfume" (Fernie 1907, 323).

Spindle whorls for spinning thread were often made of amber as a protection against the evil spirits that were known to put hexes on thread as it was spun, causing it to snarl. Amber was also used for sewing and crocheting implements as further protection against this form of witchcraft (Pęczalska 1982). The peasants from the old folk cultures of the Kurpian, Warmian, Masurian, and Kashubian regions of Poland long ago were using their spinning wheels as simple lathes to shape their oval beads. Thus, being very close to the magnetic powers of this mystical substance tossed upon their shores, they had faith

in its mysterious power "to drive away evil and bring relief in the various sufferings of man" (Grabowska 1983, 29).

A unique amber ornament typical of the folk culture of the Kurpians, called the *pajaki* or "spider," was hung from the ceiling in cottages of the rich. Amber beads were strung together to form a hanging similar to a chandelier (see Fig. 4.8a, b). Kashuba region called this decoration *Kierce* and they may have been associated with the cult of the sun. Thus, as Kosmowska-Ceranowicz states, "Faith in the healing powers of amber, of which Neolithic man had already made his talismans, has survived under thatched cottage roofs" (Kosmowska-Ceranowicz and Konart 1987, 202).

The most powerful protection against the evil eye was a phallus made of amber. This amulet was regarded as protection against any and every attack of evil spirits. Amber amulets were also made in the form of lions, dogs, rabbits, lizards, frogs, and fish in eastern Asia and were believed to add to the virility of men and the fecundity of women (Budge 1961, 356; see Fig. 4.9)

Cut into various other magical forms, amber was used widely in Baltic, Scandinavian, and Mediterranean cultures to afford similar protection against the evil eye, witchcraft, and sorcery. In the Baltic regions of Germany, Denmark, and Sweden, where gathering amber was not highly restricted, peasants carved amber hearts during long cold winter evenings while sitting by the fire. These were given to their sweethearts as a sign of their love and to keep them free from harm (see Fig. 4.10). What better way to prove one's love? Scandinavian peasants originally used amber buttons as lover's gifts, talismans against evil that might befall the loved ones. These too were fashioned by home artisans and can be found in a variety of unusual and symbolic shapes.

Some European customs decreed that amber be worn by brides to insure happiness and long life. Large amber wedding necklaces, passed down through the family from mother to daughter, were usually kept for several generations in the northern regions of Poland. The Kurpian man of honor would give amber to his daughters as part of their dowry. These necklaces were to be made by local craftsmen, and were not to be merchant amber. Well-worn tarnished wedding necklaces were typically styled in three graduated strands with the largest bead cut in a crystal pattern placed in the center. The center talisman bead was the most ornamental and up to 3 centimeters in diameter. The craftsman would often use a piece of amber with an

Figure 4.9. Baltic amber relief carving of a crocodile with finely detailed head, legs, toes, and textured surfaces. "Under the Crocodile" was a symbol for wine sellers as well as for virility. The amber hue is light cream to yellow opaque and bone amber. The back of the figure has a compact natural crust with only slight polishing. Purchased in Poland. Size: 10.9 cm x 7.1 cm x 3.8 cm, wt. 145 g. Courtesy Burgess-Jastak Collection.

Figure 4.10. Baltic amber heart with lignite inclusions called black and white opaque amber. Size: 20cm x 20 cm.

insect or plant inclusion for this center bead, giving the piece a special, age-old value as a talisman (Chętnik 1964, 107–26). Grabowska (1983, 31) relates a festive song that Kurpian wedding guests sang if a father did not provide his daughter with amber:

> Give her money for amber
>   so she'll have four sons
> Give, give, don't stint
>   Even if it costs you a rouble.

Grabowska states that if the father *had* done the right thing by his daughter, the approving wedding guests would sing:

> Oh! How many tables there were!
>   And all of linden wood!
> Oh! How many tablecloths there were!
>   And all of them from Torun!
> Oh! How many bowls there were!
>   And all of them of amber!

Illustrating amber's symbolism of courage, a Baltic folk legend called *The Torch of Happiness*, describes the enduring courage of a youth who was not deterred by seemingly impossible conditions. The tale is often exquisitely depicted in Lithuanian amber mosaics with the figure of a young boy portrayed triumphantly holding a lustrous amber torch above his head as he stands atop an amber mountain. The mosaic of The Torch of Happiness is created entirely by natural variations in shape and color of the amber material itself (see Fig. 4.11).

The Torch of Happiness legend tells of a group of very poor peasants living at the foot of a steep hill, at the top of which was the flickering Torch of Happiness. Villagers could see the torch's shining rays, but it was out of their reach. If they could only reach the torch, happiness would be theirs.

One by one, the brave villagers attempted to climb the almost vertical sides of the hill. No matter how hard they tried, the slope overcame them and they fell to the ground. When the would-be rescuers failed, they were turned to stone, each adding to the rugged climb for the next person. Finally, a young boy of great courage, knowing full well that if he did not succeed he too would be turned to stone, tried to scale the hillside. Carefully placing each footstep as he struggled over the boulders, he continued his perilous climb, and, because of his splendid effort, he was soon within view of a glowing

Figure 4.11. Baltic amber mosaic of "Torch of Happiness." Made in Lithuania. Size: Ht. 51 cm or 20 in W. 45.7 cm or 12 in. Courtesy V. Tomaslunas Collection.

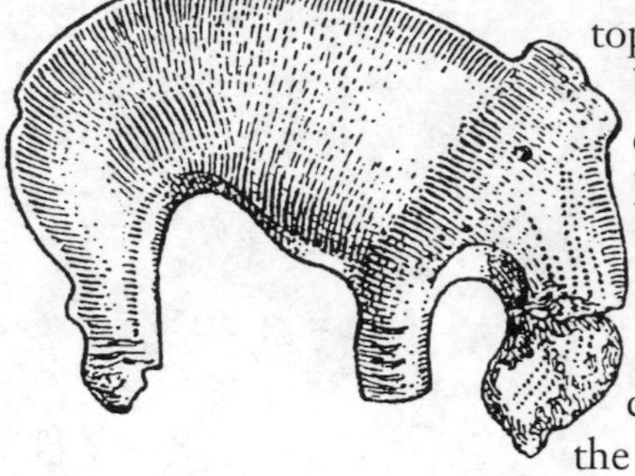

Figure 4.12. Artifact with neolithic boar drawing from Palanga Collection. Reprinted from Klebs 1880.

flame. In disbelief that he had actually succeeded in climbing to the top, he reached out for the torch. Grasping it in his hands, the boy watched with wonder as the stones of the unsuccessful climbers suddenly turned back into living persons. Surely he had reached the Torch of Happiness. Today, the amber stones found in the countryside serve as a reminder of the courage of this persistent youth.

Far to the east, amber was also used as a symbol of courage. Ancient Chinese believed it contained the "soul of the tiger," while in Buddhist paradise, the purest souls are those with bright amber-yellow faces. As the soul progresses, further merit may grow in the form of diamonds, flowers, and amber. Thus, amber was also given a religious significance (Leach 1949, 456).

The ancient amber-gatherers viewed amber as sacred to the mother goddesses because life substance was concentrated within. Tacitus, the early historian of the Germanic tribes, described inhabitants on the Baltic southern shore as worshippers of "the mother of the gods." He informs us that "the figure of a wild boar is the symbol of their [the Balts'] superstition, and he who has that emblem about him thinks himself secure even in the thickest ranks of the enemy, without any need of arms or other mode of defense." The amber boar, symbol of the mother goddess, was also the sacred animal of Celtic tribes of ancient Britain. The symbol of the boar, found on early British armor, became known in succeeding centuries as the "lucky pig" (Fielding 1945, 56; see Fig. 4.12). Mythological themes are still popular for amber carvers today (see Fig. 4.13).

The Celtic sun father, Ambres, derives his name from amber. The gemstone was used in sun worship during Neolithic times and was often carved as an emblem of the sun, with radiating circles or many small circles over the surface. Amber has been found among Cornish megaliths or stone monuments, as the central pillar representing the sun as Lord of Time (Jobes 1961, 56).

Another symbol related to sun worship was the crowing rooster, who signaled the arrival of the sun each morning. The amber egg engraved with a crowing rooster appropriately represents the many virtues attributed to amber: new birth, new dawn, a new life substance given by the sun! Etching and engraving appears to have added to the power of the amber itself, specifying the protection called for by the bearer (see Fig. 4.14).

In the Old Testament, amber was the gem of the tribe of Benjamin. Amber was burned as incense and it is believed that its pungent perfume was one of four types Moses commanded for tabernacle use (McDonald 1940, 154). The early Christian tradition included amber as a sign of the presence of God. During the crusades amber was worn in battle to protect the warriors and defend them against the enemy, just as early gladiators of the Roman coliseum had done.

It is said the former shah of Persia, the present day Iran, possessed a cube of amber that fell from heaven in Mohammed's time. The amber possessed the power to make the shah invulnerable (Fernie 1907, 323).

TALISMANIC, AMULETIC, AND FOLK MEDICINE

According to Hodges (1962, 5):

> An amulet can be a powerful healer if the owner believes with sufficient faith in what it can do; for thought is sometimes a miracle worker when the mind accepts the possibility of healing. And so faith, centered though it may be in an inanimate object, can flower to heal in a most wonderful way.
>
> Yet I think there is something more, a blending of the physical and the mystic, which makes amber a potential agent for healing as it does all precious and semi-precious stones.

Since time immemorial, amber has been worn as an amulet in Baltic as well as Mediterranean cultures. The people of these cultures believed amber was especially beneficial for protecting infants from the croup when worn as a necklace in the form of an amulet. As a charm, amber was believed to ward off fits, dysentery, and nervous afflictions. As early as Pliny's time, farmers' wives in areas where the "water had properties that would harm the throat," or was deficient in iodine, wore amber beads around their necks as a remedy against swellings or goiter.

Surprisingly, the lore attributing guardian values to amber has lasted into the twentieth century. During the late 1960s, for example, an authority announced the virtues of wearing amber beads around the throat to protect it from diseases. It was believed that the strong electrical properties of amber resulted in an electrical band forming around the throat and bringing into existence a protective power (Villiers 1973, 50). To this day, in many peasant regions of the Baltic,

Figure 4.13. Baltic amber deep-sculptured carvings in the form of mythological creatures, such as the hippocampus, giant fish, elephant, etc. The tallest figure is about 6 in or 15 cm. Made from pressed amber. Carved by Russian artist-restorers from the Amber Room Restoration Project, ca. 2003.

Figure 4.14. Amber egg engraved with symbol of crowing rooster. Amber is yellow opaque. Signed by Bryce Barker, Michigan artist.

amber earrings and necklaces are worn when one has a headache or a throat ailment.

Amber was thought to possess many other unusual curative powers. During Pliny's time, amber was powdered and mixed with honey and oil of roses for curing ear troubles; when it was mixed with honey from Attica, it was a cure for dimming eyesight. Since Hippocrates' time (about 400 B.C.) amber powder and oil of amber have been used in medicines. The powder was taken internally as a remedy for diseases of the stomach, whereas the oil, with properties resembling turpentine, was used in medicines prescribed for internal administration for asthma and whooping cough. However, amber oil was more frequently used as a liniment to be rubbed on the chest. Simply holding a ball of amber in the hands not only kept one cool during the hottest days of summer, but would reduce the temperature of a person suffering from fever.

During the Middle Ages, amber was used to ease stomach pains and goiters and was believed to act effectively against certain poisons. Even jaundice was believed to be cured by wearing amber. It was thought the unhealthy color of the skin—and with it, the sickness—was extracted magically by the powerful yellow of the stone.

In 1502 Camillus Leonardus expounded the virtues of amber as a medicine in his *Speculum Lapidum*. He stated:

Amber naturally restrains the flux of the belly; is an efficacious remedy for all disorders of the throat. It is good against poison. If laid on the breast of a wife when she is asleep, it makes her confess all her evil deeds. It fastens teeth that are loosened, and by smoke of it poisonous insects are driven away (Budge 1960, 356).

During the rule of the Teutonic knights, white amber was so highly valued for its medicinal uses that the rights for white amber were never sold when the amber monopolies were released. The order had originated in 1190 during the crusades as a charitable order in association with the German Hospital of St. Mary. The military character of the order was assumed in 1198, and the medicinal uses of amber were held with great regard by the knights (see Fig. 4.15).

The thistle was a very popular subject for Polish folk art and is seen embroidered with beads on vests for the traditional costume especially in the Zakopane region. Also, wood carvers of the mountain regions carve the thistle motif, similar to the edelweiss flower, which stands for youth, love, and health.

In the mid-1500s, Duke Albert's court physicians, Andreas Aurifaber and Gobel investigated the medicinal uses of amber. Their investigations produced a publication called the *Succini Historia*, the earliest scientific treatise relating to the specific uses of white amber. The volume was designed to promote amber as a remedy for sundry illnesses. Both external and internal administrations were recommended, especially with white amber as an ingredient. Having been used as a talisman for centuries, amber now gained scientific credibility during this mystical era (Aurifaber 1551).

Many prescriptions were written that included amber along with other gemstones. An old *Materia Medica* lists 200 medicinal stones. A dose suggested for heart disease contained white amber, red coral, crab's eyes, powdered hartshorn, pearls, and black crab claws. Such a mixture was the prized Oriental Bezoar, prescribed for different ailments. The *Bezoar* stone, a composition of powdered gems, was still accepted long after the magical qualities of gems were disregarded. The Bezoar stone, containing ground white amber as one ingredient, was also used as a component of the famous cordial medicine known as Gascion's powder. In 1778, the *Musaeum Britannicum* published by J. and A. van Rymsdyk mentions "The famous Cordial Medicine called Gascion's powder . . . which consists of Oriental Bezoar, which is the chief, White Amber, Red Coral, Crabs eyes, powdered Hartshorn, pearl, & Black Crabs Claws." Interestingly, by this date belief in the virtues of gems was nearly extinct (Evans 1976, 183).

The famous British physician Bulleyn, cousin of Ann Boleyn (one of Henry VIII's wives), wrote the following prescription:

> two drachmes of white perles; two little peeces of saphyre; jacinthe, corneline, emeraulds, granettes, of each an ounce; redded corral, amber, shaving of ivory, or each two drachmes, thin peeces of gold and sylver, of each half a scrupe.

Figure 4.15. Baltic white bone amber deep-sculptured carving in the form of a thistle. Not only was the white amber itself thought to have curative powers but the engraving, carving, and form were thought to enhance and focus those powers. Size: 5 cm x 3 cm x 2.5 cm. Purchased at Malbork Castle Amber Museum shop in 1980 in Poland.

Prescriptions of this type were commonly used against disease and poison (Schweishimer 1968, 70–1).

In 1624 Sir John Harrington, in his School of Salerne, advised:

> Alwais in your hands use either corall, or yellow amber, or some like precious stone, to be worn in a ring upon the little finger of the left hand; for, in stones, as also in herbes, there is great efficacie, and vertue; but they are not altogether perceived by us; for surely the vertue of an herb is great, but much more the vertue of a precious stone, which is very likely they are endued with occult, and hidden vertues (Fernie 1907, 293).

Matthaus Praetoius recorded in 1680 that not one amber-worker from Gdańsk, Klaipeda, Königsberg or Liepaja died of the plagues as the disease sweep across Europe.

In 1696 the *Family Dictionary*, by Dr. W. Salmon, ordered: For falling sickness, take half a drachm of choice amber, powder it very fine, and take it once a day in a quarter of a pint of white wine, for seven or eight days successively. . . . If further treatment is needed take bits of amber, and in a colsestool put them upon a chafing dish of live charcoal, over which let the patient sit, and receive the fumes (Fernie 1907, 324).

Used externally, amber was also a fumigation to cure other ailments, including tonsillitis, catarrh, or running nose and eyes. This was based on the belief that it had drying and absorbent powers. Amber was mixed with juniper berries and pine needles and burned as incense to purify the air in the room of a sick person. It was also thought that the smell of burned amber eased the pain of women in labor. The Kurpian peasants would carry small pieces of amber to church and place them in the collection tray to be burned by the priests as incense for the sanctuary.

Legend has it that in Scotland, a noted eighteenth-century smuggler, Carnochan of Galloway, had an oval amber bead, 2.2 centimeters in diameter and 1.3 centimeters thick, which he wore as a talisman hanging from a ribbon around his neck. This was particularly powerful in averting the evil eye and in curing various diseases. It is said that he stole it from some Adders who chased him on horseback across Auchencairn Bay. To be effective the amber bead had to be dipped three times in water. Then the water was given to a sick child or animal to drink. The bead was apparently lost in the smuggler's garden, but by the time it was found, it had lost its powers (Black 1893).

In the Orient, amber fumigation was accomplished by throwing powdered amber on a hot brick. The fumes, it was thought, would strengthen the individual and give him courage from the "soul of the tiger," a beast second in importance only to the dragon in Chinese mythology. Syrup of amber, a mixture of liquid acid of amber and opium, was also used in China as a sedative, anodyne, and antispasmodic drug. Powdered amber and oil of amber were prescribed as powerful diuretics.

During the eighteenth century, many critiques of the medicinal uses of stones were published. One profound critic, Dr. Slare, in an address entitled, Experiments and Observations Upon Oriental and Other Bezoar Stones, before the Royal Society in 1715, examined Bezoar stone and the other components of Gascoin Powder and attempted to prove, not only the ineffectiveness, but also the negative medicinal action of the mixture. However, the enlightened critic prescribed chalk and salt of wormwood to replace the former remedy! In

spite of these attacks, the medicinal virtues attributed to amber continued (Evans 1976, 192).

In 1770 Dr. John Cook of England, in his *The Natural History of Lac, Amber, and Myrrh; with a Plain Amount of the many excellent Virtues these three Medicinal Substances are naturally possessed of, and well adapted for the Cure of various Diseases incident to the Human Body*, wrote:

> Many are the excellent virtues of amber, especially when taken inwardly in a cold state of the brain, in catarrhs, in the headach, sleepy and convulsive disorders; in the suppression of menses, hysterical and hypochondriacal affections; and in hemorrhages, or bleedings.

Amber would "clean" any "foul inward ulcer in the lungs, kidnies or elsewhere." The dose was "60 or 80 drops for grown persons, two or three times a day, in any liquid" (Cook 1770, 16–21).

A piece of amber placed on the nose was thought to stop excessive bleeding. It is perhaps this belief in the special blood-stilling or coagulating properties of amber that explains the use of an amber handle on the ritual Jewish circumcision knife of the eighteenth century. For some time afterward, amber was thought to be useful in connection with excessive bleeding. During the early 1900s, as medical practices advanced, vessels made of compressed amber were used during blood transfusions. The amber, being a poor conductor of heat, kept a more constant blood temperature for a longer period of time than did glass or stone vessels. Thus, the blood was kept from coagulating. Today, synthetic materials are used (Gawęda 1995, 156–9).

As late as 1896 amber was still being prescribed by physicians in France, Germany, and Italy. The persistent belief that amber not only prevented infections, but acted as a charm against them, was the reason amber retained its popularity into the late 1800s and early 1900s, especially in the smoking articles industry. Its employment as a mouthpiece for cigars, cigarettes, and pipes was originally talismanic. Amber was used in the Middle East for the mouthpieces of hookahs with their many hoses because of the gemstone's supposed germicidal effect.

During the world wars, nurses from Poland and Russia would wear amber necklaces to prevent themselves from catching illnesses. An amber bead was often concealed in the inner garments of an infant to repel the illnesses in the air. The mystic qualities of the amber itself prevented invisible supernatural influences from attacking the defenseless caregivers and infants. This healing quality of amber continued in the folk cultures of Baltic countries. The folk medicine practitioner Kumuszki (1993, 27) provided the latest recipe for "Amber Medicine" in the July 1993 *Lapidary Journal* as follows:

50 grams of amber as it is found in the rough
Wash in warm water.

Put washed pieces into a .75 liter bottle and pour in straight alcohol (spirits 90 percent alcohol).

Let stand for 110 days. After 10 days it can be used.

Leave amber in bottle. Amber will not dissolve, but its essence will color alcohol amber color. After liquid is used up, take amber pieces out and grind amber up.

It can be re-used one more time, but after that it is no good.

For grippe use three drops liquid in cup of tea once a day in the morning.

If there is some body part that hurts take cotton, wet with tincture solution, and put on the spot.

For pneumonia or grippe rub on chest and on back.

As late as 1934 an official report of the U.S. Mining Bureau indicated that oil of amber was still used in pharmaceutical products (Petar 1934, 11). A volatile oil extracted from amber was used medicinally until the mid-1900s in cases of infantile convulsions.

In defense of folk remedies, recent scientific research has shown succinic acid has a positive influence on the human body and may provide immunity. Pharmaceuticals use it as an inhibitor of potassium ions and an antioxidant. Baltic amber contains 3-8% succinic acid. Thus, Poles who have an intrinsic belief in the power of amber for therapeutic purposes have developed a brand of cosmetics, created in Poland, based on amber along with ingredients such as shea butter, peach kernel extract, grape seed oil, panthenol and fructose. The product, produced by Pollena-Eva SA company, is called Bioenergetic Amber skin care. Advertisements state that these amber based products "harness the energizing, revitalizing power of amber combined with a marine complex, designed to deliver cellular energy to tired skin and visibly improve tone and texture, while reducing the appearance of wrinkles and other signs of aging and fatigue." Specialists from Pollena-Ewa's laboratories research believe that the source of energy in their products is based on the hydrating qualities of amber (see Fig. 4.16).

Other effective medicines containing succinic acid have been produced in Russia as well as in the USA. Researchers have found that the highest concentration of succinic acid is found in the amber "cortex" or the external layer of the stone. Thus, many amber workers try

Figure 4.16. Amber based cosmetics are available in Poland from the Pollena-Eva SA company. European researchers believe in amber's germicidal effect and energizing properties, stating that "the micronization of amber improves its assimilation by the stress-weakened organism of contemporary man." The natural energy of amber is able to stimulate the immune system and cellular metabolism.

to leave some of the outer crust on the nuggets when making amber goods such as bracelets and necklaces. Then raw material which is little ground may be used for therapeutic and bactericidal purposes. Polish amber workers state that when there was a plague or epidemic, they did not get sick. They also believe that treatment of amber in an autoclave destroys its aroma and also its therapeutic power.

In the book, "Amber in Medicine and Cosmetology," Dr. N. N. Moshkov, head of the institute of Amber and Regional Resources in Kaliningrad, Russia (Gierłowska, 2002), recommends massages applying amber powder to body zones:

for stimulation of hair growth and improvement in hair quality. . . .

for treatment of thyroid gland diseases . . .

for treatment of small blood vessels. The entire extremity should be massaged circularly and amber powder should be intensely rubbed in until warm.

for treatment of hemorrhoids amber suppositories are prepared by thorough mixing of 1 part of amber powder with 1 part of honey

ASTROLOGICAL SIGNIFICANCE OF AMBER

Most gems have been related to astrological signs and have significance for astrological readings. In ancient times, the celestial bodies were deemed to impart their powers to certain gems, which would then assist in exerting an influence on mortals wearing them. Amber, for example, was associated with the sign of Leo because of its golden color. People born under the sign of Leo were protected by wearing amber, but those born under the sign of Taurus would be harmed by wearing it.

It was used in reference to water signs, such as Cancer, Pisces, and Scorpio, since amber is closely associated with the sea. Therefore, amber's protective qualities would be most beneficial to those born under these water signs—the Scorpios benefitting most from reddish amber, whereas Cancers should select the white or water-clear varieties.

The earliest record indicating the relationship between amber and Cancer and Leo was in 1555 by Olaus Magnus, the priest sent to Rome by the Swedish king Gustavus Vassa, in his description of the origin of amber in *Historia de gentibus septentrionalibus*, or the *History of the Nordic People*. He describes amber as being secreted from firs or conifers—"this is particularly the case when . . . the sun . . . moves into the signs of Cancer and Leo."

Dreams that include amber signify that one will embark on a

Names of Amber

African- barnsteen;
Albanian-qehribá;
Arabic- kahramaan,
'anbar; Azeri-kæhræba;
Basque- anbar;
Belarusian-burshtin;
Breton- goularz;
Bulgarian- kehlibar;
Chinese-hupò;
Czech- jantar;
Danish- rav;
Dutch- barnsteen;
English-amber, ambre,
ambra, aumbur,
ambyr, awmer;
Carib- lamber, lambre,
lambur, lawmer, lamar,
lamour; Estonian- mere-
vaik, meriwalk,
meriphikaa;
Farsi-kahrobâ;
Finnish- meripihka;
French- ambre;
Georgian- karva/qarva;
German- Bernstein,
Ägtstein;
Greek- kechrimpari,
beronike, ilektron /êlek-
tron;
Gujarati- kerabo,
triNamaNi /truNamaNi;
Hebrew-i'nbar;
Hindi- ambara;
Hungarian-borostyán.
yanta;
Icelandic- raf, gleri;
Indonesia-batu ambar;
Irish (Gaelic)- ómra;
Italian- ambra;
Japanese-kohaku (Kanji);
Korean- ho-pak;
Kurdish- kareba;
(con't. next page)

voyage in the future. Amber is specifically associated with the proper name Anne, and those bearing that name will be protected and kept free from illness by wearing any form of the gemstone.

The ancient belief that the constellation under which one was born would influence one's fate led to the custom of wearing a stone assigned to that constellation, or a birthstone, that served as an amulet bringing good fortune and protection to its wearer. Amber is considered by some to be the alternate birthstone for November, since its color resembles that of the topaz, the regular birthstone for that month. Amber is also the gem for the tenth wedding anniversary (Kunz 1971, 55–7).

THE NAMES OF AMBER

The English name amber is based on a misconception. It is derived from the old Arabic *anbar*, a word meaning *ambergris*, which is the name of a curious waxy substance sometimes regurgitated in lumps from whales' stomachs. This material is in no way related to amber; although, like amber, it is occasionally washed ashore, since ambergris is also light enough to float on seawater. Ambergris has long been used, and to some extent is still used, to prevent the too rapid evaporation of the volatile essences in perfumes. Some authorities suggest that the Arabic *anbar* originally was applied to perfumes and incenses, and because amber also was used as an incense the word came to be applied to amber as well as to ambergris. Our present form, amber, is evidently derived from the Spanish *ambar* or *ambaeur*. Several other languages derive similar terms from the same ancient source; for example, the French word is *ambre*, the Italian, *ambra* (or somewhat amplified as *ambra gialla* (yellow amber) and the Portuguese, *alambre or ambre jaune* (Encyclopaedia Britannica 1971, s.v. "amber").

Considerably different forms appear in the Finnish *merre-kiwa* or *meri-kivi*, meaning seastone, in reference to amber's marine origin. Related forms and meanings are found in the Estonian language; *meriwalk* (seawax) or *meriphikaa* (sea resin; Williamson 1932, 50). The Russian *yantar*, the Hungarian *yanta*, and the Latvian *dzinters*, as well as the Old Prussian *gentar*, all appear to be derived from a common root. The Lithuanian term *gintaras* provides a clue to their meaning, inasmuch as *ginti* in Lithuanian means to defend, and, logically, *gintaras* means protector or defender, especially when referring to a personal amulet made from amber (Encyclopedia Lithuanica 1970, s.v. "amber"; Butenas 1973a, 110–4; 1973b, 159–64).

In Ancient Greece, the term *elektron* for amber appears in Homer's Odyssey and is derived from the word *elektor*, meaning sun's glare. Elsewhere in Greek literature, the term *Elektrides* is used syn-

onymously with Amber Islands, or the mystic places in the Far North from which this substance came. However, to the confusion of scholars of antiquity, elektron had another meaning, applied to a bright yellow alloy of gold and silver. It was not until ancient Greek graves were excavated that the presence of numerous amber beads proved that this early Mediterranean civilization not only knew amber, but prized it most highly. Thus, the translation of elektron as meaning amber was verified. As was mentioned earlier, it was Thales of Miletos who first drew attention to the attractive properties of amber sometime between 640 B.C. and 546 B.C. It was from such electrostatic properties that the ancient word elektron came to be applied to any phenomenon or device in electronics or electricity. In 1859 Franz Beckmann wrote a treatise on the origin of the name for amber in relation to these properties of amber. Curiously, the present name for amber in Greece is *beronike*, rather than electron.

Another Greek word for amber was *harpaks*, which, interestingly, means miser or to pull things toward oneself. This may relate to the observed quality of amber to attract light objects to itself when rubbed. But also observe that the Danish word *harpiks*, means resin or rosin.

Another name alluding to amber's power of attraction was used by the Persians. The Arabic *Kahroba (kah ruba)*, meaning amber, was formed by combining the two words *Kah*, meaning straw, and *ruba* meaning robber, thus describing amber as straw robber or attracter. The modern Turkish word, *Kehruba*, is obviously related to this.

The lustrous gold like appearance of amber has not gone unnoticed. The ancient European term *glaesum* or *glesum* refers to this quality. Cornelius Tacitus, in his history, *Germania*, written about A.D. 100, indicates that the source of the word was from Old Germanic languages. Tacitus wrote "the people of Germania explore the sea for amber, in their language called glese, and are the only people who gather that curious substance." A variant of this word is found in the Hebrew biblical writings of Ezekiel, as well as in St. John's Revelation, as *ghashmal*. These word forms all actually allude to the smooth shining yellow properties of amber. Therefore, the word *ghashmal* in ancient texts has caused scholars to argue over the accuracy of the translations, since the same term was applied to a pale gold metal of that period. However, the old Germanic word *glessum*, and even its Latin derivative *glesum*, both mean amber. It is from the Old German word for amber, *glaes*, that we obtain the modern word glass, but, strangely enough, *glaes* has not survived anywhere in its original meaning.

Latin- *electrum glaesum /glesum; lacrima lyncurium /langurium;*
Latvian- *dzintars, dzinters;*
Lithuanian- *gintaras;*
Macedonian-*kilibar;*
Malayalam-*kunthirukkam;* Maori-*kano pàkà;*
Mongolian- *khüw;*
Norwegian- *rav;*
Persian- *karabe;*
Phoenician- *yainitar;*
Polish-*bursztyn;*
Portuguese- *âmbar, alambre, ambre jaune;*
Prussian (Old Prussian)- *glêsîs;*
Romanian- *chihlimbarul;*
Russian-*yantar;*
Scottish-*amber,ambir &ambre; Gaelic- òmar;*
Serbian- *cilibar (ski);*
Slavic- *ambra;*
Slovak- *jantár;*
Somali- *ámbar;*
Spanish-*ámbar, ambaeur;* Swahili-*kaharabu;*
Swedish- *bärnsten;*
Syrian- *kahraba;*
Tagalog- *dagtâ;*
Tamil- *ampar;*
Thai- *ching peh;*
Tibetan- *ka su ra & spös shel;* Transylvanian-*ambrã ;*
Turkish- *kehribar;*
Ukrainian- *burshtyn, yantar; Urdu- ambar;*
Vietnamese- *h-phách;*
Welsh- *gwefr, amfer & ambr; Yiddish- burschtin*

The ability to burn is another property of amber reflected in the language of those cultures familiar with the gem. The German name *Bernstein* (*Boernstein, Borstein, or Burnsteyn*—Old German) literally means "stone that burns." The Low German word *bernen*, or *bärmen*, means "to burn or "shine." Related derivations are Dutch *Barnsteen*, Swedish *Barnsten*, Scandinavian *Bernsten*, and Polish *bursztyn*. In some early Latin writings, amber is spoken of as *lapis ardens*, which also means "a stone that can be burned" (Keferstein 1849, 61).

In contrast to the above, the Danes call amber *rav*, while in the Frisian Islands along the coast of Denmark, it is called *rov*, and in Swedish and Norwegian, *raf*. All of these names are derived from a much older Norse word, *rafr*, a name sometimes applied to gum. The *rav*, (in Swedish also means brown) is part of the name of a southern Swedish district, Ravlunda, which literally means "amber grove." Old Swedish rosaries used for counting prayers were made of amber and were called *ravband* (Dahlström and Brost 1996, 42). Some linguists suggest this root is based on the name of the island, Raunonia, mentioned by Pliny, who quoted the Roman knight sent out to search for amber and who described it as being cast ashore on Raunonia.

Buffum (1896, 13–4), in his charming study of Sicilian amber, attempts to show, through a study of *sakal*, the Egyptian word for amber, that Sicilian amber was known to the ancients. He suggests *sakal* was not an Egyptian word at all, but was adapted from its source; that is, from Sikeli, a powerful race living on the east coast of the island of Trinacria (Sicily), the area where the bulk of Sicilian amber was produced. However, the Egyptian word is also strangely similar to one used in Lithuanian and Latvian languages—*saka* or *sakai*, meaning resin or gum in Lithuanian and amber in Latvian. This word appears in names of places in the Baltic region, such as Sadatine (Sakaslina), or "Valley of Amber," a place located in Latvia north of Liepaja, and Sakasosta, or "Port of the North," a place north of Kaliningrad.

In Dalmatia, during the Roman period, amber was called *schechel*, possibly derived from the Latin *succinum*, literally, sap-stone, which comes from *succus*, meaning gum. An alternate word for amber in Spanish is *succina*, representing the only survivor of the ancient Latin term in a modern language. It is also from Latin that the mineralogical term *succinite* is derived (Lüschen 1968, 188–9). Amber sites on the internet may list many more names used for amber and their sources.

PART III

DESCRIPTIVE ASPECTS OF AMBER

SHAPES OF AMBER

*Shapes formed inside
the bark layer of tree.*

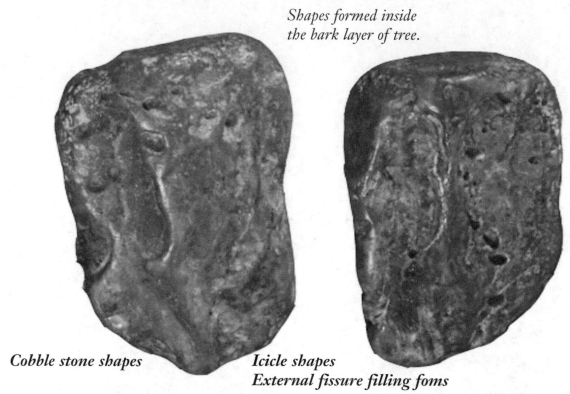

Cobble stone shapes

**Icicle shapes
External fissure filling foms**

**Dripping forms
drop shapes**

5

PHYSICAL, CHEMICAL, OPTICAL PROPERTIES WITH APPROPRIATE TESTS

Since the 1980s much scientific development has occurred that has greatly changed many of the previous ideas about the organic substance called amber and has raised many questions about international nomenclature for fossilized resins. The questions scientists from various fields and parts of the world began to ask were simply: Is amber the name that should be given only to Baltic amber because historically it was used to identify Baltic amber the longest? Or are all fossilized resins to be called amber? How long must a resin be fossilized before it is amber and not copal? Should each fossil resin from a different location continue to be named as in the past by its location or for the scientist who discovered it, or should its name be based on the source from which it was produced? More simply, is it a *stone* when it is an *organic material*? Baltic researchers developed a classification of amber based on its chemical composition containing succinic acid, which distinguished Baltic amber (succinite) from resins without succinic acid which they classified as 'retinites.' New systems have been proposed based on modern research techniques which will be discussed further in this volume.

Technically, amber is not a mineral, but rather mineraloid, along with coral, nacre or mother of pearl, jet, and pearls, due to their organic origins. All of these organic gems are used in the jewelry industry. Therefore, gemologists will come across this gem material and will be asked to identify it and verify its authenticity. Physical and chemical properties discovered with appropriate tests are important in determining if a gem is genuine amber, processed amber, a synthetic imitation, or a natural imitation, such as copal or Kauri (recent resins.)

Despite the fact that all now agree that amber should no longer be classified as a mineral but should be placed in a category called organic gemstones, gemologists sometimes continue to use the tests for minerals when attempting to identify amber jewelry. Therefore, the following information is helpful. It is important to know how these physical, chemical, and optical properties relate to amber when attempting to identify jewelry items. Most tests are destructive to amber, and should be performed with caution and not at all on finished pieces. If attempted, special care must be taken to apply the test in an inconspicuous location, such as around the bead hole, or simply take a scraping from where it will not be noticed. Heat and solubility should never be used on fashioned gems.

HARDNESS AND "TENACITY"

Hardness is considered to be of major importance in mineral gemstones; however, it is accorded lesser importance in the case of organic gem materials including amber. In all these relatively soft substances, other considerations, such as beauty and rarity, elevate them to the rank of important gem materials. To provide a scale for determining hardness and for comparing minerals with one another, the Austrian mineralogist, Friedrich Mohs, arranged ten minerals in order of their hardness—from the softest, designated as 1, to the hardest, designated as 10. These minerals were arbitrarily selected, and the differences between them in the continuum are not equal, but vary considerably. In spite of this, the Mohs scale is useful, as each mineral will scratch those lower on the numerical scale and minerals at the same hardness will scratch each other.

Mohs Scale of Hardness

1 Talc	4 Fluorite	8 Topaz
2 Gypsum	5 Apatite	9 Corundum
3 Calcite	6 Orthoclase	10 Diamond
	7 Quartz	

The hardness of Baltic amber varies from 2 to 2.5, depending on the type of specimen. Generally, amber is slightly harder than gypsum but can be scratched by calcite. Burmese amber and pressed amber may be as hard as 3, whereas Dominican amber may be as soft as 1.5 to 2. Geologically younger amber tends to be softer than amber that has been buried in the ground for a longer time. Hardness for Baltic amber using a different scale is 199–290 megapascals. (Kosmowska-Ceranowicz 1997, 3)

Because of amber's softness, it is mostly shaped into beads or pendants, as opposed to rings, because necklaces generally are not subjected to the rough wear rings receive. When amber stones are shaped for rings, they are generally cut into high-domed shapes, cabochons, or even thick natural lumps, all of which provide more bulk for the piece than a faceted shape would, thus reducing the possibility of chipping. Because of its relative softness, amber is easily carved. Artistic amber carvings continue to be made by contemporary artists working in amber.

When testing hardness, other materials than those specified in the Mohs scale may be used to give an approximate hardness. For example, the fingernail is about 2.0 to 2.5 on the scale, and this is why it is possible to slightly scratch amber with a fingernail. On the other hand, a copper penny is harder, about 3, and it easily scratches amber, and a knife blade, about 5.5, can be used to cut amber. When a steel tool is used to scrape or cut an amber specimen, a powder or small granules are produced, resulting from amber's brittleness. Fraquet (1987, 144) states this type of scrape test is useful for distinguishing amber from plastic imitations, which tend to produce small shavings rather than powder since plastics are not as brittle as amber. One will find this test used in older description of Baltic amber, also called succinite, to distinguish it from other fossil resins found in the Baltic region, such as Gedanite, which Helm (1878) called at first (1887) 'friable amber,' describing it as "less hard, splinters easily on breaking or cutting."

Despite amber's softness, it is tougher than most gemstones of similar hardness. If there were a scale of toughness from 1 to 10, amber would be placed at about 3.5. Baltic succinite tends to be tougher than fossilized resins without succinic aci?, sometimes called 'retinites.' Generally speaking, Dominican amber, which is a Class 1b amber (Anderson and Crelling 1995, xiii; see Appendix A) or a 'retinite,' is more brittle than Baltic amber and tends to break more easily when subjected to sharp blows. However, the brittle nature of various amber or fossil resins also varies with the time exposed to air and conditions of storage. Most amber breaks with a curved, conchoidal fracture, somewhat resembling that seen on broken pieces of glass. However, amber oxidizes when exposed to air, therefore, older amber artifacts are often brittle and crumbly. All amber will break and shatter when struck or dropped, so care is needed when working with museum pieces, when carving amber, and when excavating amber.

SPECIFIC GRAVITY

The specific gravity is the ratio of the density of a substance to that of water. The density of amber ranges from 0.96–1.96 grams per cubic centimeter. The specific gravity of amber is slightly higher than water, ranging from 1.04 to 1.10. However, differences may be noted even within samples taken from the same specimen. As a general rule, the clearest types or transparent ambers are more dense, whereas the osseous or bone amber varieties are less dense. A simple specific gravity test for amber serves to separate it from most of the usual plastic imitations. Because of amber's low density, it will float in a saturated solution of salt water, most conveniently made by adding four heaping teaspoons of salt to eight ounces of water. Though most amber specimens will float in this solution, the heavier Bakelite resin, for example, with a specific gravity of 1.25 to 1.55 will sink. If the sample does float, it is not necessarily definite that the piece is amber, since some imitations made of the more recent thermoplastics are lighter than amber and will even float in plain water.

In this connection, the immersion testing fluids commonly employed by gemologists for testing gems should not be used for amber because they chemically attack and dull the surface luster. Therefore, it is best to use the saltwater solution mentioned. If more accurate determinations of specific gravity are required, the water displacement method, described in any standard gemological or mineralogical text, may be used. Be cautious with salt water and rinse amber carefully after having been submerged in it because salt seeping into cracks could crystallize and cause further crazing. *Caution:* Heavy fluids sometimes used by gemologists to identify specific gravity, such as methylene iodide, bromoform or acetylene tetrabromide, should be avoided as they will attack not only amber but also most of its substitutes!

OPTICAL PROPERTIES

Amber has a resinous luster. Lacking crystalline structure, amber is amorphous and hence refracts light in a simple fashion, like glass, or is singly refractive. The degree of refraction, or refractive index, is low, ranging 1.539–1.542, considerably above that of water (1.33), but far less than that of diamond (2.42). On the other hand, it is close to the index of quartz (a double refractive mineral with indices of 1.54 and 1.55), and thus faceted amber would bend light rays to about the same degree as would amethyst, rock crystal, and other forms of quartz.

The gemological instrument used to measure refractive index, the refractometer, may be used for amber. Objects such as flat-bottom cabochons or faceted beads can easily be measured, since the instrument requires that a flat, polished surface be available for contact with the sensors. Finished amber articles set in mountings or pieces with uneven surfaces present some difficulties in measurement. However, with a little practice, refractive index readings can be taken on curved surfaces, as explained in standard gemological texts. A problem with using the refractometer arises from the fact that, in order to test gemstones, an "optical contact" must be established between the stone being tested and the small glass prism of the refractometer. For this purpose, a special fluid is used that, unfortunately, may dull the surface of amber if left on the piece. This problem can be eliminated if the measurement is done quickly and the specimen wiped off as soon as possible.

Another method of establishing the approximate refractive index of gemstones employs fluids, in which the test specimen is successively immersed until it seems to disappear from view, much in the same manner as an ice cube seems to disappear when placed in water. A close "match" means the index of the liquid is nearly the same as that of the specimen. Again, most immersion fluids are hydrocarbons or other organic compounds that more or less readily attack amber and, therefore, must be used with caution. On the whole, they should not be used, especially when amber pieces are polished, but they may be used where the effect on the surface, as in rough specimens, is not important. Needless to say, long immersion in such fluids should be avoided, since the fluids will become contaminated by the substances dissolved from the amber.

One fluid mentioned in gem literature as having a refractive index of about 1.54, similar to that of amber, and easily available from local pharmacies is oil of clove. As a word of caution, I tried using the fluid on both amber specimens and antique imitation amber beads and found to my dismay it not only dulled the surface of some specimens, but was difficult to remove, causing the surface of the beads to become sticky or tacky to the touch.

Polarized Light

In polarized light transparent amber appears light when transluminated. This is what happens when amber is placed in a jeweler's polariscope with the lens set in the dark position: a rainbow effect or iridescence appears in places where internal strains are present. As the piece is rotated between the polaroids, these iridescent spots will appear as wavering bands moving across the specimen as it is turned under the light.

Though this test is not useful for discriminating amber from plastic imitations, which may be expected to behave in a similar manner, it is helpful in identification of ambroid (pressed amber). When amber is softened and pressed, the uneven stressed markings disappear and the piece develops a flowing or roiled haziness, causing it to display an evenly lit appearance when rotated in the dark position of the polariscope.

Fluorescence

When ultraviolet light (UV) is directed on amber, it fluoresces, that is, it emits longer wavelengths than those of UV, placing them in the portion of the spectrum that is visible to the human eye. Common fluorescent colors of amber are blue and yellow, with green, orange, and white occasionally noted. Newly fractured surfaces are more flu-

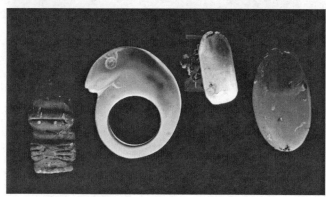

Figure 5.1. a, Baltic amber, Burmese and root amber under UV light; b, Dominican amber under UV light. Patty Rice Collection.

orescent than surfaces that have been exposed to the atmosphere. The intensity of fluorescence also varies with different varieties of amber or fossil resins. Osseous succinite is the most highly fluorescent of all, fluorescing with a bluish-white hue. Dominican amber is also highly fluorescent, fluorescing blue, green, or yellow. It is reported that yellow and mottled varieties of rumanite are also highly fluorescent, followed by simetite, burmite, clear-golden succinite, and, lastly, Siamese amber. Generally speaking, those resins with higher sulfur content are more fluorescent than are those containing less sulfur.

The characteristic fluorescent hue of clear-golden Dominican specimens is most commonly blue, but more rare yellow opaque pieces fluoresce yellow. Dominican amber also often appears to fluoresce green if viewed against a light source (back lighting), but this may result from selective absorption of light rather than from true fluorescence. Some varieties of Dominican amber and Sicilian amber not only fluoresce under a black light (UV) but also in reflected light. Dominican amber pieces possessing this quality appear as though a blue film covers the amber surface when ordinary light strikes it. Transparent pieces of Burmese amber usually fluoresce blue, but the osseous amber, or sometimes called 'root amber' if mottled with white and brown hues, from Burma (now Myanmar) may fluoresce in the orange range (see Fig. 5.1).

DIAPHANEITY AND STRUCTURE

As mentioned in Chapter 1, differences in diaphaneity or turbidity in amber specimens is caused by the enclosure of vast numbers of microscopic bubbles of varying sizes. In those areas that are cloudy, the bubbles are too small to be seen with simple lenses, but thin sections under a high-power microscope reveal minute voids distributed throughout otherwise pure clear resin. Photographs made using a scanning electron microscope show the distribution, size, and shape of the pores in various specimens. These photographs indicate that the smaller the pores, the greater the quantity. It is the number and size of these cavities that produce the different appearance in amber varieties. Early work of Bauer (1904) provided a count of the bubble enclosures in the three most prevalent amber types (see Table 5.1). This was confirmed by the work of Katinas (1988) from Lithuania and by Savkevich (1981) from Russia

Bauer reported that the bubbles are smallest and most numerous in osseous amber, having a diameter of from 0.0008 to 0.004 millimeter and a distribution of 900,000 per square millimeter (see Fig.

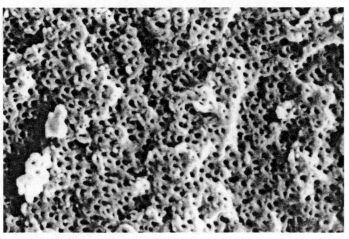

Figure 5.2. Electron microscope photo of white bone Baltic amber.

5.2). Fatty, or in Prussian or old German terms *flohmig*, amber contains the smallest number of bubbles, 600 per square millimeter, but these, with a diameter of 0.02 millimeter, have the maximum size bubble. When amber contains no bubbles, it is perfectly transparent. Cavities in the cloudy amber can be filled with a transparent substance with the same refractive index as the surrounding mass, causing the turbidity to disappear, rendering the specimen perfectly transparent. This process is sometimes used to clarify amber artificially in commercial preparations. More on this process is explained in Chapter 10.

The presence of air bubbles makes amber very porous. The size and density of the pores influence the structure of the amber, which can be differentiated into three types:

1. Compact: Slightly porous, heavy
2. Nodular, or bumpy: Moderately porous.
3. Foamy: Strongly porous, light.

The above structural classifications presently are used by Polish scientists to describe the structure of amber that influences its diaphaneity.

In the late 1980's attention was brought to the air bubbles and water droplets enclosed in amber by a national television news report that indicated that scientists were analyzing the gases enclosed in the bubbles to determine the atmospheric composition when the amber was formed. Two groups of scientists, Berner and Landis (1987, 757–62; 1988, 140–8) and Horibe and Craig (1987, 1513) attempted to crush amber in a sealed container to analyze the

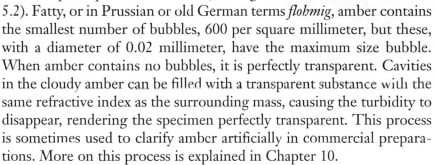
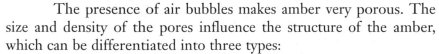

Table 5.1 Bubbles Influencing Diaphaneity		
Type of Amber	**Number of Bubbles (per mm²)**	**Size of Bubbles (mm in diameter)**
Bone (osseous)	900,000	0.0008 to 0.004
Opaque (Bastard)	2,500	0.0025 to 0.012
Fatty (flohmig)	600	0.02 and larger

enclosed gases. However, since amber has been clarified for centuries, it was apparent that it had the ability to absorb, hold, and release gases according to the general principles that apply to organic gases. Hopfenberg, Witchey, and Poinar (1988, 717–24) demonstrated scientifically that amber was not a sealant.

COLOR VERSUS RESIN SOURCE

Various suggestions have been made for the coloration of different ambers. Early studies of the coloration of amber suggested that microscopic bubbles, which interfere with the refraction of light, were the sole cause of color variations. However, current studies relating fossil resins to recent resins suggest certain colors are characteristic of certain tree sources. For example, recent pine trees, even those living today, produce resins in golden yellows from transparent to opaque, similar to those found in most common amber specimens. Such trees also produce white, ivory, and occasionally rare blue varieties of resin. The greenish shades in amber closely resemble resin exuded from a recent spruce tree. Polish Muzeum Ziemi old publications indicate reddish tints are likely in resins of deciduous trees, such as cherry and plum species, whose modern representatives exude a resin with a cherry red hue. Some researchers indicate that the hue is influenced by minerals present in the solution or compound making up the resin, such as the presence of sulfur giving the characteristic yellow hue or presence of chlorite giving a greenish hue. Also chemicals given off from the decaying plant debris in the dark brown, black, and green amber may assist in producing these darker varieties.

SOLUBILITY

Amber is insoluble in water and only slightly soluble in ether, and then only after long duration of contact. When a drop of ether is placed on amber, the ether evaporates before the amber has time to soften. Of course, one would not want to use this on their heirlom or jewelry piece as it would ruin the luster of the piece. Copal and other recent resins are more quickly attacked by ether and become sticky, providing a convenient test for differentiating amber, or fossilized resins, from recent resins such as copal (from Africa and Columbia, South America) and kauri (from New Zealand and Australia). These recent resins are discussed in Chapter 8. Crushed particles of amber are soluble in sulfuric acid, as shown in Table 5.2.

Caution: these acids require special care during use because they will attack many substances and should not be used outside of laboratory conditions. The test should be conducted in a glass test tube. The sulfuric acid will attack amber so do not use this as a test on a finished bead!

REACTIONS TO HEAT

Amber is a poor conductor of heat; therefore, it feels warm to the touch, in contrast to many other gemstones that feel cool. When rubbed briskly with a cloth or gently heated, a typical piny odor is emitted. With further heating to about 150°C to 180°C amber softens then swells and emits volatiles or essential oils composed of complicated mixtures of terpine deriva-

Table 5.2 Solubility of Amber		
Solvent	**Clear Golden Amber**	**Bone (Osseous) Amber**
Ether	18-23%	16-20%
Alcohol	20-25%	17-22%
Turpentine (oil of)	25%	
Chloroform	21-26%	
Benzene	trace	
Concentrated sulfuric acid crushed	soluble if crushed	soluble if
Hot concentrated nitric acid	soluble	soluble

tives that still remain in the fossilized resin. According to Poinar (1992, 5), most resins when first deposited are mixed with oleoresins, or volatile oils, that volatilize at normal environmental temperatures resulting in a gradual loss as the resin ages in the process of polymerization. At between 250°C and 375°C, Baltic amber decomposes, emitting white fumes with a strong pine odor, and releasing succinic acid and oil of amber. Genuine amber or fossilized resin does not actually melt, it decomposes, unlike recent resins that can be heated to a melting point and even poured into molds, wherein recent insects are often embedded! Amber ignites and burns readily, producing a bright yellow flame streaked with green and blue flashes.

Gemologists sometimes use a hot point to test amber by touching the inside of a bead hole to see its reaction to heat. The amber will slightly disintegrate with a puff of white smoke emitting a resinous, piny odor. Some gemologists scrape bits off the piece then test them with heat to avoid reactions from unknown substances.

When crushed Baltic amber is introduced into a test tube and heated, a frostlike sublimate of succinic acid crystal appears on the upper, colder portions of the tube. Filter paper moistened with a solution of lead acetate turns black when held in the tube, as a result of the presence of hydrogen sulfide in the fumes.

The residue left after volatile substances have been driven off by heating the amber to 280°C is termed *colophony* and consists principally of succinic acid and oil of amber. It is this substance that, when dissolved in turpentine and linseed oil, provides the basis for the characteristic dark-hued amber varnish used extensively in the last century. Polish scientists, after examining the barrel-shaped resin that weighed 238 kilograms that was fished out of the Baltic in 1988, determined that it was *colophony*, perhaps from some varnish company from the early Middle Ages. They report that much *colophony* is washed up on the Baltic shores as evidenced by recent research.

ELECTRICAL PROPERTIES

Amber becomes strongly charged with negative electricity, attracting small bits of paper, lint, or straw when rubbed on fur or velvet. This property of producing static electricity has been suggested as a test for identifying true amber in the past, but it is no longer useful since recent plastics possess similar qualities. Vigorous rubbing heats amber and one *may* detect the faint odor of pine after doing so, but the amber remains smooth and untouched by rubbing. On the other hand, such rubbing of nonfossilized resins, such as copal and kauri, often results in the heat softening the surface layers to the point where they become sticky or even "moving" from rubbing (as on a muslin buffing wheel), providing a convenient clue to identity.

MINERALOGICAL CLASSIFICATION

Amber is no longer classified as a mineral but as an organic material, or *mineraloid*, and for gemological purposes as an organic gemstone, along with coral, nacre, pearl, ivory, and jet. Because amber was found in the ground, it seemed logical in the past to consider it a mineral and to classify it as such among other organic materials such as bitumens, mineral waxes, and coals. Many classifications were used throughout the nineteenth and twentieth centuries. Now, with the advent of modern technology in the fields of organic chemistry, geology, petrology, and paleobotany, all these classifications were found to be inappropriate and in need of revision. Therefore, we now have confusion about what name to use (nomenclature), how to define what amber is and what it is not (definitions), and how we are to classify amber, copal, fossilized resins, recent resins, and 'resinites,' a term used by petrologists. Some say the terms are all synonymous, while others maintain that they are *not* the same and each should have a specific meaning. Polish, German, Lithuanian, and Eastern European scientists, who have been studying Baltic amber the longest and have found it is the most used commercially, have historically called Baltic amber the 'true amber' and classified fossil resins by terms of *succinite* or *retinite*, with *succinite* referring to Baltic amber based on its 3 to 8 percent succinic acid content, and retinite referring to all other fossil resins with 0-3% succinic acid content. Now scientists have found other resins that contain succinite, but generally not in as high a concentration as Baltic amber. The term *retinite*, from the late 1800's and earlier 1900's classification, is being discontinued because it does not correspond with recent more comprehensive knowledge based on a wide variety of fossil resin studies. Anderson and Crelling (1995, 1), working in the fossil fuel and petroleum industry, clarify the terms *amber* and *resinite*, indicating that the only difference is that *resinite* as a petrographic term refers to microscopic materials and *amber* refers to macroscopic materials. They suggest that the term *resin* used alone should be used only for modern samples.

Anderson and Crelling propose a new classification based on the structure of the terpene compounds and which resins have the capacity to polymerize. They suggest that scientists working in various scientific fields of study of amber need to communicate because rapid developments are taking place in the field, stating that, "the unique properties of fossil resins make them exceptional candidates as potential markers of paleoenvironments, sedimentary conditions, and geochemical processes." Anderson and Crelling's classification places Baltic amber in Class 1a or the group with *labdanoid diterpenes*, which gives them the ability to polymerize and to contain succinic acid. They place fossil resins such as Dominican amber in Class 1b or the group with *labdanoid diterpenes*, but not containing succinic acid (see Appendix A for a complete explanation). There are two classes that appear similar to Dana's *succinite* and *retinite* groups, but based on complete descriptions of the structure—which now can be determined by using infrared spectrometry and other laboratory technology such as thin-layer chromatology, gas chromatology, mass spectrometry, emission spectrometry, differential thermal analysis, and X-ray diffraction. Today's scientists look at amber as a terpene structure, organic and composed of many different elements, but the structure is what is characteristic.

CHEMISTRY

The broad composition of amber varies from specimen to specimen but is generally 61–81 percent carbon, 8.5–11 percent hydrogen, about 15 percent oxygen, and up to 0.5 percent sulfur. Some suggest it is the sulfur that gives amber its yellow color. Several different chemical formulas have been reported in the history of chemical studies from the late nineteenth century up until about 1970, but none of these are considered meaningful today. It must be recalled that early chemical analyses were based on Helm's (1877, 209) classic studies and being early research the method employed involved several sources of error. Baltic scientists, even in early studies, based identification of Baltic amber on its 5 to 8 percent succinic acid content to differentiate it from other amberlike materials.

Testing for succinic acid in the past was conducted by chemical analysis that required destructive pyrolysis, or heating to decompose the amber. The specimen was ground, then heated, with close observation of the melting process. After volatile substances were driven off, the residue was boiled in ether, then dried and weighed. After weighing, more ether was added, along with nitric acid. Deposits of crystals were examined for the presence of succinic acid. After settling for 12 hours, the remaining liquid was mixed with distilled water and separated into fractions, one of which was heated and smelled for turpentine or resinous odors, and another was filtered and diluted with sulfuric acid to indicate the presence of succinic acid deposits (Allen 1976*b*, 11). Today succinic acid and other structures can be detected more accurately by the use of the infrared spectrometer.

More significant to modern scientists is that amber is composed of isoprenoids, a particular type of hydrocarbon compound found in natural resins. The isoprene units (C^5H^8) link together, forming more complex organic molecules called *terpenoid* compounds. These are characteristically found in the resin of plants, with proportions of isoprene units varying from one plant species to another. These components, being found in both fossilized resin and in resin from recent plants, provide a basis to compare recent resins with the structure of fossil resin to determine what the source tree for the resin may have been. Thus, identifications can be made of the ancient species of amber-producing trees, giving some insight into the evolution of present-day resin-producing vegetation. However, problems may arise when these identifications are compared with living species since the "amber tree" may in fact be extinct and not extant. It has long been recognized that amber originates from resin, or the sticky pitch found on the sides of pine or other resin-producing trees—even cherry trees. Botanists describe resins as "complex mixtures of terpenoid compounds, acids, and alcohols secreted from plant *parenchyma* cells" (Langenheim 1969, 1157; see Chapter 6). Amber, or a *fossilized resin*, is no longer a *mixture*, but by polymerization, evaporation, dehydration, and effects of bacteria, its components can no longer be separated without destruction (Savkevich 1985).

In reference to providing an analysis of the chemistry of

amber, a fundamental change began to take place when amber's botanical origin was considered during the chemical investigation, such as the work of Gough and Mills (1972, 527–8). This study was based on the assumption that amber originated from conifers and past investigations, which had confirmed amber to "contain a few tenths of a percent of the monoterpene alcohol borneol, free or esterified, with about a ten percent excess of dextrorotatory over the levorotatory isomer"(Larsson 1978, 16). Based on this knowledge of its organic composition, Gough and Mills identified small amounts of nine *diterpene resin acids* with skeletons of the usual *labdane* (I), *pimarane* (II), and *abietane* (III) types in the ether-soluble fraction. Labdane is derived from agathic acids commonly found in resins of the araucarian tree *Agathis australis*, while most pine resins are characterized by high abietane derived from abietic acid.

This discovery resulted in the presumption that the bulk of succinite was derived from polymerization of *diterpenes* (which, by the way, were the major constituents of conifer resins). As stated in Chapter 1, it was in 1890 that Conwentz classified the source tree for Baltic amber as the *Pinus succinifera*. Despite many years of research having taken place previously, in 1961 Schubert (1961, 3–149) assured scientists that "without a doubt" *Pinus* was the correct genus. This inspired Rottlander, (1970, 35–51. 1974, 78–83) to put forward a theory, based on resins from modern European pines consisting mainly of *abietanes*, that amber was composed of abietic acid. As late as 1978, a German geology dictionary described amber as "flammable polyester made of abietane acid and diabietinol along with resin acids and amber acid." However, there were certain problems with this theory that were pointed out by botanists of the period. Also, Rottlander's experiments of artificially producing amber from pine resin could not be duplicated by Kucharska and Kwiatkowski (1978, 149–56; see Rottlander 1974, 78, and Larsson 1978, 16, for a complete description of this theory and its faults). This theory resulted in further confusion because even though abietanes are found in recent resins, fossil resins composed principally of abietane polymers are believed to be unknown (Langenheim 1969, 1156–69; Thomas 1970, 59–79; and Gough and Mills 1972, 527–8). Gough and Mills, by quantitative evaluation of similar experiments, estimate that 75–80 percent of the diterpene residues in Baltic amber belong to the labdane type, 15–20 percent to the pimarane type, and only 5 percent or less to the abietane type.

The definition of amber given by Wert and Weller (1988, 497) states that amber is "an amorphous polymeric glass with mechanical, dielectric, and thermal features common to other syn-

Figure 5.3. Boris Igdalov, director Tsarskoselsksaya Yantar, examining a Baltic amber with a small gecko falsified inclusion glued on a natural lump of amber and placing a plastic resin over the top.

thetic polymeric glasses." However, amber is a *natural fossilized resin* produced by ancient trees living as early as the Cretaceous period. Further investigations of the chemistry and overall terpenoid hydrocarbon organic distributions of fossil resins can be found in research that is too complicated for a detailed description in this book. However, this historical description of the changes in types of analyses clarifies the reasons for discrepancies in the literature about amber.

In 1995 the American Chemical Society sponsored a Symposium on Amber, Resinites and Fossil Resins coordinated by Ken B. Anderson and John C. Crelling, which brought together researchers to report on the most recent chemical studies, including structural characterization, isotopic composition, maturation studies, resinite-derived oils, and amino-acid distribution from aspects of the biology, geology, petrology and technology of fossil resins. The current research indicates that the ability to harden by polymerization is characteristic of resins containing a large proportion of *labdanes* with a conjugated, terminal diene unit in the side chain (see Appendix A). The suggested new classification of amber and fossil resins is based on the results of this conference and is reprinted in Appendix A. The schematic representation of the classification system for fossil resins proposed by Anderson et al. is also reprinted in Appendix A.

In Anderson's scheme, Baltic amber is placed in *Class Ia*, which includes fossilized resins containing succinic acid. European scientists have long established the succinic acid content of Baltic amber, and used it for identification in archaeological studies. Therefore, a presentation of the chemical studies related to succinic acid seems relevant. Succinite, or amber containing succinite acid, is produced in other deposits in various localities beside the Baltic, though not is such a high concentration. In the past the presence of succinic acid in an amber specimen was considered the clue indicating that the amber originated in the Baltic area. In archaeological research, when amber artifacts were found, they were tested for succinic acid to determine the place of origin. If the specimens did not contain succinic acid, the origin was considered to be other than the Baltic. Since technology has changed some of the older appraisals of amber based on the content of succinic acid, the accuracy of the old amber trade routes is being questioned. Currently researchers from Italy and Poland are collaborating in reanalyzing the trade routes.

EXAMPLES OF ANALYSIS BY MODERN TECHNOLOGY

Amber being one of the oldest gemstones, it has been subjected to study by scientists for decades. Early studies focused on chemical investigations using the analysis of the elements and the physical properties such as described previously. This variety of modern technology used for identifying the structure of amber in research studies requires sophisticated laboratory techniques to provide a "fingerprint" of resins by analysis of their constituents. Infrared spectrometry, mass spectrometry, and differential thermal analysis each measures, in a different manner, effects based on different *molecules* present in the amber.

Because amber is relatively insoluble, early work of the 1980s used techniques applied to a total solid, such as infrared spectrometry (IRS), carbon-13 nuclear magnetic resonance (C13NMR), and pyrolysis/mass spectrometry carbon-13 nuclear magnetic resonance (PC/MAS C13NMR). These techniques provide information about the carbon functionalities of the amber. Molecular analysis by gas chromatography-mass spectrometry (GC-MS) was done on the solvent-soluble portion of the amber, which has led to the recent reclassification system suggested for amber and resins.

The modern most accepted method for distinguishing succinite, or the presence of succinic acid, is to employ an IR spectrometer to record the IRS spectra and compare spectral characteristics of the resins tested with those of known specimens. Variations in the spectra are indicative of specific differences in molecular structure of various types of resin, each species having its own characteristic spectral pattern. Beck (Beck, Wilbur, and Meret 1964, 256–70) began analyzing fossil resins using an IR spectroscope as early as 1964. By 1982 the Vassar College laboratory had examined more than 5000 specimens producing analytical graphs showing IR absorption rates plotted against frequency or wavelengths. A characteristic shoulder for Baltic amber, or succinite, was established (Beck 1986, 57–110). Some caution was suggested earlier by Savkevich and Shakhs (1964, 2717–9) since different resins from the same source tree may produce different spectra and the degree of oxidation the samples have received may cause a variation within the same type of amber.

Emission spectrometry can be used to identify and quantify any inorganic elements present, such as impurities or traces of other elements. Using similar techniques, it has been found that amber occasionally contains trace elements of silver, copper, tin, lead, silicon, aluminum, magnesium, calcium, titanium, manganese, iron, nickel and others (Kuziorowska 1988, 11). Pyrite may be observed with the naked eye in some specimens.

X-ray diffraction measures the characteristic arrangement of atoms in various compounds. Thus, X-ray diffraction studies also have been used to examine amber to detect the presence of crystalline components. Though amber itself is amorphous, that is without crystalline structure, crystals of succinic acid may be detected within the amber mass. Broughton (1974, 583–94) used this technique to distinguish between New Zealand kauri gum and copal resins from Zanzibar and other African regions. This was perhaps possible because recent resins have very strong intensity value patterns.

Thin-layer chromatography was used by Frondel (1969) to link fossil resins from locations in England with those of Highgate copalite and glessites. This method separates various chemical components of a resin in solution by different speeds with which they are carried through a prepared mixture of absorbent, resulting in a chromatogram on glass plates specially prepared with a fluorescent indicator. Care must be taken in interpreting the results of even the newer technology. Savkevich (1981, 1–4) critiqued the work of Lebez (1968, 544–7), stating Kucharska and Kwiatkowski's (1979, 482–4) research indicated the former was invalid because differences obtained by thin-layer chromatography can (and in this case were) result from different degrees of oxidation and not to differences in geographic origin.

Emission spectrometry requires that the sample of amber be heated to a high temperature, and the atoms, or molecules, excited by the heat energy can be identified by wavelengths of light that they absorb or emit. Analysis of the recorded results identifies the elements present in the sample and presumably attempts to differentiate between resins from different localities. Fraquet (1987, 155) points out two major difficulties with emission spectrometry as an analysis form:

1. Amber samples may be influenced by the immediate environment where they were buried and thus show local variations rather than that of the original location.

2. Previous reference values show very large ranges. To correct for this problem, Fraquet suggests samples be taken from the interior of the fossil resin to avoid contamination.

Differential thermal analysis is generally used to identify the action of heat on resins. A sample of crushed resin would be heated in a closed tube (inert atmosphere) and weighed at each 10°C per minute increase from room temperature to 620°C or 1000°C. The reactions that produce heat are recorded in upward peaks on a graph, while those absorbing heat are shown as troughs. Fraquet (1987, 150) explains that the higher proportion of volatile components found in modern resins compared to fossil resins is graphically illustrated by an increase in trough intensity across a lower temperature range. Broughton (1974, 583–94) used thermal characteristics to differentiate between Atlantic Coastal Plain fossil resins and those from other geographic regions. Previously IR examination had not conclusively provided the framework for the geographic-botanical affinities.   Neutron activation analysis is a complex form of analysis using trace element content to identify the geologic origin of amber, but unfortunately this technology requires access to a nuclear reactor. Radioactive nuclides, formed by neutrons bombarding atoms, can be detected and identified as to the element present. In 1969, Das (1969) reported using this nondestructive activation analysis to examine samples of amber from various localities. Interestingly, he found 200 parts per billion of gold in Baltic amber and 600 parts per billion in Sicilian amber!

C13NMR spectroscopy was used by Beck, Lambert, and Frye (1986, 411–3) and Lambert, Beck, and Frye (1988, 248–63) to analyze fossil resins from other northern European locations found along with Baltic amber with some interesting results. The fossil resin referred to as beckerite often found along with Baltic amber was found to be a contaminated form of Baltic amber.* The fossil resin commonly called stantienite produced a spectrum more similar to coal than amber and was thought to be resin-impregnated wood. Gedanite, a soft resin also found along with Baltic amber, fascinatingly revealed two different materials that were called gedanite since one set of samples when analyzed provided a spectra identical to Baltic amber and the second produced a quite different spectra.** Fossil resins commonly called simetite from Sicily, rumanite from Romania, walchowite from Czechoslovakia, and delatynite from Delatyn, Russia, all produced a spectra characterized by a weak, or absent, exomethylene resonance,

Figure 5.4. Dominican amber with plant debris inclusions a, with back lighting; b, with both back lighting and surface lighting

*2006 personal conversation with Dr. Kosmowska-Ceranowicz stated, *"Beckerite is not contaminated form of Baltic amber!
**True gedanite never provided IRS identical to Baltic amber! It was badly identified Baltic amber!!"*

thus, placing them in a second category.

Andrew Ross (1998, 9) states infrared spectroscopy can indicate (as did Savkevich, Beck previously) whether the amber being examined is Baltic amber or not by looking for the characteristic plateau on the side of one of the peaks. This plateau is known as the 'Baltic shoulder.' He also indicates mass spectrometry is able to distinguish between different types of amber only that this process is not widely available and is more expensive.

These modern research instruments used in technology laboratories have provided a completely new approach to understanding amber and fossil resins. Today scientists recommend using infrared (IRS) examination of amber to verify its authenticity. However, natural succinite,* treated amber, or even tiny amber pieces pressed together produce a diagram very similar to each other and it only shows the functional groups which were not changed during any thermal treatment or pressing of the amber base.

Scientific studies in Poland, Lithuania, Germany. and Russia, where amber is found in nature continue to examine amber's natural shapes and describe bulky Baltic amber pieces as large as 3 kg in weight to icicle shaped and other dripstone forms which are composed of layered structures to small droplet forms the size of small tear drops. From these structural forms they have derived an affiliation with coniferous Paleogene trees origin which were large and produced great amounts of resin. Modern scientific analysis has suggested a coniferous tree origin from the family Pinaceae which resembled species such as today's *Cedrus* or cedar now growing in the Atlas mountain system that runs from southwestern Morocco along the Mediterranean coastline to the eastern edge of Tunisia, and *Pseudolarix* or larch tree. Today the most extensive deposits of Baltic amber are found in Eocene blue earth that occurs in the Sambian Peninsula and by the Bay of Gdańsk (Gdańsk, Chłapowo, Karwia) (Piwocki et& Olkowicz-Paprocka 1987; Kosmowska-Ceranowicz 1996).

Since commercial trade by European amber merchants has now been opened to the entire world the Commission of Experts of the International Amber Association of Poland, professional organization, has developed standard principles of classification for amber gemstones. According to Gierłowska (1998, 28), these classifications are:

Classifications of Amber Gemstones (Gierłowska, 1998)
Amber, Natural Amber, Succinite--raw material, semi-finished products or products made from Baltic amber without

natural Baltic amber as well as Ukrainian and Saxony amber

any changes to its natural physical and chemical properties.

Treated Amber (Succinite)--semi-finished products made from Baltic amber with artificially changed i.e. through temperature or pressure, physical and chemical properties.

Doublet, Triplet or Multi-element Amber--consists of two, three or more layers, at least one of which must be natural or treated amber. It is permitted to put a colorful layer made from a durable material between amber elements or place a formed amber piece on the base of another material.

Reconstructed (Pressed) Amber--material from tiny pieces of Baltic amber or amber meal pressed together in high temperature and pressure.(Gierlowska, 2003, 23.)

The International Amber Association refrains from the use of reconstructed amber in the manufacture of jewelry, however, they traditionally use it for manufacturing larger objects such as boxes and other ornaments. They suggest trade companies at each stage of sale contain general information on what processing was used if any.

Even though the technology is not available to the amber connoisseur, collector, or fossil collector, the contributions made by the scientists in various fields such as geology, botany, petrology, and ancient DNA research add to the intrigue of the resin that was viewed as having magical powers by early humans.

Figure 5.5. Modern amber jewelry displayed in Russia at St. Petersburg shop in 2004. Both Polish and Russian amber artists produce a variety of amber objects using a variety of techniques. They have learned modern techniques to enhance the raw material.

Wing of inclusion in Bitterfield amber. Courtesy Krumbiegel.

6

SCIENTIFIC STUDIES OF AMBER

If thou couldst but speak, little fly,
How much more would we know about the past?
—Kant

HISTORICAL SPECULATIONS
ON THE ORIGIN OF AMBER

Amber has interested scientists for centuries. The mystery of its origin stimulated inquiry and speculation as early as 500 B.C. by Aeschylus (525–456 B.C.) and Sophocles (490–406 B.C.). Ancient myths associated amber with tears, apparently because of the specific, droplike form of the pieces. Amazingly, even the ancient tale of Phaëthon's sisters turning into poplar trees and shedding tears of amber hints at the organic origin of amber.

Pliny's *Natural History*, written about A.D. 79, about the origin of amber dismissed views expressed by his predecessors as groundless, stating emphatically that amber was of vegetable origin, formed from a coniferous tree gum that hardened with time. As it hardened, it fell into the waves of the sea, later to be deposited on the shore. As proof, he pointed to such properties of amber as its capacity to give off the odor of resin upon rubbing, the smoky flame resembling that of pine resin when burned, and the presence of insects within amber (see Fig. 6.1).

Strangely, this essentially accurate information of the origin of amber was ignored or forgotten. It was not until the Renaissance evoked a new interest in amber that the question was reexamined. Three outstanding scientific authorities of that time all concluded that amber was of inorganic origin: namely, Athanasius Kircher, S. J. Philippus Aureolus Bombastus Theophrastus von Hohenheim (who called himself "Paracelsus"), and Georg Bauer (who called himself "Agricola"). The latter, Georgius Agricola (1494–1555), the leading authority on mining and processing of useful minerals, believed amber

Figure 6.1. Primitive fly (Diptera) in Baltic amber Eocene Age.

Figure 6.2. Modern day fakes available in Poland photographed at a dealer's booth in the Tucson Gem and Mineral Show 2004 for sale at a price of $100.

was formed by solidification in air of liquid nonorganic bituminous matter flowing from the "bowels of the earth." In fact, he thought petroleum was merely liquid amber, and, as proof, he noted that amber often contained liquid-filled bubbles or uncongealed "petroleum." He believed petroleum, as it seeped from the bottom of the Baltic, rose to the surface in liquid form and embedded small gnats and flies at the surface.

Adherents of the "petroleum theory" lived in Italy, Switzerland, and southwestern Germany, never having traveled to the Baltic to see amber as it was actually found. Yet they described "oil slicks" on the Baltic Sea and told of amber being washed up on the shore while still in a "tar-like" form. Some of them even speculated that oil was seeping out of the sands of Samland's coast. However, no Baltic fisherman had ever seen oil slicks or found an amber lump in a tarlike condition.

Athanasius Kircher (1601–1680) supported the liquid petroleum theory based on finding small fish imbedded in amber specimens in his collection; he suggested they were trapped in the oil as it rose to the surface. Unfortunately, since about a.d. 1500, fake fish, lizards, and frog inclusions have been placed in amber specimens by hollowing out the piece, inserting the small animal, filling the interstices with rapeseed oil, and resealing. No fish has yet been found naturally imbedded in amber; recently, however, small geckos have appeared. Daniel Herman in 1583 and Nathanael Sendel in 1742 published the first volumes exposing these fakes in amber (see Fig. 6.2 a, b, c.).

The belief in amber's inorganic origin predominated for several hundred years, but other, equally unfounded theories appeared that attributed amber's origin to living creatures. For example, some authorities thought amber was congealed honey of wild bees, and others believed it to be a secretion of ants. Still others were sure it was excreted by whales, thus confusing amber with ambergris, a regurgitation of whales, even though at that time the distinction between these two products was already well known.

About 1750 George Louis Leclerc, Comte de Buffon, the eminent French naturalist, proved that coal and some other minerals, including petroleum, were of organic or vegetable origin. Buffon asserted that these minerals were remains of old forests, and if this were true, amber, even if it were "solidified petroleum," also would be an organic product.

About the same time, Karl von Linne (Carolus Linnaeus), the Swedish naturalist and founder of modern systematic botany, collected facts about amber that convinced him it was organic in origin. By now, the assembled knowledge about the nature of amber beds, the mineral remains found in it, and the amber pieces found in chinks and under the bark of fossilized trees forced the reexamination of Agricola's petroleum theory.

Mikhail Vasilievitch Lomonosov (1954, 389), a Russian scientist, was the first to denounce the petroleum theory in its entirety. While compiling the catalog of the Mineral Collection of the Russian Academy of Sciences in 1741, Lomonosov closely studied amber and found evidences confirming amber's vegetable origin. Emphasizing its unique physical and chemical properties, he noted the nearness of the specific gravity of amber to that of pine resin and pointed out that "water separated from amber by chemical means

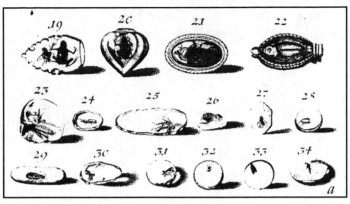

Figure 6.2. First illustrations of fakes in amber; a, by Sendelio in 1742; b, Title page of Munster's and c. of Herman's 1583 publications of fakes in amber.

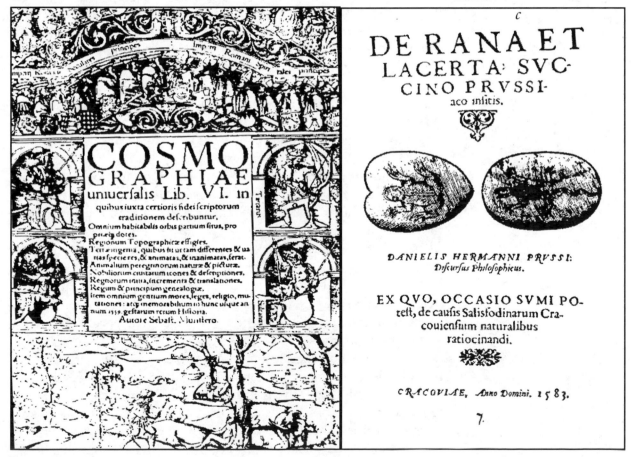

smells of succinic acid, a property of growing things." In a speech delivered in 1757 to the academy in St. Petersburg, he denounced the petroleum theory and stated that amber had to be fossilized resin of some ancient tree and that the prevailing explanation of the presence of insects in amber was unfounded.

Ten years after Lomonosov's speech, Friedrich Samuel Bock (1767), a great expert on amber from Königsberg, confirmed Lomonosov's findings in a monograph whose title is freely translated as *Attempt at a Short Natural History of Prussian Amber*. Bock said everything found naturally imbedded in amber related to a forest environment; for example, flies, gnats, and ants, all of which crawl along trunks of pine trees. He also noted that amber was found deep in soil associated with a blue loam. Based on this observation, Bock reasoned there may have existed during Pliny's time islands between Sweden and Samland covered with forests that produced resin, but these islands had sunk during intervening centuries. It was from these forests, Bock said, that amber was continually being washed ashore, especially during storms that churned the seawater.

Elaborating on the studies and logic of Bock, scientists began placing the time of the existence of amber forests further and further into the past. Buffon became convinced that the origin of the Earth preceded recorded history by about 80,000 years, and plants and animals during this period grew or lived in places where they no longer exist, with some species now extinct. By 1811 there no longer was any doubt that amber was the product of pinelike trees, but because no living tree could be found producing amber, it had to be resin from a prehistoric tree.

During the 1830s, geologists began studies of deposits in which amber was found, discovering fossilized plants and leaf prints in lignite coals dating to the Paleogene period. The expert on Tertiary (now designated by Paleogene) fossils, Oswald Heer (1859, 1869), of Switzerland, identified the remains of swamp cypress and sequoia similar to North American varieties, but it was not until 1850 that George Zaddach (1867, 85–97) found that amber was fossilized long before the first tree of the lignite forest grew. Baltic amber is now thought to have formed during the early Paleogene period, or at least 40 million years ago.

Since amber occurs only in secondary deposits, having been transported from its original source by geologic processes resulting in changes in land formations, glaciation, and changing action of sea currents, scientists have not identified the exact epochs of the Paleogene period in which amber forests existed. However, stratigraphic investigations of the present amber-bearing deposits of the Baltic region indicate

Upper Eocene. This means the primitive forest must have come into existence no later than 40 million years ago, and may have existed before then, but an exact time span of these forests has not been definitely established. Some of the richest Baltic deposits of amber also occur in the Upper Eocene strata, suggesting that the forests still existed during that epoch.

The areas of Europe covered by these forests are still only approximately defined, and, as mentioned in Chapter 1, extensive continental changes took place between the Cambrian and Paleogene periods in northeastern regions of Eurasia or the area called Fennoscandia. During the Paleocene, or the earliest epoch of the Paleogene, Fennoscandia is believed to have covered the entire Baltic Sea area and more. Early nineteenth-century scientists offered several descriptions of the boundaries of this primeval forest, but these earliest descriptions limited its size, with Berendt (1845, 61–124) erroneously believing, as Bock had earlier, that the forests had existed on an island near Sweden that became submerged and was at the bottom of the Baltic Sea. Berendt theorized that amber deposited by these sunken forests washed up on the Baltic shore from Klaipeda to Gdańsk. It was Berendt who placed the location of the forest at latitude 55 degrees and longitude 19 to 20 degrees, obviously much smaller than what is currently believed. By the end of the nineteenth century, scientists such as Heer (1869, 1–4) and Conwentz (1890*b*, 81–151) expanded the size of amber-bearing forests to cover the area now comprising Finland, the Baltic Sea, and large portions of Scandinavia, as well as northern regions of Latvia, Lithuania, Poland, and Germany.

CURRENT GEOLOGIC THEORIES ON THE ORIGIN OF BALTIC AMBER

Currently, it is believed that amber-bearing forests extended far beyond the boundaries of Fennoscandia, covering much greater regions than imagined by earlier authors. In 1943, Komarow (Zalewska 1974, 82) described it as a vast wooded area covering much of the northern Eurasian continent, reaching almost as far south as the Black Sea. He included the area from the Scandinavian mountains to the northern Urals, northern regions of Pomerania, Lithuania, Latvia, portions of Belorussia, and the Ukraine.

Using radioactive methods for dating rocks, scientists now calculate that the Upper Eocene began approximately 40 million years ago, while the Lower Oligocene ended around 34 million years ago. This makes it possible that amber-bearing forests existed over a span of 5 million years, as confirmed by finds of amber deposits with-

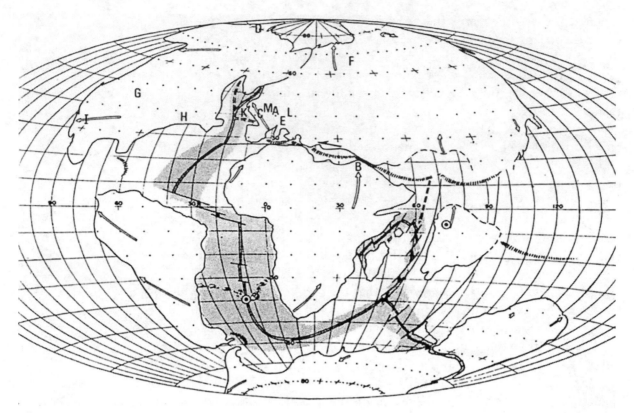

Figure 6.3. Position of the continents at the end of the Cretaceous, with locations of amber as old as or older than Baltic amber.
Legend: A, Bornholm, Denmark (Jurassic); B, Lebanon (earliest Cretaceous); C, Northwestern France (early Cretaceous); D, Alaskan Arctic Coastal Plain (early Cretaceous); E, Central Europe (Cretaceous of different ages); F, Central North Siberia (Upper Cretaceous); G, Cedar Lake, Manitoba, Canada (late Cretaceous); H, Atlantic Coastal Plain (late Cretaceous); I, Simi Valley, California (Eocene); K, Southeast Coast of England (Eocene); L, Baltic Coast (Eo-Oligocene); M, Rhine Valley (Eo-Miocene). Based on data from Larsson 1978.

in strata dated to these periods. During its span of existence, boundaries of the forest changed, with the range both increasing and decreasing with the passage of time, its limits generally expanding toward the south. The continents were not completely separated during the Cretaceous age and were still separating during this period (see Fig. 6.3). Northern Eurasia was subjected to repeated upheavals of the Earth's crust during the forest's existence, with major changes in boundaries as prehistoric seas invaded the region, then regressed, causing continuous changes in the configuration of coastlines. At the same time, slow but continuous changes in climate may have altered the composition of the flora of the forests, yet trees continued to supply large quantities of amber-producing resin. Fossil resins have been produced in various geologic times and localities, but never in such vast quantities as are found in the Baltic area, especially before the upper Eocene epoch.

Knowing that amber was a product of forests and therefore of land origin, geologists were puzzled to find amber deposits occurring in marine sediments. Further geologic studies provided reliable evidence that vertical movements of the Earth's crust took place near the end of the Eocene and during the Oligocene in central and eastern European areas, which caused repeated marine transgressions and regressions. As a result, marine deposits were laid over land deposits, which, in turn, could have again been elevated to support a new forest and further deposits of amber. In 1911 Kaunhoven (1913, 1–80) found evidence that the boundaries of marine and continental areas in Sambia during the Paleogene progressed through as many as nineteen changes (Zalewska 1974, 57). Today scientists know that this very process of being carried by the saltwater seas, being cast up on the shores, or carried by rivers into shallows of its delta sediments, then covered with silts and clays prevented oxygen in the atmosphere from eroding the surface away. This process gave the amber-producing resin the appropriate burial, allowing it to continue the process of polymerizing into amber.

The oldest Baltic amber deposits occurring in the lower blue earth layer were transported from nearby lands by the Eocene Sea. Streams flowing from amber-bearing forests carried the first deposits from the forest floor with the current as it emptied into the sea. Kosmowska-Ceranowicz (1995, 200) suggests that during the Upper Eocene a river flowing from the north had a fingerlike branching delta that extended as far as the western shores of the Sambian Peninsula and to the east of Gdańsk as far as Karwia on the Polish coast. It was this theoretical river that carried the amber that is now deposited in the sediments that are called blue earth (see Fig. 6.4). Currently, this digital-shaped delta is the region called the Chłapowo-Sambian Delta, the amber-rich sediments between the Sambian Peninsula and the western shore of the Hel Peninsula. According to this theory, the delta was laid down in the shallow, northern coastal zone of the ancient intercontinental Northwest European Eocene Basin about 40 million years ago. Poetically, Kosmowska-Ceranowicz relates this to Pliny's myth, naming the river the Eridanus. Three core drill samplings in the Chłapowo region indicated that amber seams in the western part of the delta lie deeper than in its eastern part. The Poland area deposits, as mentioned in Chapter 3, lie at a depth of about 120 meters, which makes exploitation impractical at

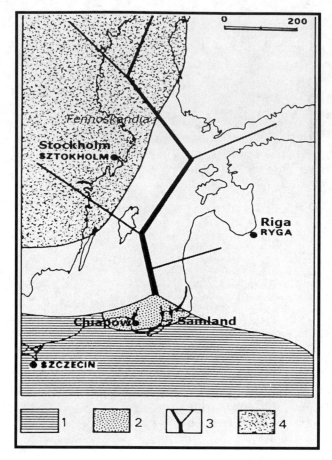

Figure 6.4. Map of the hypothetical amber-bearing river and its delta, carrying the richest deposits of Baltic amber. By K. Jaworowski, 1984; reprinted from Kosmowska-Ceranowicz 1989.

the present time (Kosmowska-Ceranowicz 1996a, 5). The first stage of amber's transport from the amber-bearing forests was through the delta of the "Eridanus River" (see Fig. 6.5). As regions covered by amber-bearing forests were flooded during major marine transgressions at the close of the Eocene, soil containing amber deposits were eroded. Masses of amber, buoyant in agitated salt water, were carried by wave action and deposited on the sea bottom with sands, clay, and remains of sea fauna (see Fig. 6.6). Although amber occurs in a secondary bed, geologists believe Eocene amber-bearing deposits are only slightly younger than the major amber lumps themselves because the greater part of the amber was being moved with the seas before it was deposited in the successive strata. The secondary beds, however, are richer in amber than were the original beds of resin in the amber-bearing forest. Because sea waves continued to wash over the forest ground cover and soil, loosening and transporting lumps of resin to new marine beds, the secondary deposits continued to build up into thicker and greater amounts.

The upper blue earth deposits of the Baltic were formed in a similar manner during the Lower Oligocene, but scientists are not certain if the primeval amber-bearing forests still existed or if they continued to produce resin during this epoch. However, younger, geologically speaking, amber is found in the Baltic region along with the Baltic succinite.

Successive sea transgressions washed out amber resin and transported it to new areas, and, beginning with the middle Oligocene, the invading sea attacked exposed fragments of amber from previously formed deposits and carried them to new places. For this reason, traces of amber are now located in several strata in the Paleogene sequence from the Eocene to the Miocene. It is even found in brown coal deposits that were formed just prior to the Eocene and are associated with beds of lignite.

During the Ice Age, further transportation of amber took place, particularly during the Pleistocene epoch, when glaciers extended over southern Baltic regions near Sambia, cutting off considerable portions of the amber-bearing beds. Amber, along with the remains of the blue earth, was picked up by the swift currents of the glacial waters. Then, amid the mass of stony mud, it was transported southward. This kind of transport differed from the transport from the mother forest, as it was now mixed with more turbulent waters as well as mud and stones, causing the lumps to be rolled over and over, grinding away their original forms and rounding the edges of the pieces. Moraines, glacial rivers, and sea currents carried deposits over the European area and scattered them on sea bottoms and in land sediments of glacial origin. Therefore, glacial amber deposits appear in Quaternary sedi-

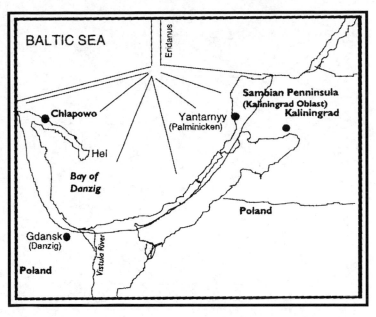

ments, because amber from Paleogene deposits was carried over considerable distances by glacial, fluvial, or fluvioglacial action resulting in a distribution across Poland, Belorussia, Lithuania, Latvia, and Germany; up to the east coast of England, over Jutland and the southern coast of Scandinavia (Kosmowska-Ceranowicz 1997, 6; [see Fig. 6.7]). The last type of transportation continues to take place today as amber continues to be deposited in new sites by the action of sea waves eroding seabed amber deposits. Shore currents pick up loosened amber and toss it about in stormy periods, depositing it on the beaches. The most complete postwar studies of the Sambian deposits (Kaliningrad Oblast and environs) were published by Savkevich (1980, 17–28) from Russia and by Katinas (1988) from Kaliningrad region, Russia. The map of Paleogene deposits of Baltic amber in Figure 6.7 diagrams the distribution of sediments of Upper Eocene, Oligocene, and Paleocene sites in the Sambian region.

Figure 6.5. The amber-bearing Chlapowo-Sambian Delta laid down by the hypothetical river Eridanus in the northern coastal zone of the Eocene basin as in Kosmowska-Ceranowicz 1997.

Katinas also presented at the six meeting on amber in Warsaw 1988 an interesting theory of the process of the amber-producing resin turning into amber after it was separated from the tree. He believed the diagenesis, that is, the breaking up of the radicals and recombining of molecules by polymerization, took place in three stages. First, as the resin was exposed to the air and airborne bacteria, volatiles escaped, terpenes evaporated, and oxidation and isomerization of the resin followed. This process continued as the resin rested on the forest floor and in the soil, continuing to be exposed to microorganisms. This progression was stage two with polymerization of the molecules forming long chains of molecules and continued

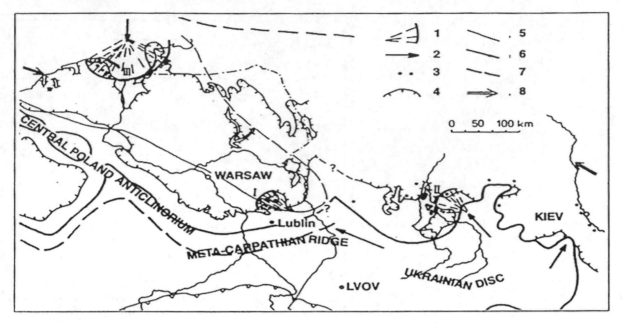

Diagram of deposit formation in the ancient seas – c. 40 million years ago
1 – amber-bearing deltas: I – Parczew, II – Klesov, III – Sambian-Chłapowo; 2 – direction from
which material was borne; 3 – amber sites; 4 – limit of Upper Eocene deposits; 5-7 – extent of North-
-West Eocene Basin; 8 – direction of transgression

Figure 6.6. Diagram of deposit formation in the ancient seas about 40 million years ago showing extent and three deltas. Reprinted from Kosmowska-Ceranowicz 1997.

"auto-oxidation." The resin continued to harden as it was transported underwater in the salt waters of the Paleogene seas, the salt water itself having a preserving quality for the amber-producing resin. During this time period Katinas described a restoration of the resin as it turned into amber and later was deposited in the delta locations of the Paleolithic river (Katinas 1971, the lecture in 1988). This process is still the topic of today's scientists as well as an attempt to compare fossil resins to recent resins to determine the source tree from which the amber came.

A historical chart of the exploitation of amber deposits in Poland in the past was published in 1974. However, it must be noted that exploitation has, of course, continued over the last 20 years, with new recovery methods such as the use of hydraulic hoses, forcing water under pressure to wash out fine-grained sands along with amber near the Baltic shores and marshy regions, and dredging in the lagoons and lakes, but the chart provides information that illustrates how extensively amber was laid down by the Paleogene seas and the glaciers during the Quaternary period (see Table 6.1).

Basically, amber comes from Paleogene deposits on the Sambian Peninsula in Russia or from accumulations found in Quaternary sediments exploited in Poland (Kosmowska-Ceranowicz 1995, 200). From the map in the diagram of Figure 6.6, one can log-

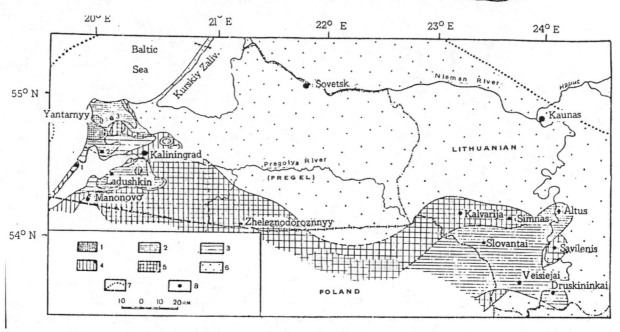

Figure 6.7. Map of distribution of Paleogene deposits of Baltic amber in Kaliningrad region. (Katinas 1971, page 69) Reprinted from Savkevich 1988.

ically see that Baltic succinite may have been deposited in various locations covered by the Paleogene seas. Today, succinite outside the Baltic region is found where it was laid down as a Paleogene deposit along the southern shores of the Northwest European Eocene Basin. Kosmowska-Ceranowicz (1995, 200) describes two other known amber-bearing deltas: (1) The Parczew Delta in Poland, located southwest of Warsaw, is considered to be Middle and Upper Eocene sediments; (2) The Klesov Delta, in the Ukraine, reaches as far as Gatcha in Belorussia.

Kosmowska-Ceranowicz (1997, 5) estimates the stores of amber deposits in the Parczew Delta south of Warsaw in the Lublin region to be about 6911 tons (Strzelczyk 1991, 1–8). These deposits are relatively easy to extract since they are rather shallow.

UKRAINIAN AMBER—PART OF THE BALTIC FORMATION

Ukrainian amber from the Klesov Delta has been extracted since about 1978. In 1993 Koshil, Vasilishin, and Panchenko 1993, 34–7) estimated the stores of amber to reach up to 420 grams per cubic meter with an average of 57 grams per cubic meter. At that time, the largest lump found at Klesov weighed 1040 grams. The Muzeum Ziemi collection contains pieces of Ukrainian succinite in various hues (see Figure 6.8, a,b). These specimens illustrate the rounded corners of the pieces that were smoothed during transportation by the Paleogenc seas. The museum's largest Ukrainian

Figure 6.8. Ukrainian amber specimens a, with dark smooth crust and yellow interior b, indicating transportation to a secondary location. Courtesy Muzeum Ziemi, Warsaw, Poland.

amber specimen weighs 800 grams. Kosmowska-Ceranowicz (1997, 6) also reports succinite being found in the Miocene sediments of Poland and Ukrainian Carpathian Foredeep. She believes this fossilized resin actually originated from ancient neighboring forests during the Miocene period. Core samples were taken in the Pikule section, south of Janów Lubelski, finding amber at a depth of 97.3 meters and in the sulfur mine in Machów in Sarmatian loam deposits. A large amber cobblestone weighing 880 grams was found in the sulfur mine area. However, in the Ukrainian region of the Carpathian Foredeep a greater concentration of amber is found (Srebrodolskiy 1980*a*, 1980*b* 184–6). In Yazov amber was extracted from clay and sandstone deposits in a calciferous binding and was dated as late Miocene. This amber is transparent and has a yellowish brown color. Its hardness is 1.7 on Mohs scale. A study of the amber from the Carpathian region using infrared (IR) spectrometry produced an absorption spectra (taken from the Yazov amber) that was similar to that of succinite (Srebrodolskiy and Glebovskaya 1971, 1081–3). Analysis by IR spectrometry of fossil resins from various localities is used most extensively to compare the fossil resin's con-

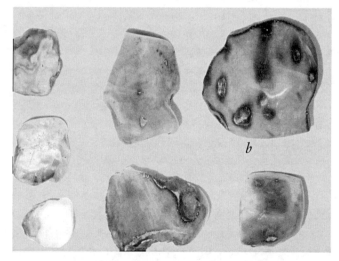

b

stituent compounds to those of Baltic succinite, which has a characteristic Baltic shoulder in the spectra. This comparison is used to determine if the fossil resin is succinite or not, or to place it in the classification suggested by Anderson (see Appendix A). Also, the spectra thus obtained are examined to see if absorption peaks in amber can be matched to modern resins, and thus suggest a common botanical origin.

In spite of its similarities, the succinite from the Carpathian region differs from Baltic succinite by its higher content of sulfur, which ranges from 3.21 percent to 4.89 percent. In the Baltic region, the average sulfur content is only 0.03 to 0.05 percent. In the older deposits located in the Ukraine, the sulfur content reaches up to 3 percent. Kosmowska-Ceranowicz's investigations indicate that the presence of sulfur is a result of diffusion of hydrogen sulfate that, in combination with oxygen, constitutes one of the components of the amber macromolecule (Kosmowska-Ceranowicz 1997, 6).

BITTERFELD AMBER FROM THE BALTIC REGION IN GERMANY

In 1974 amber (succinite) was found in central Germany at the Goitsche pit, where Paleogene brown coal was being mined. Interestingly, the Bitterfeld amber appears in upper Oligocene - lower Miocene deposits, which are millions of years younger than Baltic and Ukrainian amber-bearing deposits. This indicates that the amber-producing trees in the Baltic grew over a long period of time. Many and varied fossils were collected during the time the mine was open and this extensive collection is now at the Humboldt Museum in Berlin. Over 48 species of flora and fauna have been identified in the Bitterfeld amber. Bitterfeld is a city in the province of Saxony Anhalt. The mine is in the Halle-on-Saale region where the amber was systematically extracted from 1975 onward. Krumbiegel reported on this amber and the fossil inclusions found within it at the 1988 Symposium on Amber in Warsaw (Krumbiegel and Krumbiegel 1996). Based on IR spectrometry, Barthel and Hetzer (1982, 314–36) indicated that the resin came from an ancient conifer *Cupressospermum saxonicum* Mai, which was widespread in central Europe during the Paleogene. However, other investigators find the spectra similar to Baltic amber. Thus some authors claim Bitterfeld amber to be redeposited Baltic amber, and others claim the amber-producing forests continued to grow over the span of 18 million years. The annual yields of Bitterfeld amber reached up to 50 tons. Dry mining was conducted up until 1990. However, at that time, the mine was flooded with water as part of an environmental protection scheme. Then, underwater methods involving the use of a dredge continued exploiting the area until 1993. From as early as 1980, the Bitterfeld amber was used commercially by the factory in Ribnitz-Damgarten (the Ostseeschmuck Company, previously the Fischlandschmuck Company) producing amber jewelry and other ornamental material (Kosmowska-Ceranowicz and Krumbiegel 1988, 17; see Fig. 6.9).

Figure 6.9. Bitterfeld amber series: a, Brown coal mine at Goitschew pit, 1970, Bitterfeld, Germany; b, Goitschew pit in 1988; c, Extracting amber from Bitterfeld lower Miocene deposits with mechanical scoop; d, Barbara Kosmowska-Ceranowicz examining the Bitterfeld Miocene deposits at Goitschew pit; e, Variations in Bitterfeld succinite, a Baltic amber younger than that of Poland; f, Detail of wing venation in a fossil insect in Bitterfeld amber; g, Detail of the antennae of a fossil insect in Bitterfeld amber. Photos by Günter Krumbiegel, Halle/Saale, Germany.

Table 6.1 Exploitation of Amber Deposits in Poland

Geological Age	Lithological Description of Bed	Important Exploitation Sites	Mode of Exploitation	Most Active Period of Exploitation	Remarks
Holocene	Muddy-salty delta deposits and sandy beaches	Delta of the Vistula, the Vistula sandbar, Hel Peninsula promontory, banks of Kurpie rivers	Collection of beach amber and fishing offshore after stormy weather	Neolithic to the present	Yield to 4 ton/year. Mostly from the Vistula bar between Sobieszewo and Krynica Morska. "The Amber Shore"
Pleistocene	Gravel, sands, and moraine clays, also river sands and gravels (alluvioglacial deposits)	Bursztynowa Góra near Gdańsk, Braniewo and its vicinity (region of Sambia), the Narew Basin (Puszcsz (forest) Kurpie)	Primitive open-cut pits of the "box" type	18th and 19th centuries, small scale to present	At Bursztynowa Góra signs of mining date to some hundreds of years ago. In the Puszcsz (forest) Kurpie over 100 small pits with a good output were worked during the 19th century
Pliocene	Sandy-silty deposits of the brown coal sediments	Bory Tucholskie - Tuchola Forest	Open cuts in shallow depressions where brown coal had been mined (about 60 mines	1835 to 1865	"Nests" of amber were of such output that over 200 workmen were employed at one time
Miocene	Deposits of brown coal: fine-grained quartz sands, mostly with a high admixture of mica, also sandy silts	Region of Słupsk (Możdżanowo, Starkowo, Ugoszcz), Brętowo near Gdańsk-Wrzeszcz, Braniewo (region of Sambia), Massurian Lake District, Narew Basin, Puszcsz (forest) Kurpie	In Pomerania, small pits varying in depth with shafts provided with brattice; in the Masurian Lake District and the Narew Basin, small open pits of the "box" type (60 mines in 1835-1865)	18th and 19th centuries, small scale to present	In Słupsk region, the output of amber was abundant. Over 100 workmen used to be at work at one time. At Brętowo three pits worked until 1940
Lower Oligocene	Glauconitic sands of the amber formation (upper "blue earth")	Możdżanowo near Słupsk (western Pomerania)	Deep boring (the amber-bearing beds occur at a depth between 100 and 105 m)	1950 to 1956	Deposits rich but costs of exploitation high
Upper Eocene	Deep under water, near Chłapowo, Baltic shelf	Offshore belt of the Bay of Gdańsk, under water and on shore, Stogi near Gdańsk	Below the bottom of the Bay of Gdańsk by dredging; on shore by extractor pumps	1970 to 1972, core drillings in 1987 to present, extraction continues	The accumulate mass of amber is an extension to the west of the Sambian beds, laid down in the Tertiary river delta

Source: Adapted from Z. Zalewska, *Amber in Poland, A Guide to the Exhibition* (Warsaw: Wsydawwnictwa Geologiczne, 1974)

AGE OF AMBER

The age of amber is sometimes difficult to determine. The geologic method of identifying the index fossils in the layer of earth where the amber deposit rests is often misleading. This is because of the continued transportation amber experienced during the Paleogene period by the river and seas and in the Quaternary period by the glacial and fluvial movements. Also the use of chemical analysis is questionable in determining the age of fossil resins. According to Anderson,

> It is certainly true that definite and measurable transformations occur in the chemical structures of fossil resins over time. . . . The reason for this is that all chemical transformations, except nuclear decay proceed at a rate which is determined by (i) the rate constant of the particular reaction and (ii), the temperature at which the reaction is allowed to proceed. Normally, a reaction will proceed faster as the reaction temperature is increased. Consequently, the vast majority of geochemists rely on the concept of "maturity," which takes into account both age and thermal history. To put this concept in context, two identical resins, deposited at the same time and buried at the same rate in sediments with different geothermal gradients, will have different maturities due to acceleration of the chemical changes in the resin which has experienced higher temperatures. That is, even though the samples will be the same age at any given point in time, they will have different maturities (Anderson and Crelling 1996, xv).

With this in mind, geologists still attempt to establish the age of the geologic strata by studying the fossilized materials associated with the amber, as well as spore and pollen assemblages occurring within the amber when new fossiliferous amber deposits are uncovered outside the Baltic region. Details of the source and age have been established for some of the well-documented fossilized resin locations, such as the Simojovel Formation in Chiapas, Mexico (Oligocene-Miocene), the beach alluvium of Cedar Lake in Manitoba, Canada (reworked Cretaceous), and numerous sites in the Cretaceous strata of Alaska.

AMBER DEPOSITS OF RUSSIA AND FORMER COUNTRIES OF THE SOVIET UNION

As a result of the ancient rivers, seas, and glacial action transported over the Baltic regions, amber deposits and amberlike fossil resins are found in numerous areas of Russia, and northern regions of Lithuania, Latvia, and the Kaliningrad region where amber has been recovered since antiquity. (The term "amberlike" fossil resins is used here as in Russian literature, which reserves the name "amber" for Baltic succinite. Other fossil resins found even in the Baltic area are considered "amberlike" fossil resins, thus given other names.) Historically, as early as the 1700s, amber was discovered in the northern regions of Siberia (Yushkin and Sergeyeva 1974, 152–3).

Cretaceous amber was reported as early as 1730 from the Taimyr Peninsula in northern Russia or Siberia. This amber is found in four main deposits. The amber deposit at Romanikha, eastern Taimyr, the region where the northernmost forest of larches grows, is located in the Khatanga Depression and has been estimated to be 80 million years old. This amber is found as "lens" in lignite embedded in the sand and clay wall on the side of the mountains at Yantardakh, a Russian name meaning Amber Mountain. The pieces often contain inclusions. In the center of the Taimyr Peninsula amber is found on the edge of Lake Taimyr in Baikura-neru Bay. Most of the fossiliferous amber has been recovered from this area. The deposits from both the central and western regions of Taimyr are considered to be 100 million years old from the Cenomanian epoch, Dolganian and Begichev formations. The fourth deposit is located just off the west coast of Taimyr, at Arctic Institute Island. The tiny inclusions in this Siberian amber have been studied by scientists from the Paleontological Institute in Moscow for many years (Grimaldi 1996a, 55). Cretaceous amber has fascinated scientists who study the evolution of insects because amber's unusual quality of preserving insect inclusions by "embalming" conserves them almost intact. Cretaceous age insects often are distinctly different from recent insects, whereas insects were quite well developed by the Paleogene.

In the middle of the nineteenth century, early geologic field trips discovered amberlike fossil resins in eastern and southern regions of the Soviet Union. Reports describe grain size, color, and transparency, but not the quantity of amber contained in the deposits. Interestingly, amber was found along the shore of the Arctic Ocean and was reported to be used as a substitute for labdanum (ladanum), a dark-colored brittle resinous substance from Old World plants known as rock rose (genus, *Cistus*), which was used in flavorings and perfumes.

In 1936 and 1937 extensive work was begun in Transcaucasia for extraction of fossil resins found near the village of Shasha, but the small amounts found made the work unprofitable. During this same period, amber was found in the far east of Russia while processing coal. The amber was separated by hand during the treatment of the coal. Unfortunately, only 8 to 10 percent of the total material mined was amber, therefore, it was not considered useful for the jewelry industry. However, a fossil resin found on the eastern coast of southern Sakhalin and on the beach near the villages of Vzmor'ye, Firsovo, and Starodubskoye is both distinctive and unusually decorative (Savkevich 1975, 919–23; Zherichin and Sukacheva 1973, 3–48). From the the late 1800s to about the 1980s, when new fossil resins were discovered, investigations focused on whether they were succinite or not (retinite). Often, even though many were not succinite, they were still used in the jewelry industry. Investigations of this era found that most amberlike fossil resins were not succinite. In fact, in each region different types of fossil resins were found and named by the investigator. Today's scientists, focusing research on finding the botanical affinity or source tree for the fossil resin, have considered this nomenclature confusing. Nevertheless, these names have been used throughout a hundred years of research and seem to identify the area where the fossil resin was found. For example, in the Sambian, or Kaliningrad district of Russia, six types of fossil resins occur: succinite, gedanite, glessite, stantienite, beckerite, and krantzite. The succinite, or the classic Baltic amber, makes up 90 percent of the total yield. In the Carpathian area, rumanite, schraufite, and delatynite are found along with succinite.

OTHER FOSSIL RESINS ASSOCIATED WITH BALTIC AMBER

Baltic amber deposits often are found in conjunction with smaller quantities, perhaps as much as 2 to 3 percent, of other fossil resins. Research continues to determine if these fossil resins were formed from source trees other than the amber-producing trees from which the typical Baltic amber (succinite) was produced. Or are they the result of differing geochemical conditions in which the resin found itself during the first stages of hardening, or, as Katinas (1971) described—"diagenesis"? Keep in mind that most research was conducted by Eastern European scientists on Baltic succinite, who labeled this the "true" amber. Then, as paleo-botanists began studying all the fossil resins and their chemical composition, it was decided to classify amber based on those that contained the identifying characteristics of succinic acid content. Those that did not contain from 3 to 8 percent succinic acid were assigned to a group labeled retinites. Today's researchers using IR spectrometry and other sophisticated technology to study the structure of fossil resins have discarded the use of retinite as a label. Anderson's classifications are based on the structure of the fossil resins as the determinant for placing a fossil resin into related categories (see Appendix A). This information is important because much research was completed long before the 1995 classification system was suggested.

Because several other fossil resins have been known since mining of Baltic amber began, the ones commonly occurring in association with Baltic amber deposits will be described here. In the past, some called these fossil resins "pseudoambers" since most were not considered suitable for use in the jewelry industry. However, the term fossil resin will be used to describe the research findings related to each. In the 1980s, Lambert and his colleagues, analyzing other types of fossil resin from northern Europe using the carbon-13 nuclear magnetic resonance (C13NMR) spectroscopy, found results that differed from earlier chemical and geologic studies. The results of their studies will be included in the information about each fossil resin (Lambert and Frye 1982, 55–7; Lambert, Beck, and Frye 1988, 248–63; Beck, Lambert, and Frye 1986, 411–3).

Gedanite, a fossil resin found associated with Baltic amber, is pale yellow transparent, and when first uncovered appears to be coated with a white powder. The name gedanite comes from the old name for Gdańsk—Gedanu (Kosmowska-Ceranowicz 1996*a*, 7). It is known to miners as brittle amber because it splinters easily. Its hardness is only 1 to 1.5 on Mohs scale because of its brittleness, therefore, it has limited use in jewelry. The fracture is conchoidal, with a glassy luster, and a very resinous luster when polished. The specific gravity is 1.06 to 1.066, with a density of 1 to 2. When heated to 140–180°C, it softens and inflates, and, with continued heating, it melts. Gedanite is soluble in linseed oil and alcohol. It contains much less succinic acid than succinite. A chemical analysis of gedanite yields: carbon 81.01 percent, hydrogen 12.41 percent, oxygen 7.33 percent, and sulfur 0.25 percent. Enclosures are rare, but botanical debris, such as small fragments of a pinelike wood, decomposed leaves, and a few insects, have been found. Gedanite is thought to be the fossil resin of an extinct pine species, *Pinites stroboides*, that resembles a five-needle white pine species (Zalewska, 1974, 33). More recent studies by Beck, Lambert, and Frye (1986, 411–3) and Lambert, Beck, and Frye (1988, 248–63) found conflicting information, or perhaps two different materials may be called gedanite, because one sample yielded a spectra identical to Baltic amber and the second quite a different spectra. Perhaps the specemin was balltic amber and had been misidentified as Gedanite previously.

Beckerite, named after one of the members of the firm of Stantien and Becker, developers of the Baltic amber industry, is another fossil resin usually found in lumpy, opaque, or cloudy masses, only the edges of which are slightly transparent. Some report the fossil resin to be similar in appearance to a piece of charcoal. Specimens remain opaque even after grinding and polishing. Beckerite is tough and difficult to pulverize, but when crushed, it produces a gray-brown powder. It is denser than Baltic succinite and contains only traces of succinic acid. It was thought to be the fossil resin from a leguminous tree, but C13NMR studies show that the material belongs to the Baltic amber group and is a contaminated form of Baltic amber. Among the miners of the 1800s, beckerite was called Braunharz or "brown resin." Krumbiegel reports that the resin was first discovered by Kunow, a conservator from the Institute of Zoology of Königsberg's Albert University, in 1867 located in the blue earth at Gorosskuren (Primore) and Kleinkuren (Filino). Krumbiegel reports further discoveries made in 1871 at Warnicken (Lesnoe) and in 1872 at Palmniken (Yantarnyy). The German amber deposit near Bitterfeld also has occurances of beckerite alongside the succinite amber in the Upper Oligocene to late Miocene Bitterfeld micaceous sands at the bottom of the brown coal layer. It appears in the form of pieces ranging from 3 to 7 centimeters in diameter. It is a hard, brittle material with spikey protrusions and fissures (Krumbiegel 1997, 38–9).

Krumbiegel (1997, 39) also points out that his analyses do not confirm the theory by Beck, Lambert, and Frye (1986, 411–3) that beckerite is a contaminated form of succinite, but that the question of identifying which resiniferous tree beckerite derived from remains unanswered.

Stantienite, named after Stantien of the Stantien and Becker amber firm, is found mainly in mining of the blue earth near Yantarnyy (Palmnicken). It resembles beckerite in its dark color and complete opacity, and presents a dull, dark black fracture surface. Specimens become strongly glittering after polishing. It is extremely brittle and more easily powdered than beckerite, producing a cinnamon-brown powder with less succinic acid than beckerite. Early miners called it Schwarzharz, or "black resin," after its color. The C13NMR studies produced a spectrum more reminiscent of coal than amber. Researchers speculate it may actually be resin-impregnated wood.

Glessite is a light brown, almost opaque fossil resin that breaks with a conchoidal fracture displaying a greasy luster. Glessite's name is derived from the ancient name of amber, *gles* or *glez*. Its hardness is about 2 on Mohs scale. The fossil resin lacks fluorescence. No plant or animal inclusions are found. In 1968 Frondel, one of the early pioneers in using modern technology for identifying amber, studied this fossil resin (1967, 1411–3, 1968, 381–2). Using IR spectrometry and X-ray diffraction, she found the presence of alpha-amyrin, a crystalline triterpenoid alcohol, characteristic of angiosperms in both recent and fossil resins. Therefore, she attributed the material to have originated from a tree in the genus *Burseraceae* that produces an aromatic resin often used as incense. During the past 30 years more sophisticated technology has been developed and surprising results have caused many researchers to reexamine earlier studies. In 1995 at the American Chemical Society's *Amber, Resinites and Fossil Resins* symposium, Langenheim (1995, 9) emphasized that resinous materials from one location may have been derived from several botanical sources and analyses of several samples are necessary to draw conclusions since various samples may give different results. Langenheim cites C13NMR and pyrolysis-gas chromatography-mass spectrometry (py-GC-MS) data all indicating a similarity between glessite and succinite (Anderson and Botto 1993, 1027–38; Kosmowska-Ceranowicz and Krumbiegel 1994, 394–400).

Krantzite is a soft fossil resin found in the coal areas of Saxony, as well as in the Baltic Kaliningrad area.

Schraufite is a reddish resin with the generalized chemical formula of $C^{11}H^{16}O^{12}$. It is usually found associated with lignite and jet in the Carpathian area of the Bukovina Romania (former Soviet Union) and in certain sandstones of Austria (Srebrodol'skiy 1980). The C13NMR and mass spectrometry studies place this material in a category along with delatynite and other fossil resins characterized by weak or absent exomethylene resonances.

Delatynite, found in Delatyn, Russia, and other areas of the Baltic and Carpathian regions, is a fossil resin mentioned in Soviet scientific literature without further details as to its appearance, size, or properties.

RELATIONSHIPS TO BALTIC AMBER'S BOTANICAL ORIGIN

Historically, investigations into the organic inclusions and botanical imprints on the forms of amber began in a systematic manner when German naturalist Berendt (1830) examined over 2000 specimens. Göppert joined him in 1845. As a result of their combined work in 1853, Göppert (1853, 450–76) reported to the Berlin Academy 163 different species of vegetable remains, divided into 24 families and 64 genera.

In 1890, Conwentz (1890*b*) published a large volume on the results of his studies on inclusions in amber. Being the first to examine the rare pine enclosures such as needles, fragments of wood, and pollen grains, he concluded that at least four different species contributed to the formation of Baltic amber. Because so few fragments of pine are preserved in amber, and distinction of the particles now detached from parent trees is so difficult, it was, and still is, almost impossible to identify exact species. In view of these difficulties, Conwentz recommended that all prehistoric conifers that produced amber resin be described with one collective name, *Pinus succinifera*, or amber-bearing pine, and this name was adopted by other scientists and is still used. However, scientists using modern technology, focusing attention on comparing fossilized resin with various recent resins using IR spectrometry measures, continue in their attempts to identify the exact source tree based on characteristics of today's trees.

Realizing that the small fragments of plant tissue were engulfed and preserved by the fresh sticky resin produced by amber-yielding trees 40 million years ago, just before the tissue entered the first stages of decay, paleobotanists continue to study the flora, searching for evidence of plant evolution. Small microscopic inclusions such as fungi, lichen, mosses, and liverworts are rare but are intriguing since the entire organism is preserved. The simplest forms of life in amber, such as bacteria, are being studied by microbiologists such as Raul Cano. Owing to the difficulty of the use of the instruments necessary to study microorganisms in amber, few studies have been done. The first published literature was in 1929 by Blunck (1929, 554–5). Using a method of partially dissolving the amber to obtain the bacterial cells and a light microscope with a magnification of 800x, he found pollen, fungus spores, and bacteria. In 1983 Katinas reported finding clusters of spherical bacterial cells in Baltic amber. In May 1995 Cano reported finding bacteria in amber, and to have extracted and cultivated the ancient DNA from the bacteria extracted from the stomach of a fossil stingless bee embalmed within the amber (Gerhardt 1995).

The first verified slime mold from Baltic amber was identified by Domke (1952, 152–62). Fungi are common in amber, being found growing on the remains of arthropods and plants. Fungi were reported by Conwentz (1890*b*), Göppert (1853, 450–76), Czeczott (1961, 119–45), and Katinas (1983). These simple plant fossils—the mosses and liverworts—are rare and according to Poinar (1992, 12) best preserved in Baltic amber. Czeczott (1961) presented a summary of bryophytes, the mosses, and liverworts in Baltic amber, giving descriptions of six different species of mosses. Though fern fossils are rare in amber, Czeczott also reported two different species of fern. It is also possible to identify plants by constituent parts preserved in amber, such as flowers, seeds, needles, or leaves, as well as resin-saturated wood.

According to Poinar (1992, 74) between the years 1830 and 1937 up to 700 species of spore- and seed-producing plants were described from either impressions or inclusions in Baltic amber. However, the Polish study by Czeczott in 1961 (450–76) based on morphological analysis delineated only 216 species that could be validated as growing in the amber-producing forest. At the time of her study, she included within these validated plant species 63 cryptogams, that is, bacteria, slime molds, fungi, lichens, liverworts, moss, and ferns, which now are placed in separate kingdoms other than the plant kingdom. Her study reported 52 gymnosperms and 101 angiosperms identified from flowers, fruits, or seeds and some by fragments of leaves and twigs.

Though not appearing in fossil records until the early Cretaceous, angiosperm plants dominated by the end of the Cretaceous period in the Baltic region and were encountered abundantly in amber in the form of palms, cinnamon, magnolias, oaks, and others in studies by Caspary and Klebs (1907, 182). The most abundant angiosperm plant remains in Baltic amber have been identified as tufts of tiny stellate hairs, produced by the protective outer casing of oak tree buds. Poinar (1992, 77) indicates that these hairs resemble those found today on oak bud scales, but recent analysis found five major trichome, or varieties of hair types, in Baltic amber. All occur today among representatives of the family Fagaceae, which not only includes oaks, but also chestnut and beech.

The two classic studies presented by Conwentz (1890*a*) and Schubert (1961) describing the Baltic amber-producing tree based on morphological analysis included illustrations of leaves, cones, and fine detailed drawings of wood remains in amber (see Fig. 6.10). Numerous conifer plant remains were identified such as cypress, tuja, sequoias, and other pines. The fresh wood chips, interestingly, were

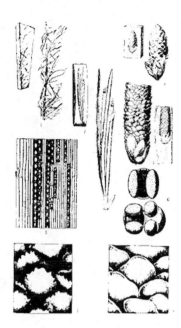

Figure 6.10. Drawings of botanical inclusions in amber. Reprinted from Conwentz 1890a.

Figure 6.11. Drawing of flowering plants (angiosperms) from Baltic amber. Reprinted from Conwentz 1890a.

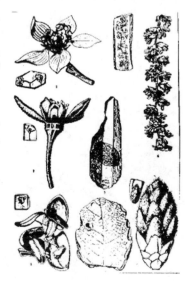

shaped as if they were serrated by small incisor teeth, indicating the action of small rodents.

Among the angiosperms identified in amber, 23 percent were tropical, with 5 families (12 percent) generally limited to growing in the temperate zone and the rest, about 20 families, or about 60 percent, were cosmopolitan or have discontinuous distribution (see Fig. 6.11).

MORPHOLOGY OF AMBER AS FOUND IN NATURE

Another morphological study, providing scientists with insights into the origins of amber, is the form and shape of the raw amber pieces found in nature. By examining various lumps of raw amber, scientists discovered that the shapes were not accidental but depended on whether the resin was formed inside or outside the tree trunk. The teardrop shapes of amber were recognized as having been produced from a tree "weeping tears" of amber even in ancient times. Scientists today explain away the myths, noting resin dripping from an injury in a limb on the tree becomes viscous and sticky, forming tear-shaped droplets as it hardens during the first stage of diagenesis while losing volatiles and exposure to the atmosphere. As evaporation and dehydration take place, the droplets wither. Today these pieces may be found with a wrinkly crust over the surface (see Fig. 6.12).

Iciclelike forms, or stalactites, of raw amber are frequently found in various sizes. These also were formed outside the tree by dripping liquid resin. As the resin continued to exude and streak down the outside of the tree trunk, the new resin formed layers over the previous layers. The iciclelike forms are most likely to contain small insects or fragments of plants as inclusions. As the resin dripped from the tree, small flying insects or organisms living on the bark of the tree would be trapped in the sticky surface. When more resin exuded, it coated over the first layer already partially dried. The fresh resin would quickly embalm the organism, trapping it within. Amber has an unusual preservation property; as it excludes oxygen and dries out the tissues preserving them intact for millions of years.

Since these icicleslike shapes were repeatedly covered over with fresh resin, each layer being exposed to atmospheric conditions, they produced a layered, loosely adhered amber that sometimes splits to form very thin plates, exposing perfectly visible fossil insect inclusions. This amber is usually transparent. Amber pieces are often found showing these concentric rings from the layering of the resin (see Figs. 1.5 and 1.13). Also, interestingly, pieces were formed when resin exuded over a previously formed icicle or droplet, completely covering the semidried resin and resulting in a transparent shape with a clear view of the amber formed inside. Polish scientists called this formation "amber within amber" (see Fig. 6.13 a,b).

Another form of amber is the "cobblestone"* that is produced by a large exudation of amber-producing resin. "Cobblestone" are fromed by dripping resin similar to stalactite forms. These pieces usually have a flat-shaped side with perhaps imprints of the debris they rested on. There may be a raised peak where the sticky resin dripped and piled up from the continuous flow from the tree above. If the resin was produced in a large quantity at one time, the resulting form may be opaque or even foamy.

Amber nodules formed within the tree were a result of the resin filling up a crack in the interior or in a split between the bark and the growth layer. These pieces form a mold of the pocket, with a flat top formed as a result of the surface of the liquid resin as it filled up the fissure, thus the resulting amber form tapers along the sides and bottom of the pocket. These pieces are usually curved, forming to the circumference of the tree. By measuring the angle of the curve, scientists can estimate the size of the tree that produced the amber-yielding resin. These lumps are usually opaque. This internal natural mold may have the imprint of the tree's bark on one side or holes where the growth layer imprinted the resin. Thus, the shape of the surface of the amber lump indicates where it was formed—in an underbark fissure, inside the bark, or in a so-called resin pocket between annual growth rings.

Another surface feature of amber that provides information about the environment is a covering of barnacles or other skeletons of colonial crustaceans found on amber that rested for some time on the seafloor. Sedentary barnacles, *Balanus improvisus*, living in the offshore zone of the seafloor at that time, settled on the amber lumps and became so firmly attached that they must be ground off before cutting and polishing (see Fig. 6.14).

Botanists have found that the amber-producing trees were notable for their low ratio of wood to cork production. The resin was produced in the epithelial cells of resin canals and also in the parechymal cells of the live, inner bark. Various reasons have been suggested for why the trees in the Fennoscandian region produced such an abundance of resin. Conwentz (1890*a*) attributed it to a disease; however, botanists today identify some varieties of trees that normally produce volumes of resin, such as the *Agathis*, which produces the

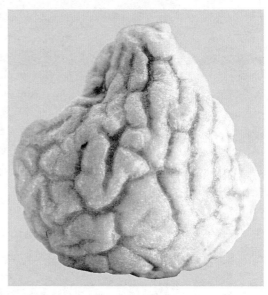

Figure 6.12. Teardrop-shaped amber with wrinkly crust formed during dehydration. Courtesy Muzeum Ziemi, Warsaw, Poland.

a

Figure 6.13.a, Icicle-like forms in Baltic amber formed by external fissure fillings. An iciclelike form of resin covered by a later exudation of amber-producing resin, resulting in "amber within amber" such as; b, "amber within amber" Dominican amber specimen.

b

kauri gum. Some botanists suggest that resins exude to attract or repel insects. Also Polish scientists point to the influence of the climate. Scientific tests demonstrate that the quantity of resin produced in forests growing in various zones all over the globe increases toward the equator. Also, the flux of resin increases when the weather is warm and humid. During the early Eocene there appears to have been rapidly rising temperatures in the European territory as observed by the sediments. Also there was an increase in the humidity during the Paleocene period. These abnormal climatic conditions for the flora growing at the time may have provoked an excessive formation of resin that was not part of the trees' normal production. The warm air changed the delimitation of the climatic zones and instigated a migration of subtropical flora to northern regions, thus extending the zone of the forests (Katinas, lecture 1988).

BOTANICAL ORIGINS OF BALTIC AMBER AND ITS BOTANICAL AFFINITY TO RECENT TREES

Today, the antiquity of the trees in the amber forest has been more accurately determined than in Conwentz's time, but the exact type of the Baltic amber source tree is still in question. For over 50 years scientists followed Conwentz's determination that amber was from the *Pinus succinifera* or a coniferous source. Frondel (1968, 381), using X-ray diffraction to detect and identify crystalline components in organic complexes, detected a substance called alpha-amyrin in several amber samples, including Baltic amber. Alpha-amyrin is a major component in resins from some species of angiosperms (flowering trees). Since Baltic amber had been considered a product of gymnosperms, this new data cast an entirely new light on the amber source of the primeval amber forest.

Up to the mid-1960s, scientific studies of amber mainly confined investigations to identification and classification of insect and plant inclusions. However, new analytical methods and new approaches by botanists to define biological affinities to recent resins provided for a renaissance in the substance of amber rather than its inclusions. Old chemical composition studies were looked upon as meaningless, and botanists announced that the complex terpenoid compounds produced by plants were the significant structures that should be studied. These terpenoid compounds vary in proportion from one plant species to another. Studies comparing the composition of amber to recent resins produced by present-day plants delineated the possible evolution of amber-producing plants.

Some researchers using IR spectrometry now find that Baltic amber more closely resembles the Araucariaceae family, like the kauri tree *Agathis australica* (Poinar and Haverkamp 1985, 210–21; Langenheim 1969, 1156–69). Others revealed certain similarities between the absorption level of some beams of light in both Baltic amber and resin from modern-day cypress *Cedrus atlantica* (Moskwa 1987; Katinas , lecture 1988; Kosmowska-Ceranowicz and Konart 1995). Since studies of the botanical inclusions and wood fragments preserved in amber were similar to pine and not araucarian, the pine origin is still considered. Some believe there may have been a tree that no longer exists that had characteristics similar to the araucarian and the pine since no araucarian leaves are found embedded in amber (Pielińska 1997, 11). Larsson (1978, 21) suggests that the source tree may have been a primitive type, representing an early stage of development for the *Pinaceae* with a chemistry common to the araucarian. He compares it to the *Pinus lambertiana*, which still produces labdane-type resin rather than abietic acid that is produced by most pines today. *Pinus lambertiana*, or sugar pine, grows in the northwestern part of North America, and its resin's IR spectrum reveals the Baltic shoulder characteristic of Baltic amber (Poinar 1992, 28). Grimaldi (1996*a*, 54) reports a recent discovery of some living trees in the pine family, belonging to the genera *Keteleeria* and *Pseudolarix*, that do indeed produce succinic acid. He theorizes that, since pine resin is characterized by abietic acid and Baltic amber is distinguished by succinic acid, the question may be resolved based on this discovery. Fossilized resin, in 40-million-year-old *Pseudolarix* cones from Axel Heiburg Island in the Canadian arctic, has been found that also contains succinic acid. As of yet common agreement has not proven the actual tree from which Baltic amber was derived, and scientists continue their investigations.

Figure 6.14. Baltic amber coated with barnicals.

Langenheim (1995, 23), botanist from the University of California, gives a detailed description of the botanical production of resins, the trees that normally produce an abundance of resin, the resins that will polymerize and the resins that do not contain the terpenoids necessary for polymerization (labdanoids), and the possible source trees for Baltic amber. She summarized the pros and cons regarding the Agathis as being the plant source of Baltic succinite as follows:

Pro *Agathis* Involvement
1. Amber chemistry predominantly similar to extant
    A. australica (Araucariaceae)
2. Massive accumulation with many large pieces

similar to extant *A. australica.*
3. Much of associated amber flora similar to that in some Pacific subtropical-tropical forests where *Agathis* occurs today.

Con *Agathis* Involvement
1. Absence of succinic acid.
2. Only resin-producing plants included in amber or associated with it, from other coniferous sources.
3. Wood with enclosed succinite considered to be "pinaceous."

Langenheim describes resins from coniferous families like Araucariaceae, Taxodiaceae, and Cupressaceae as containing labdanoid compounds that readily set under the correct conditions of oxidation and light. These resins become solid when they are exposed to the atmosphere soon after having flowed from the tree. Labdane-type resins are also found from the angiosperm family Leguminosae, particularly the *Hymenaea* and *Copaifera*. The Pinaceae, which were historically identified as the source of the amber-producing resin, contain low or no labdanoid compounds but are characterized by abietic acid. Thus, their resin does not set as quickly and is not found in a solid state in large quantities. Continued research will provide a definite answer to this question.

Much research to identify source trees or their botanical affinity to recent trees has been published using amber samples from the different fossil resins found in locations throughout the world. This research is summarized in Table 6.2 adapted from Poinar (1992). Studies comparing the composition of amber to recent resins produced by present-day plants delineated the possible evolution of amber-producing plants.

INSECT INCLUSIONS IN BALTIC AMBER
Runge (1868, 1–70) found 174 different species of flies, ants, beetles, and moths, 73 species of spiders, and some species of centipedes in amber. Such imbedded specimens of extinct life are often found with details such as compound eyes, wing veins, scales on insects, and hairs on the legs of spiders clearly visible (see Fig. 6.15). It is speculated that insects were attracted to the sticky, sweet surface of the resin as it flowed over the bark in its liquid state. When these insects became engulfed in the viscous material, their bodies were preserved just as they were without serious damage. In some specimens, only fragments of insects, such as antennae, wings, or legs, are found, which

Table 6.2 Plants Considered as the Source of Different Amber Deposits

Amber Deposit	Proposed Plant Species or Genus	Plant Family	Type of Examination	Reference
Baltic	*Pinus sp.*	Pinaceae	Morphological-anatomical	Aycke 1835
Baltic	*Pinites succinifer* (Göppert & Haczewski)	Pinaceae	Morphological-anatomical	Göppert 1836
Baltic	*Abies bituminosa* (Göppert & Haczewski)	Pinaceae	Morphological-anatomical	Haczewski 1838
Baltic	*Pinites succinifer* (Göppert & Berendt)	Pinaceae	Morphological-anatomical	Göppert and Berendt 1845
Baltic	*Taxoxylum electrochyton* (Menge)	Pinaceae	Morphological-anatomical	Menge 1858
Baltic	*Pityoxylon succiniferum* (Krasu)	Pinaceae	Morphological-anatomical	Schimper 1870, 1872
Baltic	*Picea succinifera* (Conwentz)	Pinaceae	Morphological-anatomical	Conwentz 1886a
Baltic	*Pinus succinifera* (Conwentz & Göppert)	Pinaceae	Morphological-anatomical	Conwentz 1890
Baltic	*Pinus succinifera* ((Conwentz & Schubert)	Pinaceae	Morphological-anatomical	Schubert 1961
Baltic	*Agathis*	Araucariaceae	Chemical-morphological	Kostyniuk 1961
Baltic	*Pinus*	Pinaceae	Chemical	Röttlander 1970
Baltic	*Agathis*	Araucariaceae	IR spectra	Langenheim 1969
Baltic	*Agathis*	Araucariaceae	IR spectra	Thomas 1969
Baltic	*Agathis*	Araucariaceae	IR spectra, chemical analysis	Gough and Mills 1972
Baltic	*Pinus halepensis* (Miller)	Pinaceae	Resin analysis	Mosini and Samperi 1985
Baltic	*Agathis*	Araucariaceae	Pyrolysis mass spectrometry	Poinar and Haverkamp 1985
Baltic	*Cedrus atlantica*	Cypress	IR spectra, chemical analysis	Katinas 1988
Mexican	*Hymenaea*	Leguminoseae	IR spectra, inclusions	Langenheim and Beck 1968
Dominican	*Hymenaea*	Leguminoseae	IR spectra, inclusions, Pyrolysis mass spectrometry, NMR spectroscopy	Langenheim and Beck 1968; Poinar and Haverkamp 1985; Lambert, Frye, and Poinar 1985
Dominican	*Hymenaea protera* (Poinar)	Leguminoseae	Morphological	Poinar 1991b
Alaskan	*Sequoiadendron, Metasequoia, Taxodium*	Taxodiaceae	Plant fossils in amber beds	Langenheim, Smiley, and Gray 1960; Langenheim 1969; Poinar and Haverkamp 1985
Alaskan	*Agathis*-like	Araucariaceae	Pyrolysis mass spectrometry	Poinar and Haverkamp 1985
Alaskan	*Agathis*	Araucariaceae	NMR spectroscopy	Lambert, Frye, and Poinar 1985
Canadian	*Agathis*	Araucariaceae	IR spectra, pyrolysis mass spectrometry	Langenheim and Beck 1965, 1968; Poinar and Haverkamp 1985
Bitterfeld (Glessite)		Burseraceae	NA	Kosmowska-Ceranowicz and Krumbiegel 1989
Bitterfeld (Zygburgite)	*Liquidambar*	Hamamelidaceae	NA	Kosmowska-Ceranowicz and Krumbiegel 1988
Borneo Sarawak	*Shorea*	Dipterocapaceae	NA	Kosmowska-Ceranowicz 1994
Canadian	*Agathis*	Araucariaceae	IR spectra, pyrolysis mass spectrometry	Langenheim and Beck 1965, 1968; Poinar and Haverkamp 1985
Canadian	*Agathis*	Araucariaceae	NMR spectroscopy	Lambert, Frye, and Poinar 1985
Jordanian (Middle East)	*Agathis*	Araucariaceae	IR spectra, mass spectrometry, thin layer chromatography	Bandel and Vavrà 1981
Lebanese (Middle East)	Araucarian	Araucariaceae	IR spectra, mass spectrometry, thin layer chromatography	Bandel and Vavrà 1981
Romanian	*Abies*	Pinaceae	Associated fossils	Protescu 1937
Romanian	*Sequoioxylon gypsaceum* (Göppert)	Taxodiaceae	Plant tissue in amber, silicified wood, fragments in amber strata	Ghiurca 1988
Atlantic Coastal Plain (New Jersey)	*Cupressinoxylon bibbinsi* (Knowlton)	Coniferales	Wood anatomy	Knowlton 1896
Atlantic Coastal Plain (New Jersey)	*Liquidambar*	Hamamelidaceae	IR spectra	Langenheim 1969
Atlantic Coastal Plain (N.J.)	*Sequoiadendron, Metasequoia*	Taxodiaceae	IR spectra	Langenheim in Wilson, Carpenter, And Brown, 1967
Atlantic Coastal Plain (N.J.)	*Agathis*	Araucariaceae	NMR spectroscopy	Lambert, Frye, and Poinar 1985; Langenheim in Wilson, Carpenter, and Brown, 1967
Arkansas	*Shorea*	Dipterocarpaceae	IR spectra	Saunders et al. 1974

Source: Adapted from G. O. Poinar, Jr., *Life in Amber* (Palo Alto: Stanford University Press, 1992), 26-7.

suggests they were torn off as the insects attempted to extricate themselves from the sticky mass. It is noteworthy that insects are most often found in shelly amber, the kind that formed from numerous layers of resin, corresponding to successive flows of resin as described earlier.

Microscopic examination of many of the insects with tough cuticular shells preserved in amber shows hollow body cavities where internal organs have decayed. In these cases, what is actually observed is a mold in the amber, lined with a pigment composed of metamorphosed and carbonized material from the tough horny outer covering or exoskeleton of the insect (Larsson 1978, 9–11).

However, as early as 1903, Kornilovitch (1903, 198–206), a Russian scientist, identified the presence of striated muscles in the limbs of insects found in amber. Unfortunately, the existence of deoxyribonucleic acid (DNA) was not yet established. About 30 years later, a German researcher, named Keilbach (1937, 398–400), confirmed the presence of striated muscles and widely publicized the findings. But it was not until 1982 that the American press created a news media blitz when Poinar and Hess (1982) in California succeeded in describing musculature in the legs and internal anatomy of a sectioned fly enclosed in Baltic amber. Interestingly, they found the digestive tract and brain of the insect preserved in place, the lungs—in reality air sacs or membranous tracheoles (fine tubes) that deliver oxygen to the organs—still intact. The complicated musculature of the insect was preserved with the fine striations previously described by Kornilovitch, but now more fully understood than before. These striations result from the process of muscle contraction. In insects, muscle filaments of actin and myosin proteins slide past each other during contraction and where the ends of the filaments align, bands form. Using a 20,000x magnification, much greater than Kornilovitch had access to in 1903, the bands were distinctly seen and validated. (Grimaldi reports that Baltic amber does not preserve internal tissues as well as Dominican or Mexican amber. In Dominican amber there is often virtually no shrinkage of soft tissues and no traces of decomposition, presumably because of its unique chemistry; Grimaldi 1996a, 122).

In 1981 and 1982 Poinar and Hess made a major discovery in isolating tissue from a female fungus gnat *(Mycetophilidae: Diptera)* preserved in Baltic amber. Examination with a transmission electron microscope identified cell structures that correspond to muscle fibers, nuclei, ribosomes (the cellular factories that assemble proteins), endoplasmic reticulum (membranes that transport substances within the cell), and lipid droplets or fat globules. Between the fine bundles of muscle tissue, the mitochondria were located, which surprisingly remained intact for over 40 million years (Bada et al. 1994, 3131–5). Wilson, a molecular biologist at Berkeley, continued to study the chromatin, the chromosome material in which DNA or fingerprint patterns of the tissue is embedded and attempted to sequence the DNA from fossil inclusions in Dominican amber (Wilson 1985, 164–73). This idea was the embryo on which *Jurassic Park* was fabricated! Poinar believes that comparison of the tissue with those of recent descendants will indicate how quickly evolution works on a cellular level. Now amber has its place in the field of molecular genetics, the area of the most recent research carried out on amber inclusions.

Amber has a remarkable property that preserves insect inclusions in a virtually unaltered state in three-dimensional form, the only difference being the original color of these ancient fossils very rarely survives. This extraordinary preservation of biological materials included within amber is undoubtedly due to the properties of amber. Though the exact process is not clear, it is likely related in part to rapid dehydration of the soft tissues by the resin. Electron microscopic studies show muscle fiber and delicate internal organs preserved and to some extent the molecular structure as well (Wang et al. 1995, 255–62).

Earlier research focused on insect inclusions in order to recreate the scene of the Eocene forest by studying the habitat in which recent descendants of similar varieties live today. Insect inclusions in Baltic amber indicate that the climate 40 to 50 million years ago was much warmer than at the present time. Many of the species found no longer inhabit the cool Baltic region, but resemble those now found only in the temperate climates of Europe and North America. Very few of these species exist in their exact form today, though most survive in modern variants.

Of special interest to scientists are the ants found in Baltic amber because they give much information about the evolution of insects. Not only were they abundant 40 million years ago, but some types are now extinct or are no longer found in the Baltic area. One species, now found only in Sri Lanka, weaves leaves together with fine threads to form its nest. Having no silk-producing organs, the ants hold their larvae between their mandibles, gently squeezing the larvae to encourage spinning. Other ants hold the leaves in their mandibles until the threads dry and the leaves are secured in place. These curious ants occur in Baltic amber and one was even found still holding a larva. The ant most abundantly found in Baltic amber, however, is similar to the most common ants in Europe and North America today, the mound-building black ant, *Formica fusca* (Brues 1951, 56).

Though flies make up to 54 percent of all the insects trapped in amber, today they account for a considerably larger portion of the total insect biomass than they did 40 million years ago. Most examples found entombed in Baltic amber are of small varieties, possibly because it was more difficult for them to pull themselves free from the sticky mass than it was for larger insects (Brues 1951, 58).   In 1910 Klebs (38–52), after collecting insect inclusion specimens of amber for 40 years, surveyed the extensive Stantien and Becker collection at the Königsberg University Geological Institute Museum and estimated the proportions of enclosed insects as follows:

Diptera (two-wing flies) 50.0%
Pseudoneuroptera (termites, mayflies) 10.7%
Rhynchota (lice, gnats) 7.1%
Neuroptera (caddis) 5.6%
Hymenoptera (bees, wasps, ants) 5.1%
Arachnoidea (spiders, mites, scorpions) 4.5%
Coleoptera (beetles) 4.5%
Orthoptera (cockroach, grasshopper) 0.5%
Microlepidoptera (little moths) 0.1%
Various 1.1%

The collection studied by Klebs included about 120,000 of the finest amber specimens that contained inclusions of flora and fauna selected during amber mining. At least 70,000 arthropod inclusions were found among these specimens. It was thought that this entire collection was destroyed by fire during World War II, but, fortunately, reliable sources report that this is not true. In 1965 Hennig, a German scientist, studying Diptera in Baltic amber, indicated that the collection was still available for study. According to him, it was moved during the war for safekeeping to the Geological-Paleontological Institute of the University of Göttingen, Germany, and is still there. (Ritzkowski [1996, 297] also reports that in 1958 the collection was placed at the University of Göttingen.) That portion of the University of Königsberg collection assembled by Berendt is now housed in the Paleontological Museum of Humbolt University, Berlin, while some specimens are reported to be in the British Museum of Natural History, Department of Paleontology. Thus, the collection appears not to be lost forever, but is available for future generations. Other major collections of fossiliferous Baltic amber are located in the British Museum of Natural History, London; Muzeum Ziemi, Warsaw; the Palaeontology Museum, at Humboldt University, Berlin; Museum of Comparative Zoology at Harvard University, Cambridge, Mass.; Institute for Geology and

Figure 6.15. Baltic amber inclusions from one necklace. Necklaces of this type were thought to bring good luck. Each bead contained an insect:
a, primitive mosquito;
b, immature two-wing fly, (Diptera);
c, primitive wasp with white coating;
d, primitive fly;
e, mite (Acari);
f, primitive mosquito;
g, primitive larvae.

Paleontology, Göttingen; the Zoological Museum, Copenhagen; and the Geological Institute, Moscow. Major North American collections of Baltic amber are located in the Museum of Comparative Zoology at Harvard University and in the Natural History Museum in Chicago. These collections were obtained from William A. Haren, who had a private collection from Germany that was purchased by the Museum of Comparative Zoology. This museum also purchased the Herman Hagen collection of approximately 8000 pieces. After coming to Harvard University, Hagen, who was born in Königsberg, continued his interest in fossils in amber. (See Appendix B for a listing of locations of fossil collections.)

A recent report by Kulicka (1996*a*, 11–3) indicates that research into the fauna present in Baltic amber found that 93 percent of the inclusions consisted of insects. The remainder were arachnids, myriapods, and a small proportion of microscopic life-forms. Her findings indicate that approximately 70 percent of insects belong to the suborder Diptera, with the family Nematocera (primitive three-horned flies) being the most common, followed by ants, wasps, and bees (Hymenoptera); beetles (Coleoptera); caddis flies (Trichoptera), and bugs (Heteroptera.)

The abundance of Diptera of the Mycetophidae family indicates that the forest environment was rich in humidity and hydro reservoirs. Also, larvae of these insects live in fungi, which need humidity. Thus, the numerous inclusions of larvae of aquatic insects such as caddis flies (Trichoptera), mayflies (Ephemerida), and net-winged insects (Neuroptera), testify to a great number of water reservoirs. Interestingly, there are both types of insects—those that live on stagnant water and those that live on fast torrents. The latter forms indicate that some of the forests were growing in mountainous regions.

Fossil inclusions of butterflies (Microlepidoptera) suggest that there were clearings in the forest. Though butterflies are rare in amber, the Adelidae, which *is* found, is a species attracted to light and would be found near clearings. Butterflies in Baltic amber were systematically listed by Skalski (1973, 153–60).

Other insects that thrive in warm climates and are found in amber are crickets, mantids, bristle tails, silverfish, and firebrats (Thysanura). Spiders and termites are also found in amber and may be an indication of a warmer climate. The presence of pseudoscorpians in Baltic amber gives further evidence of a warmer climate, since many of these are found in subtropical zones today. Some genera and families found in Baltic amber now are living only in Australia, New Zealand, and in the tropical zone of Africa.

Surprising among the fossils are insects typical of the temperate climate as well. For example, craneflies (Tipulidae) and springtails

(Collembola) living in the amber-producing forests still exist in Poland (Stach 1972, 416–20). The arthropods present in Baltic amber show a disproportion in the rhythm of the evolution of arthropods in the ancient forests. Some fossil arthropods are extinct, with no living relatives today; for example, aphids of the family Elektraphilidae and Hymenoptera Pelecinopteridae. At the same time, there are fossil arthropods that have forms of the same types still living today. By analyzing the arthropods in Baltic amber, Kozlowski (1951, 446–57) found the majority to be foreign to the fauna living today in Central Europe, with many species belonging to tropical fauna found in the Americas. There were also species now living only in Australia, New Zealand, and the tropical zone of Africa.

Fossil insect inclusions found in Baltic amber often are surrounded by a milky white substance. Electron microscope studies on organic inclusions in amber by the Muzeum Ziemi demonstrated that this milky white halo was related to foam formed during the process of decomposition of the body of the insect when numerous small gas bubbles are produced during putrefaction. Moisture from the insect's body also assists in decomposition; thus, preservation processes of rapid dehydration are effected by the volume of moisture present at the time the insect is embalmed. The quantity of these small gas bubbles, the density of their arrangement, and the change in their diameter indicate the intensity of the process of decomposition in the first stages of fossilization (Mierzejewski 1978, 79–84).

CONCLUSIONS DRAWN FROM ORGANIC INCLUSIONS FOUND IN BALTIC AMBER

From the flora and fauna inclusions in amber scientists have pieced together a picture of the primeval amber-producing forests. Since amber occurs in secondary deposits, it is impossible to know if that picture is exact. Also, not all forms of life in existence during the resin-producing period were trapped in amber. As mentioned, about 200 species of spore-producing and seed-producing flowering plants have been identified from remains enclosed in amber, which suggests the vegetation was both rich and varied. The identified spore-producing enclosures in amber include firs, cypresses, junipers, pine, spruce and Arbor vitae, while the seed-producing flowering plants were represented by the oaks, magnolia, and cinnamon, as previously mentioned, but also by beech, maple, and chestnut. Remains of palms with fan-shaped as well as pinnate leaves have been identified too. Numerous ferns, mosses, and flowering herbaceous plants formed a ground cover in the ancient forests. Having knowledge of the present habitats and natural environments of various species of plants and ani-

mals similar to those found in amber, scientists are able to describe the environment of the amber-producing forest. Fungi inclusions suggest humid areas comprised portions of the forest, though some ants and other insects indicate dry regions also existed. The presence of some beetles indicates that some amber-producing trees grew on mountainsides, as these insects are characteristically found in areas of swift-running water. Stagnant ponds were also present in some locations, as evidenced by gnats and their larvae. Fleas and gadflies indirectly imply early mammals, since these insects are predators on mammals. The presence of wood-boring insects indicate particular kinds of trees were present in the amber forest (Zalewska 1974, 74). For example, the beetles often found in Baltic amber are from the family Anobiidae. About 40 different species have been described, the most common belonging to the genus *Anobius.* Today there are about 1200 species of Anobiidae, most living in dry wood, bark of trees, and cones of conifers, and in tree pith in tropical zones. Most live in communities and lead a commensal life. Also termites and other insects, along with palms and tropical shrubs, suggest the climate was subtropical to tropical. Much of the vegetation debris preserved in amber belongs to species that grew in a subtropical climate or a warm temperate zone. On the other hand, numerous plant inclusions characteristic of the temperate climate zones are found, too, such as the pine, larch, and maple, suggesting that the forest also may have covered a mountainous region where the climate was cooler as a result of the elevation. Since a wide variety of flora normally growing in different climates are all found in amber, scientists reason that they may have grown at different altitudes.

Some lumps of fossil resins also bear surface impressions of leaves, preserving details of vein and cell structure. In other specimens, surface textures provide clues to the environment in which the amber came to rest. For, example, lumps found in sandy, dry, well-aerated areas, such as just below the surface of a dune, tend to be thick-crusted. Since sand provides such lumps with free access to the atmosphere, oxida-

Figure 6.16. Flaky red oxidation crust on baltic amber The red crust is often found on pieces from sand dune areas.

tion occurs, resulting in the formation of a thick flaky crust that deepens in color with age, gradually thickening to the point where it readily flakes off (see Fig. 6.17.) Amber masses from diluvial deposits laid down during the Pleistocene may show striations or cracks formed as the pieces were transported by glacial debris. Amber lumps with natural holes in them are especially interesting, apparently having been formed by amber-producing resin dripping onto, and surrounding, fragments of twigs. Such twigs became partially enclosed by resin and in time decayed and crumbled away, leaving tubular voids in the amber.

Of the numerous animals present in the Eocene subtropical forests, only certain specimens were trapped in the sticky resin. It is clear that the ancient forests were inhabited by numerous species of vertebrates, as a verified vertebrate inclusion was found in the early 19th century in Königsberg, but disappeared. This unique, *authentic* specimen was lost during World War II. However, it is now in Göttingen (Ritzkowski 1999). It is a small lizard about 43 millimeters long without a tail. This type of lizard belongs to the species of *Nacras*, now living in Africa. It is speculated that it was covered with amber-producing resin when it lay dead on the succiniferous forest floor. However, at the Gdańsk amber symposium in 1998 a new find of a lizard in Baltic amber was reported by Kosmowska-Ceranowicz. Though rare, small lizards have been identified in Dominican amber.

The presence of birds in the succiniferous forests is indicated by the few remains of fleas, parasites, and feathers. Occasionally, footprints of birds are found imprinted on the surface of a piece of amber. Generally, the feather remains were identified as those from woodpeckers and jays. Tufts of hair from mammals are found in Baltic amber. Studies of this hair found that it belonged to rodents or marsupials. Other proofs that mammals were present in the ancient forests are fossil horse flies, fleas, and paw marks.

In 1978 an excellent summary of 150 years of research on Baltic amber formation was done by Larsson in his *Baltic Amber: A Palaeobiological Study* in which he describes in detail the fossil inclusions in Baltic amber. Another book by Schlee (1980), but in German, describes fossil inclusions in Baltic and in Dominican amber. Poinar's 1992 book, *Life in Amber*, provides a treasure chest of information for the entomologist interested in ancient fossil arthropod inclusions in amber. Poinar has publish many findings of unusual life in Baltic amber; such as, nematodes emerging from an ant and one from a planthopper, a wasp larva emerging from its host ant, a rare scorpion, and a rare fly (see Fig. 6.17a, b, c, d, e. See appendix for complete references of first fossil records.)

Figure 6.17. Rare inclusions in Baltic amber identified by G. Poinar.
a, first fossil record of nematode parasitism of ants - mermithid nematode emerging from an ant;
b, mermithid nematode emerging from a planthopper;
c, first fossil record of endoparasitism of adult ants -braconid wasp larva emerging from its host ant;
d, rare wind scorpion;
e, rare tanyderid fly, <u>Macrochile spectrum.</u>
(See appendix for references of works published regarding first fossil records).
Photos by G. Poinar. Courtesy G. Poinar.

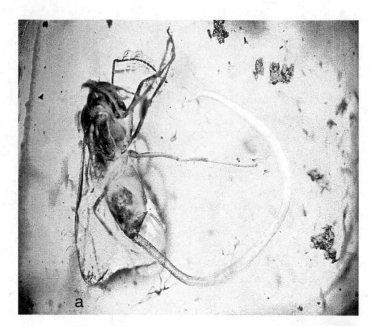

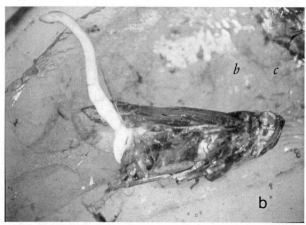

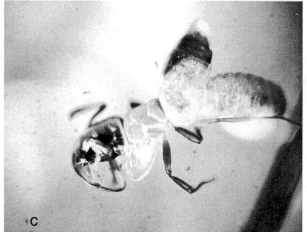

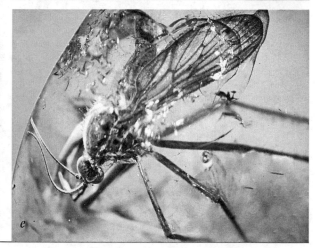

ANALYSIS OF BALTIC AMBER IN
ARCHAEOLOGICAL FINDS

The composition of amber not only yields useful information to botanists, but is valuable to archaeologists in determining the origin of amber artifacts. Amber containing succinic acid was considered to be of Baltic origin, whereas amber without succinic acid or classified as a retinite, was determined not to be from the Baltic source. It was Helm (1885, 234–9) who believed that Baltic amber was uniquely distinguished by its succinic acid content, thus providing a historical milestone in archaeology. However, there are limits to its usefulness. Though Helm was correct in claiming all Baltic amber contains from 3 to 8 percent succinic acid, this does not mean that other ambers contain less or no succinic acid. In fact, ambers from other localities may contain comparable amounts. On the other hand, the test for succinic acid did rule out Baltic origin if no succinic acid was found in an unknown specimen. Another drawback to testing for succinic acid is that it requires destruction of more than a gram of the amber artifact because of the chemical analytical method used, requiring heating the amber until it decomposes.

Beck (1971, 235) summarizes how modern laboratory methods can be useful in amber studies, stating:

> The success of modern instrumental analysis to decide questions of archaeological provenance by means which are either nondestructive or which use very small samples has opened new approaches to the amber problem. Among the options, infrared spectroscopy has proven both decisive and convenient. Infrared spectra do not distinguish unequivocally among all of the many non-Baltic European resins, probably because many of them have the same botanical sources. But they do permit the positive identification of Baltic amber and thus allow the direct recognition of imports from the north among the large number of amber finds in European, and particularly in Mediterranean, archaeology.

Because archaeological samples are often contaminated with other substances, IR peaks are sometimes obscured, making identification impossible. An archaeological study group, directed by Beck at Vassar College, developed a computer technique to analyze IR spectra to aid in identification. This technique was successfully applied to amber artifacts collected from various Greek sites, including Mycenae, and provided evidence that as early as 1550 B.C., Baltic amber originating in the northern regions of Europe was used in Greece in the Mediterranean region.

Beck's experiments tested other analytical methods, which could aid in determining the provenance of fossil resins (Beck 1965, 272–6). Besides IR spectroscopy, pyrolysis followed by gas chromatography and analysis for trace elements proved useful. Specimens with too much oxidation of the surface do not produce characteristic spectra as a result of the chemical changes that have taken place. When amber is exposed to oxygen over time, it develops an opaque crust that is found on the raw material and with continued oxidation the surface will crumble away. For this reason, other methods are necessary for determining the type of resin (see Fig. 6.16).

Beck's IR spectroscopy methods of analyzing amber are used extensively by Polish researchers when studying amber finds throughout Poland to identify true Baltic amber from other fossil resins. Russian researchers also use these techniques since many different fossil resins are found in the Eastern European regions. These sophisticated techniques have led to some question as to the accuracy of the determination of the old trade routes. As mentioned earlier, Polish and Italian scientists are reexamining the artifacts previously studied to verify and correct any differences.

Since 1985 Shedrinsky, of the Conservation Center of New York University, has applied analytical pyrolysis in several variations, such as pyrolysis-gas chromatograph (Py-GC), pryolysis mass-spectrometry (Py-MS), and pyrolysis gas chromatograph/mass spectrometry (Py-GC/MS), with successful results for analyzing amber materials in art and archaeology. He states that "a general picture of the chemical structure of amber and plant resins which after close comparison by multivariate (statistical) analysis, can be used for differentiating classification and identification purposes." He recommends using the combined approach of Py-MS and Py-GC/MS to provide more information to characterize ambers, copals, and other resins and to provide a fingerprint of each material (Shedrinsky and Baer 1995, 138–42).

AMBER VARIETIES—FOSSIL RESINS OTHER THAN BALTIC SUCCINITE

CRETACEOUS AGE AMBERS

The age of a fossil resin is an important question because the commercial value of amber depends on its rarity and age. At the present time no consensus or standard has been accepted as to the exact point in polymerization when the resin changes from "recent" to "fossilized." Some amber dealers currently sell copal for "young amber," often deliberately confusing the public as to the difference. Dealers can get a much higher price if the buyer believes that the "fossil" is in "fossilized resin," as opposed to a recent specimen in recent resin. Current researchers in the field of botanical inquiry focus studies on amber, seeking its relationship with recent resins. They also attempt to determine the age of these ancient resins in order to study the paleoecology and evolution of plants.

There are various approaches to grouping fossil resins into ages for study. One approach may be to investigate the amber deposits' geologic occurrence, another may be to use modern technology for determining when the amber-producing resin originated. Another field of science that has a different interest in amber is Petrology. Petrologists, in their search for fossil fuels, have come across resinite materials in chinks in the coals, some even as old as the Triassic period or even older. Platt (1997-2005) lists the oldest fossil resin recorded as being from the Paleozoic Age, Carboniferous Period in deposits from Northumberland, United Kingdom, this would date the resin from 280 million years ago (Ma) to 345 Ma. He also lists resin from the Permian Period (225– 280 Ma) coming from the Cekarda River, Ural Mountains in Russia. The Mesozoic Era, which consists of the Cretaceous, Jurassic and Triassic, has produced amber but in smaller and scarcer quantities. One problem associated with Mesozoic amber is the level of degradation it undergoes since ancient fossil resin is usually badly affected by oxidization, erosion,

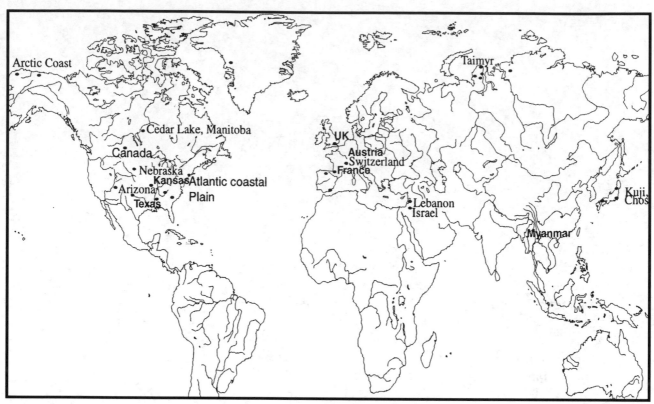

Labels on map: Arctic Coast · Cedar Lake, Manitoba · Canada · Nebraska · Kansas · Atlantic coastal Plain · Arizona · Texas · UK · Austria · Switzerland · France · Lebanon · Israel · Taimyr · Myanmar · Kuji, Chos

Figure 7.1. Distribution of Cretaceous Age Amber.

excessive heat and pressure. Usually, due to exposure to temperature fluctuations, this aged amber, being fragile and highly fractured, is found in small pieces no bigger than a few centimeters (Poinar 2001, 15).

Platt (1997-2005) states the majority of amber that has been discovered and studied originates in the Cenozoic Era and Late Cretaceous. For example, fossil resin found in Arizona in the Chinle formation of Triassic age, about 225 million years old, differs from younger amber. Recent reseaerch has shown this amber is Cretaceous resin (Kosmowska-Ceranowicz, personal communicatiuon, 2006) These fossil resins are dark red, very brittle, and are thought to have originated from an extinct plant similar to the Cycads (palmlike plants). They are perhaps true amber, depending on how the term is defined, but they are chemically different. Only microscopic fossil organisms have been found in these resinites. A Triassic age fossil resin from Bavaria was found to include primitive bacteria, protozoa, fungal spores, and what appeared to be plant spores. Also, Poinar (1992, 68–71) reported algae and protozoan amoebalike organisms in this 225-million-year-old Bavarian amber. At the present time these are the oldest fossil inclusions reported in amber.

Though there are fewer representatives of Cretaceous period

fossil resins than Paleogene fossil resins, Cretaceous ambers contain fossiliferous inclusions that provide unique information on the evolution of insects and other organisms. For example, Cretaceous ambers from New Jersey and Lebanon have been the source recent discoveries conducted by microbiologists related to fossil inclusions and ancient DNA, the constituent part of the nucleus of a cell that contains the genetic code of the species from which it came (see Chap. 11). (Figure 7.1 shows the distribution of Cretaceous amber.

During the Cretaceous period, from 65 to 140 million years ago, angiosperms or flowering plants began to spread and insects rapidly developed. According to Grimaldi (1996a, 21), who conducted a study of the largest Cretaceous amber deposits on the Atlantic Coastal Plain in Maryland and New Jersey, the angiosperms and insects comprised three-quarters of all life forms during this period. It appears evolution of flowering plants and insects affected each other and assisted in the apparent "explosive radiation" of these life forms. Grimaldi (2000; 1995) found the earliest evidence of "true ants" with a social order, earliest gilled mushrooms, and even feathers in Cretaceous amber (Hikbbett, Grimaldi, Donoghue 1995). Thus, fossils in Cretaceous ambers provide a window into the development of early insects—many of which are extinct today.

Gymnosperms, such as Cycads and other conifers, predominated during the Cretaceous even though angiosperms were diversifying. Grimaldi (1996a, 21) points out most Cretaceous amber deposits currently studied are chemically related to the Araucariaceae, which is one of six families in the *Coniferae* group. Conifers have been known from fossils that are more than 290 million years old, and fossils of *Araucaria* have been found in locations around the world from the Jurassic to the Paleogene. (Table 7.1 lists the four major conifer groups found during the Cretaceous period.)

Araucarians produce tremendous amounts of resin that hardens after having been exposed to the air and that is insoluble in water.

Table 7.1 Table of Four of the Six Families of Conifer Trees		
Coniferae Phylum	**Family**	**Common names**
Pinophyta	**Araucariaceae**	**Kauri pine, monkey puzzle, Norfolk Island pine**
	Pinaceae	**Pines, larches, spruces, hemlocks**
	Cupressacreae	**Cedars, cypresses, junipers**
	Taxodiaceae	**Sequoias, Bald cypresses**

Cretaceous ambers are very brittle and fracture easily. Since Cretaceous amber becomes crumbly when exposed to the atmosphere after many years, museums are preserving rare Cretaceous fossils in amber by embedding them in a synthetic resin (Grimaldi 1996*a*, 24).

Though many Cretaceous ambers are thought to have originated from araucarian, others have been determined to be from different sources such as the fossil resin from San Juan Basin, New Mexico. This fossil resin is found embedded in stumps of an ancient Taxodiaceae tree. Also amber from Mississippi, found in the Upper McShan Formation, is associated with Taxodiaceae, Cupressaceae, and Pinaceae fossilized wood that is determined to be about 90 million years old. Fungal spores and hyphae have been identified in this amber but as yet no insects (Grimaldi 1996*a*, 27). Grimaldi's team also reported Taxodiaceae the source indicatd in Wyoming amber, thus they concluded all Cretaceous amber was not formed by members of the Araucariaceae as some recent proposal suggested (Grimaldi, Lillegraven, Wampler, Bookwalter & Shedrinsky 2002, 163).

Another unique Cretaceous amber is found in both Austria and France. This amber is similar in composition to New Jersey amber with similar fossil inclusions. But, interestingly, the amber, having been formed near pyrite deposits, is found with the mineral seeped into microscopic cracks and bubbles and even into some of the insect inclusions. Both Poinar and Grimaldi point out that pyrite allows for high-resolution X-raying since the mineral is much denser than the surrounding amber and has in some cases replaced the original insect in detail. In the French amber, Grimaldi (1996*a*, 25) reports only about 20 insects and insect parts in 1 pound of raw amber. This yield is relatively small compared to Cretaceous amber from Canada that produces twice as many inclusions.

A Cretaceous age fossil resin is found in Choshi, Japan. The unique contribution to scientific study provided by this early Cretaceous amber was the inclusion of portions of a bird feather that dropped from a bird that flew in the primeval forest 120 million years ago. Another location of Cretaceous amber from Kuji, Japan, was discovered in shore deposits. The amber was dated as originating during the late Cretaceous period based on the marine fossils that occur along with it. The pieces are a caramel color and other shades of yellow, with some pieces appearing opaque. The Japanese carve objects from this amber because it is often found in large pieces, the largest being 44 pounds (20 kg) found in 1927 and another weighing 35 pounds (16 kg) found in 1941. These pieces are now in the National Science Museum Tokyo (Grimaldi 1996*a*, 32).

Canadian Amber: Chemawinite or Cedarite

Canadian amber, considered to be about 80 million years old, has been known for over a century, but little has been collected or used in ornamentation. The best known and most widely studied source is Cedar Lake, Manitoba. In 1889 local Indians brought specimens to W. C. King, the officer at the

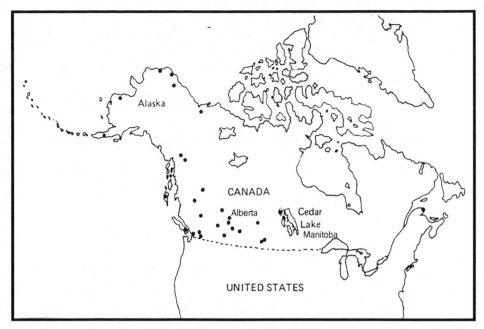

Figure 7.2. Map of Canadian amber deposits.

Chemawin Trading Post. In 1891, this beautiful amber or fossil resin was named Chemawinite after the Indian name for the Hudson Bay Company post that is near the deposits (McAlpine and Martin 1969, 819). O. J. Klotz, of the Geological Survey of Canada, attempted to find other deposits in the Cedar Lake region, but results were disappointing. Only traces of amber were found at several places along the Saskatchewan River flowing into Cedar Lake, and none of the deposits were rich enough to support a commercial mining venture. In 1893 Tyrrell (30A–1A), of the Geological Survey of Canada, also obtained and studied the amber, while surveying the northern Manitoba area. He reported amber deposits extending to a depth of 1 meter for a distance of about 2 kilometers along the lake shore, and extending back about 25 to 36 meters away from the shore. His estimate of the amount of amber in the area was about 600 metric tons. In 1897 Klebs (199–230) described this amber and used the name of Cedarite, but the name Chemawinite is more often used. Cedar Lake deposits are thought to be carried into the lake by the easterly flowing Saskatchewan River, as it runs through lignite beds in Alberta and south Saskatchewan (McAlpine and Martin 1969, 819; see Fig. 7.2).

Chemawinite is found as small nodules of brownish color in sands and gravels along the southwest edge of Cedar Lake and is considered to be Cretaceous in age (60 to 130 million years old). It contains no succinic acid. Langenheim (1969, 1157–69) relates the resin to that from the Araucariaceae. In1985 Poinar and Haverkamp (210–21), analyzing the resin's infrared (IR) spectra and pyrolysis mass spectrometry, indicated an affinity to the *Agathis australis,* similar to New Zealand pine trees that produce kauri resin. Lambert, Frye, and Poinar (1990, 43–52) reported similar results using carbon-13 nuclear magnetic resonance (C13NMR or shortened to NMR) spectroscopy. Analysis of the NMR spectroscopy showed Canadian amber to be quite similar to Atlantic Coastal Plain amber and that found in Washington and in Alaska.

Though Canadian amber is most abundant at Cedar Lake, where it is found in alluvial deposits, it also occurs elsewhere in low-grade coal and lignite beds and in carbonaceous sediments. In 1933 the Craig Amber Mining Company at Cedar Lake supplied a small quantity of gem material, but in all, only about 1 ton of amber was produced from this area between 1895 and 1937. Much of the amber was sold to varnish manufacturers. In still another attempt to exploit the deposits, the firm of Native Minerals Limited attempted to obtain amber from the region for commercial purposes in the mid-1950s. During the winter amber-bearing sand was scraped up from the shore and shipped to Winnipeg for sorting. This venture also proved to be unprofitable because the amber was found mainly in the form of very small granules. Some of these specimens were donated to the Canadian National Collection of Insects.

Amber resembling Cedar Lake deposits is found in coal deposits near Medicine Hat, Alberta, and is often called "Grassy Lake amber." The coal deposits belong to the Foremost Formation and are overlaid by Upper Cretaceous soft claylike rocks, called bentonites, indicating an age of 75 to 78 million years by radiometric dating (Folinsbee et al. 1964, 525). Poinar (1992, 50–2) points out that this corresponds to the minimum age of the Cedar Lake amber deposits laid down before the close of the late Cretaceous period, making the amber about 70 to 80 million years old.

Chemawinite is important scientifically for its inclusions of well-preserved insects, spiders, and mites, among the best to be found anywhere. Walker (1934, 5–10) was the first to report fossils found in Canadian amber. It also contains pollen grains, spores, and fragments of plants from the Upper Cretaceous period, some 70 million years ago, as well as many unusual insects that, though predominant during the epoch, are now extinct.

In 1937 Carpenter (7–62), a paleontologist from Harvard University, organized an expedition to Cedar Lake to collect Chemawinite fossiliferous specimens. Working for three months, he obtained 400 pounds (181.4 kg) of amber. Later attempts to collect specimens for scientific study yielded smaller and smaller quantities. Some of this collection of yet unexamined amber is now housed at the

Museum of Comparative Zoology at Harvard University. In 1950 and again in 1963, for example, Brown and Bird (McAlpine and Martin 1969, 819), collecting for the Canadian National Collection of Insects, found only 2.5 pounds (1.13 kg) of amber nodules, ranging from the size of a pea to that of a robin's egg. The amber was found along the sandy shores of the Cedar Lake mouth of the Saskatchewan River, mingled with shells, coal fragments, and organic debris. In 1969 McAlpine and Martin (819) reported the site no longer accessible because a dam at the foot of Cedar Lake at Grand Rapids had raised the water level to cover the amber-collecting areas. However, Poinar (1992, 50–2) indicates wave action still washes some amber up on newly formed higher beach areas, where it may still be collected (see Fig. 7.3).

Figure 7.3. Nodules of Chemawinite from Cedar Lake, Manitoba. The pieces are a reddish semi-transparent amber. Size: about 2.3 cm Courtesy of the Canadian Museum of Natural History, Toronto, Canada.

During their studies of inclusions in Chemawinite, McAlpine and Martin (1969) examined 470 pieces with identifiable inclusions of fauna. About 300 different species were represented, but only 30 of these were completely described at that time. Interestingly, the specimens studied did not readily fit into present orders, as is the case with those species found in Baltic amber, which can be assigned to modern genera or related categories. At that time too little was known about Cretaceous amber insects to draw conclusions about their paleoecology and environment, but Canadian amber provides certain useful clues. First, no species—and very few genera—occurring in Cretaceous amber occur either in Baltic amber or any other place in the world today, which suggests that these evolutionary lines suffered extinction either in late Cretaceous or early Paleogene times. Second, some species found in Canadian amber, such as aphids, may be directly associated with coniferous trees. Since 40 percent of the inclusions in Canadian amber are Diptera, a rich, moist environment is indicated. The arthropod fauna suggest climatic and ecological conditions similar to present day Florida. The study by McAlpine and Martin revealed that the species in Canadian amber are more closely allied to species now living in Australia, New Zealand, and South Africa than to any others.

Other than the Cedar Lake occurrence, there are 32 locations in Canada where amber has been reported, none of which is a signifi-

Figure 7.4. A piece of Upper Cretaceous walchowite, from Moravia, Czech Republic, in matrix of carbonaceous sandstone, as exhibited in the Muzeum Ziemi, Warsaw, Poland. Photo by Patty Rice.

cant producer. These locations are shown on the map in Figure 7.2.

In 1969 Langenheim (1962) reported that some of the Manitoban amber was similar chemically to the fossil resin from Moravia, .Czech Republic. This amber was named "walchowite," dated as Upper Cretaceous, and thought to be derived from the *Dammara* phylum or other araucariaceous sources. She points out that the spectra of ambrite from Auckland, Manitoba, amber, and walchowite from the Moravian region all are similar to those of resin from living *Agathis australis* and *Agathis labillardier*, but not other species of *Agathis* and *Araucaria*. It seems Araucarians disappeared from the Northern Hemisphere in the early Paleogene and are distributed today only in the Southern Hemisphere. Analysis by NMR spectroscopy (Lambert and Frye 1982, 55–7; Lambert, Beck, and Frye 1988, 248–63; Beck 1986, 57–110) indicated that walchowite and chemawinite were characterized by a weak or absent exomethylene resonance, placing them in a group of fossil resins showing similar NMR patterns, such as simetite from Sicily, rumanite from Romania,   from Bukovina, and delatynite from Delatyn (former USSR; see Fig. 7.4).

CRETACEOUS AMBER FROM THE UNITED STATES

Tennessee. The first fossil insect discovery from North American amber was identified by Cockerell in 1916 as a caddis fly, *Dolophilus praemissus*, in amber recovered from Coffee Sand, Tennessee. This amber is found in small pellets occurring in beds of lignite north of Newman Cemetery in the northwest section of Hardin County (Cockerell 1916, 89; Jewell 1931, 94–5).

Mississippi. As mentioned earlier, 90-million-year-old amber was found in Mississippi associated with fossilized wood from Taxodiaceae, Cupressaceae, and Pinaceae. The amber was found in small pieces up to 3.5 centimeters in diameter and ranged from yellow to dark brown in color. Some pieces included fungal spores and hyphae. Lambert (et al 1996) dated this amber as 80-90 Ma using NMR assigning the source tree as *Agathis*.

Texas. Another fragile fossil resin is found at Terlingua Creek, in chunks of soft coal. Fossil resin deposits are found in Cretaceous coals of Maverick and Brewster counties in Texas (see Fig. 7.5). Some pieces were translucent and of a quality suitable for lapidary purposes, but they were much too small. A poor quality brown amber has been found in Paleogene deposits along the Gulf Coastal Plain.

North Carolina. Amber from North Carolina was found in lignite beds near Goldsboro in the Black Creek Formation. Small pellets of Cretaceous amber, which are a transparent, clear yellow, have been collected by local hobbyists. The amber is considered to be about 75 million years old.

Kansas. A dark opaque amber, Cretaceous age fossil resin, discovered by George Jelinek, occurs in lignite beds along the Smoky Hill River in Ellsworth County, but the beds are no longer accessible because of the back waters of the Kanopolis Dam. Smoky Hills are capped with sandstone to limestone and rocks that were deposited on or near a sea floor during the Cretaceous Period. Less than 50 pounds of the amber were found, but it was distributed to various museums. Buddhue (1938a) initially proposed the resin be called kansasite (p. 8). He later renamed the fossil resin jelinite, in honor of the collector George Jelinek (Buddhue, 1938b, p. 9). Aber and Kosmowska-Ceranowicz (2001, 32) describe this rare amber as being translucent to opaque, dark orangish-brown to yellow, with a resinous to waxy luster. It is extremely brittle with a thin gray crust. Their work, submitting Jelinite to infrared spectrometry, gave an IR spectrum similar to Cretaceous resins from Wyoming, USA, and the Canadian Cedar Lake, Manitoba, and Grassy Lake, Alberta. Thus they believe it could be included in the Cedarite group of fossil resins. These IRS data correlate to other fossil resins found in sediments in Asia and Europe, and to resin of the living *Agathis australis* or Kauri pine growing extensively in New Zealand. Aber and Kosmowska-Ceranowicz conclude the data suggests Kansas amber originated from the Araucariaccae even though these conifers are restricted to the Southern Hemisphere today. To read more about Kansas amber go to the website Amber World by Susan Ward Aber, (1996-2005) Emporia State University.

Wyoming. A very brittle fossil resin was described by Kosmowska-Ceranowicz, Giertych, and Miller in 2001 as being reddish-yellow in color. The amber was found embedded in the Lance Formation, a compact, lime-free grey loam in Upper Cretaceous deposits. Kosmowska-Ceranowicz, et. al., classified it in the same group as the jelinite from Kansas, the cedarite group of fossil resins. According to Aber (1996-2004), Steve Levine, a geologist, found another amber in

Figure 7.5. Samples of amber from Terlingua Creek, Texas. Very fragile lignite and carbonaceous material with lenses of amber embedded. Crumbles easily.

Wyoming in the mid to late 1970s that came from the Battle Spring Formation, a carbonaceous un-altered arkose, a sandstone with quartz and abundant feldspar, Eocene in age. It was a dark colored nodule, shattered from blasting at Western Nuclear's Seismic Mine at Jeffrey City, Wyoming. Grimaldi and his team examined amber from south central Wyoming on the Hanna Basin and found evidence based on scales of conifer cones containing resin which suggested Taxodiaceae as the common source, yet the amber chemistry also suggested Pinaceae.(Grimaldi, Lillegraven, Wampler, Bookwalter & Shedrinsky 2002, 163)

Nebraska. A little know amber, listed by Grimaldi (1996, 20) as being Upper Cretaceous age from Seward, Nebraska, and also by by G. Platt, on his Amber Home Web site, was reported without any description. Correspondence in 2005 on the Nebraska Geological website elicited a response from Robert M. Joeckel who stated;

> I have found amber in two places in Nebraska, both in the Dakota Formation, which straddles the boundary between the Lower Cretaceous and Upper Cretaceous: (1) near Pleasant Dale, west of Lincoln, in Seward County (south and slightly east of the city of Seward); and (2) due south of Fairbury, Jefferson County, Nebraska, about 1.5 mi north of the Kansas border. In both cases, the finds were small (5 mm or less), irregular masses of poor-quality amber--but amber nonetheless; the host rocks are relatively dark, carbonaceous shales that contain other fossil plant material such as angiosperm leaves, conifer shoots, wood fragments, etc. I suspect that amber is more widespread in the Dakota Formation and that, perhaps, larger pieces may yet be found.

The Kansas amber and Wyoming amber are both from the Dakota Formation.

Alaska. Cretaceous age amber was reported in Alaska as early as 1870 by Dall as a result of his survey of Alaskan resources. Amber was found in lignite beds on the Alaskan peninsula, in the alluvium of the Yukon delta, and near the coal deposits of Fox Island. In 1957 University of California, Berkeley, investigators reported Eskimos being aware of amber deposits in the Aleutian Island chain. Interestingly, the Eskimos described "spreading walrus skins between two boats at the base of a cliff and pulling down debris along with amber onto the skin" (Dall 1870, 625). The University of California, Berkeley, investigators explored the area at Smith Bay in the Alaskan Arctic, north of the Brooks Range, finding exposed cliffs with coal and lignite with embedded amber. The amber was brittle. They found

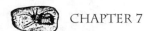

pea-size pieces in gravel banks. However, at the mouth of the Omalik River, amber occurred more abundantly. Also, amber was found at Kuk Inlet near the junction of the Avalik, Ketic, and Kaolak rivers (Poinar 1992, 53–6). Langenheim (1960, 1345–56) found amber in layers of Cretaceous rocks occurring in the Arctic Coastal Plain and Arctic foothills. It was considered to be deposited originally in thin coal beds or carbonaceous shale layers and was dated as late Cretaceous.

Based on a study of the fossils associated in the amber-bearing strata, Langenheim considered Alaskan amber to have been produced by a representative of the Taxodiaceae family; but later, using IR spectrometry, found this was not substantiated. Poinar and Haverkamp (1985, 210–21), on the basis of pyrolysis mass spectrometry, produced a similarity to recent *Agathis* resin and suggested an araucarian origin. Lambert, Frye, and Poinar (1990, 43–52; 1996) analyzed Alaskan amber using NMR spectroscopy, finding support for the *Agathis* origin (Poinar 2001,18).

Atlantic Coastal Plain, New Jersey, and Maryland. The Atlantic Coastal Plain has received the most study of its fossiliferous amber, having been first reported in 1821, thus known for more than 175 years. Today attention is focused on the New Jersey amber currently being studied by fossil collectors in the paleontology society of the American Museum of Natural History. *The New York Times* reported in January and February 1996 issues:

> In an exciting discovery, scientists on an expedition from the American Museum of Natural History stumbled upon one of the richest deposits of amber ever found. The deposits, found in an undisclosed area of New Jersey, contain fossils of insects and plants previously unknown.

Scientists are attempting to keep the exact location of the deposits secret to preserve the ancient fossils. The news that made the pages of the *New York Times* was described as "the world's oldest mosquito fossil, the oldest mushroom, the oldest moth, the oldest bee, and 'a feather that is the oldest record of a terrestrial bird in North America.'" Philip Hilts described the feather in the January 30, 1996 *New York Times*, while Paul Nash (February 8, 1996) reported on a specimen containing a cluster of flowers from a primitive oaklike tree, also claimed to be the oldest flowers preserved in amber, dating to the era when flowering plants first began to proliferate.

Earlier reports describe amber found in clay deposits along the shores at Cape Sable, Maryland, that was considered to be of Upper Cretaceous age. One particular amber piece was described as containing "a gall made by scale insects." Other amber finds were reported on Staten Island when mining in clay pits for brick manufacturing. Grimaldi (1996*b*, 84–91) reports that the workers would store the amber in barrels and burn it to keep warm in the winter! Grimaldi, having conducted extensive investigations of the largest Cretaceous age amber deposits in North America, namely, these New Jersey locations, describes amber also being found in abandoned clay pits. The source of the fossil resin is chemically similar to the araucarian. However, many botanical fossil inclusions, such as twigs, and the microscopic structure of lignite associated with the amber indicate Cupressaceae. The amber has been dated as 65 to 95 million years old. The amber nodules are clear red to yellow, with about 70 percent being turbid, some with small bubbles and particles of plant debris (Grimaldi 1996*a*, 27–30).

This ancient amber has revealed new insights into the development of flowering plants and insects of the Cretaceous period. Two of the most important discoveries of fossil insects were found in New Jersey amber. The first was discovered by Harvard entomologists, Frey and Beck, in 1966 and reported by Carpenter and Brown in 1967 (1038–40), as a primitive ant that provided a link between the tiphid wasps and the most primitive known living ants (Walker 1971, 31). Though older fossil ants have been described since then, this remains the oldest definitive ant, called "Frey's wasp ant," or *Sphecomyrma freyi*. This insect possessed both wasplike and antlike characteristics, thus providing the link to the ants that today, as a group, include over 14,000 living species (see Fig. 7.6). In 1998 Grimaldi secured its status as an ant by discovering a well preserved New Jersey specimen of *S. frey* identifying their netapleural glands, which only ants have.

Another unique fossil is a stingless bee, *Tigona prisca*, discovered by Grimaldi (1996*b*, 28), belonging to an evolutionarily recent group of stingless bees. Grimaldi found the fossil in an old museum collection that was authenticated as New Jersey amber about 65 to 80 million years old. He explains its significance because bees forage on pollen and nectar and advanced bees such as the one he found in Cretaceous amber would mean angiosperms had to correspond in age to the insect. This changed current thought regarding the evolution of flowering plants during the era and also raised much controversy among the paleoecologists. Some scientists remain

Figure 7.6. "Frey's wasp" the famous ant, Sphecomyrma freyi, fossilized in New Jersey amber, providing a link between wasps and ants. In 1998 Grimaldi secured its status as an ant. Photo courtesy Harvard University Museum of Comparative Zoology.

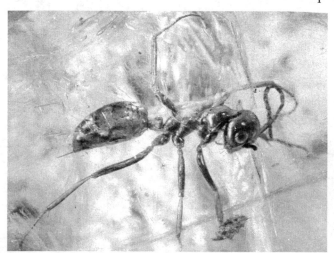

"skeptical about the authenticity of the specimen" (Grimaldi 1996*a*, 28).

However, not to be deterred Grimaldi, in the 1990s, examined new excavations of the Cretaceous deposits in New Jersey yielding much amber and a stunning array of 90-million-year-old flowers preserved in clay. Grimaldi (1996*b*, 29) describes the flowers as being "tiny made entirely of carbon preserved perfectly with stamens, anthers, pollen, stigmas, petals, glands and even the cells visible." Since many of these flowers were surprisingly advanced evolutionarily, belonging to tropical families such as laurels, tiny magnolialike flowers, heaths, and witch hazel families, the stingless bee fossil is more easily explained and the New Jersey amber find more feasible as an authentic specimen. These flowers had to be pollinated by insects, since the flowers produce a viscous substance that makes clumps of pollen adhere to the hairs of the insect such as the ancient *Tigona prisa*.

Fascinatingly, this new batch of ancient fossil insects revealed that some of today's insects had close relatives extending back nearly 100 million years. Much research is continuing on the inclusions in the New Jersey amber. Another ancient botanical fossil of interest is a fully developed mushroom found in East Brunswick, New Jersey, Cretaceous amber (Grimaldi 1996*a*, 28–30).

CRETACEOUS AMBER FROM THE MIDDLE EAST

Fossiliferous amber has been found in the mountain areas of Lebanon, Israel, and Jordan. This is the oldest extensive fossiliferous amber known, having been deposited some 135 to 120 million years ago in the early Cretaceous period. The amber was first reported by Fraas (1878, 257–391), a German paleontologist over a hundred years ago, who took his amber collections back to the Stuttgart Museum of Natural History. However, study of the fossiliferous amber was not conducted until 1970 by Schlee and Dietrich (40–50) and in 1978 by Schlee and Glöckner. In 2001 Poinar and Milki provided a complete study of Lebanese amber, with on site visits by Milki, identifying it as having the oldest insect ecosystem in fossilized resin. They identify the source amber-producing tree as an extinct Araucarean. Amber from Jordan was also attributed to this source tree (Lambert et al. 1996; Shinaq & Bandel 1998). Interestingly, a new species, *Agathis levantensis*, was described by Poinar (2001, 17) based on leaf remains and is considered the source tree of Early Cretaceous amber of Lebanon and surrounding areas.

The amber in Lebanon is described by Schlee (1990) as occurring in primary deposits of the early Cretaceous as well as in second-

Figure 7.7. Lebanese amber weevil Lebanorhinus succinus. Poinar Amber Collection. Photo by G. Poinar.

ary deposits in the district of Jazzine, about 23 kilometers east of Saida. In the 1970s - 1980s amber was collected by Acra (1972, 76–7) from deposits in Dar al-Baidha, between Damascus and Beirut. The Acra collection, over a thousand fossiliferous pieces selected over several years, was given to the American Museum of Natural History for study by Grimaldi and his team. This collection, valued for its unique ancient fossils, has provide some of the oldest record we have today of certain arthropods - the most common of which are male scale insects, parasitic wasps, midges, and bark lice. Moths embalmed in this ancient amber have preserved even the microscopic scales on their wings, which indicates that they are true lepidopterans, well developed at this early era, and are the oldest definitive moths. A well-preserved caterpillar with the tiny spinneret still remaining on the tip of its head was identified as the oldest known caterpillar found in amber. Entomologists studying amber find the rarest organisms are the pseudoscorpions, millipedes, and stinging wasps (Grimaldi 1996a, 36).

A piece of Lebanese amber contains the oldest known example of external parasitism by including a biting fly being bitten by a mite frozen in action in the ancient amber resin. Also the oldest example of internal parasitism is displayed by Lebanese amber containing a nematode visible within the abdomen of a midge. Poinar (2004) describes the parasite, mermithid nematodes, within the body of its host, assigning it to a new genus, *Cretaciomermis*. Nematodes have been reported in amber from other locations also. Studies of Cretaceous amber continue to reveal new species of extant animalia.

In 1993 Poinar succeeded in sequencing DNA from the oldest amber-embalmed fossil—a weevil trapped in Lebanese amber dating back to the early Cretaceous period over 130 million years ago (Cano et al. 1993, 536–8; see Fig. 7.7). Of course, this again raised questions from competing researchers, "Should these rare old fossils be destroyed?" (Lewin 1996, 201).

In Israel amber is found on the southern slopes of Mt. Hermon, on Naftali Mountains in northern Israel, and in Kokhov and Barboor areas in southern Israel. Amber from Israel and Lebanon is considered to be of the same deposit. It is found in the form of droplets and nodules ranging in size from very small to several centimeters in diameter. The color ranges from transparent yellow to brown and honey colored. The amber is frequently associated with lignite, fossil wood, and carbonized plant remains. *Agathis*-like plant fossils were found associated in the amber-bearing strata and using IR spectrometry similarity was noted with resins from the Araucarian by Bandel and Vavrà of Austria (1981, 19–33).

A new amber sample, recovered in the Persian Gulf by an oil worker during drilling, came from a mile down in Cretaceous deposits. It was analyzed by Poinar's team, who concluded it was formed by an araucarian, possibly a new species, *Agathis levantensis* sp. n. They used the amber as a bio-marker to identify the coal which contributed, in part, to the oil production. This is the only known amber sample to be analyzed from such a depth (Poinar, 2004, 207-209).

CRETACEOUS AMBER FROM MYANMAR OR BURMITE, BURMESE AMBER

Burmese amber usually occurs in small pieces and varies in color from burnt orange to reddish shades with swirls of color. It was originally described as Miocene age by Noetling in 1892, then later Eocene age (Stuart 1922; Chibber 1934) or only about 45 Ma. However, more recent studies by Poinar (2003; Zherichin & Ross (2000) assign Cretaceous based on primitive insect types found in the amber. Burmite is now thought to be at least 100 Ma. Burmite was unavailable for many years with the mines closed, but were reopened in about 1988, officially in 1995-1997 and 2000 (Cruickshank and Ko Ko 2003, 444). In 2002 Cruickshank and KoKo revisited the site to verify the source of the amber and obtain information of the geology and age of the host rocks. These researchers suggest the amber-bearing strata they witnessed as Upper Albian to Lower Cenomanian, with the preponderance of the amber as Upper Albian. The amber is typically reddish brown in color with various shades of yellow to red; transparent to translucent to opaque; with thin white calcite veins commonly observed in the amber. Cruickshank describes the amber mine site during his visit as resembling a small open pit mine all excavated by manual methods with 60 men present during his visit. Amber was produced from unweathered rock as they stripped overburden. Sedimentary rocks including limestone beds, coal, carbonaceous material and some shales were exposed (a complete description may be found in Cruckshank and Ko Ko 2003, 447). The Cretaceous age "determination is significant for the study of insect evolution . . . indicating the oldest known definitived ants have been identified in this amber from Myanmar." After their verification as Cretaceous age for Burmite, fossils from former Burma now in Museum archives, one historical collection maintained at the Natural History Museum, London, are being restudied and reidentified (Rasnitsyn and Ross 2000; Grimaldi, Engel & Nascimbene 2002). New subfamily, genus and species are being found in this Cretaceous amber. Also, fossiliferous amber from the new mines was sent in 1999 and 2000 to Grimaldi

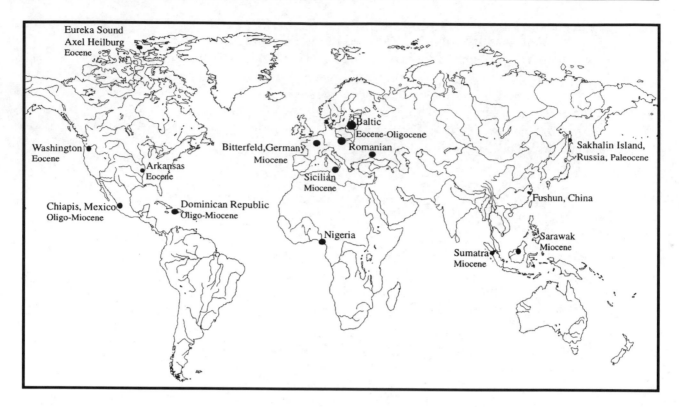

Figure 7.8.
Distribution of
Paleogene Age
Amber.

and his team and they found numerous insect inclusions to study. After which this team proposed that the age is as early as Turonian-Cenomanian (90-98 Ma) (Grimaldi et al. 2002) Even though the amber from Myanmar is now considered to be Cretaceous age, its history will be described later along with other Asian amber sources.

PALEOGENE AGE AMBER

Dominican Amber from the Dominican Republic
The Dominican Republic has been a known source of amber in North America longer than any other amber deposit in the Western Hemisphere. Columbus, in his account in 1496 of his second voyage to the New World, stated that amber was found on the island, then known as Hispaniola, and described a mining region near the Tower of Conception, a fortress located on the border of the country ruled by Guanahanis or Guarionexius (Rice 1979, 1804). Amber ear ornaments were found in Indian tombs, but were somewhat disregarded without much interest paid to these finds. Thus, we know amber was known to the Taino Indians and other peoples living on the island before Columbus arrived.

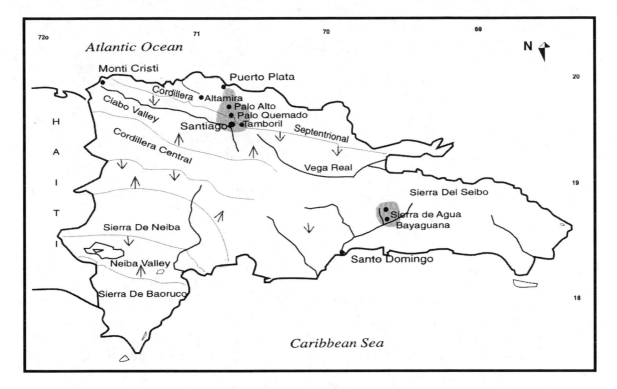

Hispaniola is the second largest island in the Greater Antilles Archipelago of the Caribbean Sea. It is generally believed to have been formed by volcanic activity as the Atlantic plate subducted under the Caribbean plate. The islands are actually peaks of a submerged volcanic mountain range that belched fire and smoke during the late Cretaceous age. Deep trenches border the island and form the deepest portion of the Atlantic Ocean—more than 4690 meters deep at the Puerto Rico Trough, located just northeast of the Dominican Republic. The first geologic survey in 1949 was conducted by an Italian geologist, Renato Zoppis de Seno, who described three subsurface structures or arcs; the arc of Jamaica, the arc of Cuba, and the arc of the Bahamas, whose axes tend northwest-southeast, paralleling the long coastal direction. The arcs on Hispaniola form three valleys and four mountain ranges running almost parallel to each other. In 1966 Brown (11–84) described four structural trends converging on the island; the main axes of the island arc, the southeastern part of the Bahamas, the swell extending from Central America to southwestern Hispaniola, and the Beata Ridge. The oldest dated rock formations on the island are plutonic tonalite of Cretaceous age (see fig. 7.8 and 7.9).

Tremendous tectonic changes occurred during the Paleogene period with islands appearing and then disappearing as they were again covered by the sea. During the Ice Ages, sea level continually

Figure 7.9. Dominican Republic mining area in Cordillera Septentrional and in Bayaguana Basin.

rose and fell, causing new areas to be inundated by the sea and leaving behind sedimentary deposits with each regression. Thus, much of the surface rock occurrences are sedimentary, dating to after volcanic activity ceased in what became the island of Hispaniola. According to Pompilio Brouwer, Director General de Minas y Petroleo during the Trujillo administration, amber is found in the arc of the Bahamas, which stretches across the northern coastal regions of the islands, currently called the Cordillera Septentrional or Northern Range (Brouwer 1959, 1). The amber is generally associated with marine deposits and materials laid down in freshwater lagoons, associated with lignite and carbonaceous deposits in horizontal layers of sedimentation. However, as a result of faulting and folding of the Earth's crust in these mountainous regions, the amber strata may slope as much as 20 to 30 degrees southward. In the Cordillera Septentrional, in the Monte Cristi Range, stratifications indicating repetitive submersion under water during the Upper Eocene-Lower Oligocene are apparent. Flysch deposits, with alternate layers of shales or siltstone and sandstone formed by turbidite flows, are exposed on the faces of steep ridges in route to the amber mining area. These marine deposits originally were at beach level, but as upheaval occurred, the strata took on the southward slope (Rice 1980*b*, 70–2, 1981*b*, 145–52). Some time during the Paleogene period, the land became covered with resin-producing trees. As traumatic climatic changes took place, tropical trees growing on the island produced an abundance of resin that dripped down their trunks, gathering in chinks and crevices, often entrapping small insects living on the tree bark or attracted to the sweet aroma of the viscous resin. This material hardened and lumps fell to the ground. As land areas flooded with water, the insoluble resinous masses, being light and buoyant, were swept away by currents. Just as in the Baltic area, these repetitive inundations of the land with water resulted in amber being placed in secondary deposits. This, of course, adds to the difficulty in precisely determining the age of the amber. Interestingly, Grimaldi reports that the amber from La Toca shows little sign of reworking, as it still retains irregular and angular shapes of the pieces, a result of low-energy water action.

Figure 7.10. A native uses his pack burro for transporting rough amber to the market in Santiago.

"It must be stressed that these deposits are marine, so the amber has been redeposited from its original terrestrial origin" (Grimaldi 1995, 211).

Although this amber was mentioned in the literature in the nineteenth century, it was not exported commercially. In 1891 Haddow noted:

> At Santiago, in San Domingo, in the valley of the brook Acaqua, amber pieces, some as large as the egg of a goose, reward the explorer. The acaqua brook carries away the amber from hills of marl which is rich in petrifications, and bears a near resemblance to Miocene clay of the Vienna basin (Haddow 1891, 8).

In 1905 amber containing inclusions of decayed twigs was described as originating from an amber-bearing formation in the Cordillera Septentrional (Samples 1905, 250), but despite early knowledge of the source, Dominican amber did not begin to be exported on a commercial scale until about the 1970s.

The main Dominican deposits are located in the Cordillera Septentrional along the northern border of the Vega Real central plains area, about 10 to 30 kilometers northeast of Santiago, between Altamira and Canca (see Fig. 7.10). The original and largest mining site is northwest of Tamboril in the Pena region. The location is between two gorges, Los Meninos and Perez, of the Arroyo Capancho tributary of the Rio Gurabo. The original amber location was found at an approximate elevation of 1240 meters, and mountainous terrain made mining difficult (Sanderson and Farr 1960, 1313). The second site was located below Pico Diego de Ocampo, near Pedro Garcia in the Palo Alto de la Cumbre region, and the sloped outcrop here was approximately 300 meters in length. This mountainous location was opened for commercial mining in 1949 by the government under the direction of Pompilio Brouwer. However, large scale operations did not last long, because of the unstable condition of the sandstone. This mountainous terrain consisted of Paleogene sandstones and shales along with a variety of eroded conglomerates and big limestone beds. The overburden sandstones were composed of quartz, feldspars, hornblende, and small amounts of magnetite and limonite, deposited in layers, often compacted forming quartz diorite or a variety of tonalite. The amber-bearing strata consisted of dark carbonaceous material about 45 centimeters thick.

Conditions of the overburden that decomposed easily when wet made large-scale underground tunneling impossible because galleries could not be supported with wood beams (Brouwer, personal communication, 1980). Therefore, early mining by the government was an open operation with undercutting along the amber-bearing strata. This resulted in a shelf of unstable sandstone protruding from the mountainside as extraction of the amber-bearing strata progressed. During the rainy season, the sandstone would decompose "like China clay," causing landslides. Eventually, the weakened protrusion of sandstone collapsed and production on a large scale was abandoned. (Rice and Rice 1980, 39). The mountains consist mainly of Paleogene marine sandstones and shales, along with a variety of eroded conglomerates. Wherever carbonaceous material and lignite occur in the sandstones, amber is likely to be present. In 1960 Sanderson and Farr (1960, 1313) described a geologic section of the mining area as follows:

> The uppermost layer [varies] in thickness up to 15 m. Below this layer in this section was a soft layer of clay shale varying from 1/2 to 2 m. in thickness, followed by a harder layer of silty shale 2 to 2.5 m. thick. Below the latter was a fourth layer of unknown thickness, of grey sandstone in which the amber occurred. Contacts between these units are gradational. The amber, which is confined to a thin bed, is removed by breaking chunks of sandstone, first removed with a pick, then by hand and with a heavy knife.

The age of Dominican amber was not definitely established for many years. Since Dominican amber is very fossiliferous, dealers and collectors alike desired their collections to be as old as possible, because age added rarity and value to their collections. In 1959 the first excavations at Palo Alto were identified based on the strata where the amber was located and the index fossils in that strata. Brouwer (1959, 16) tentatively assigned it to the Oligocene. The early report in 1960 by Sanderson and Farr (1313) also dated the amber as Oligocene age, based on strata and index fossils associated with it. However, studies in 1979 by Hueber, paleobotanist of the Smithsonian Institution, showed that amber occurred in Lower Miocene formations similar to those at Chiapas, Mexico (Hueber and Langenheim 1986, 8–10). Later reports by Baroni-Urbani and Saunders (1980, 213–23), at the Ninth Proceedings of the Caribbean Geological Conference, placed the age at Lower Miocene or about 23 million years old, while at the same conference Eberle et al. (1980, 619–32) placed the age at late Eocene or about 40 million years old.

Different qualities of amber occur in the Dominican Republic, and Brouwer points out that these are in secondary deposits (Rice, personal communication, 1980). Therefore, age of the amber cannot be definitely determined by the formation in which the amber occurs, but modern technology has provided a more precise method of determining the age of fossil resins.

Table 7.2 Characteristics of Dominican Republic Amber Mines

Mines in the Cordillera Septentrional

Mine (Slide Area, Excavation Sites, Pits, or Tunnels Following Veins Containing Amber)	Amber Features	Age Estimated from C13NMR Data (millions of years)	Age of Surrounding Sedimentary Rock on Basis of Nannofossils (millions of years)	Age
La Toca and La Pena (personal visit)	Hard, clear, yellow to dark red amber. La Pena mine has produced vertebrate fossils	30-40	30-40	Early Miocene (ca. 23 million years) to Mid-Oligocene (ca. 30 million years)
La Cumbre (personal visit)	Hard, yellow to red amber; few fossils			
Tamboril and Las Cacaos	Hard, transparent, yellow to red, but known for blue amber	30-40		23-30 million years
Palo Quemado (personal visit)	Hard, mostly red to yellow amber, some blue			
Los Higos	Hard, transparent, yellow to red amber, some green; few fossil insects		12-23	
La Bucara	Hard, transparent, yellow found in rock, to red found in soil; some blue		20-40	
Palo Alto (personal visit)	Transparent, yellow to red amber; many fossils of invertebrates	20-30	23-30	23 million years
La Valle La Medita Ya Nigua	Hard, brittle, yellow to red amber; light blue. Fossils include pseuodscorpions and vertebrates. Largest lump of amber found in the Dominican Republic is from La Valle mine.	20-30		
Los Aquitos	Similar to Palo Alto with some invertebrate fossils	20-23		
Pescado Bobo	Similar to Las Cacaos with some dark blue amber; few fossils			
El Naranjo	Transparent, yellow amber, some with bluish hue, some with metallic deposits; few fossils			
Juan de Nina	Pale yellow, fairly soft amber, with metallic inclusions associated with organic material			
Las Auyamas	Yellow amber; some fossils			
El Arroyo	Yellow and yellow-blue amber; few fossils			
Carlos Diaz	Yellow to red amber; some vertebrate fossils			
Villa Trina	Yellow to red amber			
Aquacate	Yellow to red amber; some vertebrate fossils			

Mines in the Bayaguana Region, Foothills in Eastern Part of Dominican Republic

Sierra de Aqua (personal visit)	Hard, brittle, pale yellow to red amber; few fossils			
Comatillo (personal visit)	Yellow-grayish, fairly soft with some hard amber; filled with decayed wood and organic matter	15-17		

Mines in Valley between Cordillera Septentrional and Cordillera Central

Cotui	Clear "amber," pale champagne color, some with yellow tinge, very soft; melts easily; reacts visibly to solvents	15-17 (some younger)		Some as young as 300 years, carbon-14 data

Sources: Amber Features data adapted from G. O. Poinar, Jr., *Life in Amber* (Palo Alto: Stanford University Press, 1992), 32-33. Also based on personal visit in 1980 by Patty C. Rice. Age Estimated from C13NMR Data adapted from J. B. Lambert, J. S. Frye, and G. O. Poinar, Jr., "Amber from the Dominical Republic: Analysis of the Nuclear Magnetic Resonance Spectroscopy," *Archaeometry* 27 (1985): 43-51. Age of Surrounding Sedimentary Rock data adapted from C. Baroni-Urbani and J. B. Saunders, "The Fauna of the Dominican Republic Amber: The Present Status of Knowledge," in *Proceedings of the Ninth Caribbean Geological Conference* (Santo Domingo, August 1980), 213-23; and W. Eberle, W. Hirdes, R. Muff, and M. Pelaez, "The Geology of the Cordillera Septentrional," in *Proceedings of the Ninth Caribbean Geological Conference* (Santo Domingo, August 1980), 619-32. Age based on D. A. Grimaldi, 1994. "The Age of Dominican Amber." In *Amber, Resinite and Fossil Resins*, K. B. Anderson and J. C. Crelling, eds. (Washington, D.C.: American Chemical Society), 203-217.

Baroni-Urbani (1980, 213–23) attempted to clarify the discrepancies regarding the ages reported for Dominican amber based on C13NMR by studying the maturation of exomethylene resonance. The C13NMR study used exomethylene as an indicator because recent resins tend to show more exomethylene. Grimaldi, disagreeing with the older ages, attempted to refute these findings by comparing fossils found in Dominican amber with those found in Baltic amber, which is agreed to be at least 40 million years old. He pointed out that fossils in Baltic amber include more extinct species than do those in Dominican amber, indicating a younger age than 40 million years for the Dominican fossil resin. Grimaldi (1995, 203–17) reported his findings at the American Chemical Society's symposium on Amber, Resinite, and Fossil Resins in 1995, suggesting differing ages for amber recovered from different mines. Unfortunately, Grimaldi's report focused on refuting the work of Eberle, Baroni-Urbani and Saunders, and Poinar, but neglected to provide an age for amber from each mining area. In general, he suggests an age ranging from about 23 to 28

Figure 7.11. Miner removing amber with a machete from the sandstone matrix at Palo Alto mine, 1980.

million years for mines in the Cordillera Septentrional. Surprisingly, based on Carbon 14 dating of Cotui sample resins, he indicated an age of as few as 300 years for some pieces recovered in the Cotui region, yet there are hard pieces of amber recovered in the Bayaguana regions. It seems amber extracted from the Bayaguana region varies in age, depending on the area from where it was recovered (see Table 7.2).

Currently, numerous small mines are scattered among the hills in the Northern Mountain Range, or Cordillera Septentrional, and are usually named after the small villages located nearby, such as Palo Alto, Palo Quemado, La Toca, La Valle, La Tamboril, La Aguita, Las Aquacate, La Bucara, Los Higos, and El Arroyo, with approximately eleven mines in the area.

Dominican mines vary from shallow cauldron-shaped diggings in most of the areas to larger pits found in a few amber regions. Many mines in Cordillera Septentrional are almost inaccessible and can be reached only by hiking along steep mountain paths or by pack burro. The amber is then transported in bags on the pack burros down the mountainside to the cities for sorting and working into jewelry, which is sold in the shops in Santo Domingo and Santiago (see Fig. 7.10).

Figure 7.12. Mountainous terrain in La Toca mines area, the only region in the Dominican Republic where actual tunneling is used for retrieving the amber.
Figure 7.13. La Toca and La Pena tunnels viewed in the distance on the steep mountainside in the Cordillera Septentrional.

Palo Alto, where two qualities of amber are found, one appearing to be softer and perhaps younger than the other, is one of the earliest locations where amber mining was attempted commercially. Even though large-scale commercial excavations have been discontinued, amber is still extracted from the Palo Alto region by miners working with picks, shovels, and machetes to dig small, shallow cauldrons in outcroppings of the amber-bearing strata.

Figure 7.14. La Toca mine tunnel opening with a miner guarding the entrance along with several Dominican children who assisted at the mine during the summer of 1982.

The amber is removed by hand from lumps of black sandstone or deeply embedded pieces are loosened from the matrix with a strike of the machete. Mining at Palo Alto occurs in several strata of black carbonaceous clay alternating with sandstone. Approximately 5 meters from the upper surface a deposit of red amber is found. The lower vein at 30 to 33 meters produces a hard, brittle golden amber. The amber from the upper vein tends to be more fragile and is thought to be of more recent origin (see Fig. 7.11).

Red amber is found in various locations of the Dominican Republic, but particularly in the Palo Quemado mines, associated with rich deposits of limonite, a reddish-orange iron oxide mineral. This may affect the coloration, producing the red surface of the amber, or the coloration may be caused by oxidation as a result of decomposing organic materials within the amber resin during the fossilization process. Often the red coloration simply covers the surface and does not extend throughout the piece of amber.

La Toca, where a hard, fine-quality golden amber is found, has produced many well studied fossils. La Toca is the only mine in the Dominican Republic that actually has tunnels following the dark gray carbonaceous amber-bearing strata (see Fig. 7.12). This mine is located about 3 kilometers north of Palo Quemado and approximately 3 kilometers east of Palo Alto de La Cumbre on the north side of the mountain peak. The La Toca mines actually are narrow tunnels that follow a vein into the mountainside for 25 to 30 meters (see Figs. 7.13) . These tunnels are only about 1 meter in diameter and the miners must descend on hands and knees carrying their lights and machetes to loosen the amber nodules (see 7.14). All the material is removed by hand using picks, chisels, and mallets. The La Pena mine, located in the La Toca region, is where vertebrate fossils have been recovered.

The terrain at La Valle, is similar to Palo Alto, with the base composed of a layer of sandstone that varies from 4 to 12 meters in width. Pieces of lignite and carbonaceous material occur in this sandstone, and as the sandstone lessens, beds of silt, silty clay, and clay appear. Amber is again found in heavily carbonaceous upper layers of clay. La Valle produces hard brittle amber from yellow to red in hue. Transparent yellow amber with a light blue hue over the surface in reflected light is produced here. Many unusual fossils have come from this mine region, which also includes pits called La Medita and Ya Nigua. The largest natural lump of rough amber recovered in the Dominican Republic to date was from La Valle, one of the small mines located in the mountainous Palo Alto region north of Santiago The piece measures 18 inches (45.7 cm) across the largest portion and is 7 inches (17.8 cm) thick and weighs a total of 17.5 pounds (8 kg) (See Fig. 7.15). Like most amber mined from sandstone locations, the rough surface is covered with a grayish crust. The gigantic amber specimen is now located in the Colon Gift Shop Museum in the Plaza Criolla in Santo Domingo. In 1987 Fraquet (122) reported that this large amber lump actually was recovered from the Yaniqua region in Sabana de la Mar County, rather than the La Valle mine as was indicated during my visit to the Colon Gift Shop in 1980. The unusual lump is described the same and was found in March 1979. It is oval-shaped and a dense deep brown color. Sometimes it is very difficult to get the true story from dealers!

A unique type of blue amber is mined at the Los Cacaos mine in a strata composed of blue glauconite clay dated as a Miocene formation. Many varied explanations prevail to account for the unusual coloration. The "blue" Dominican amber is an unusually fluorescent

Figure 7.15. Author, in 1980, holding the largest piece of amber ever found in the Dominican Republic.

a

Figure 7.16. Green
amber from Dominican
Republic: a, green amber
snuff bottle; b, Dominican
amber snuff bottle with
insect inclusion. Purchased
from Daniel McAuley.
Author's Collection.

b

variety that in transmitted light appears to be clear yellow to a yellow-brown color. But in reflected light, a blue surface hue appears as if produced by an oily film. Blue amber is highly prized and commands a steep price on the market. There is also a variety of blue amber that has many lignite inclusions in a deeper brown color with a darker blue hue over the surface. These pieces may lose their blue fluorescence after long exposure to sunlight (see Fig. 7.16a & b).

Amber deposits running underground are mined using a pick to break loose the sandstone. Mining must be done without blasting or heavy pounding, as the impact would shatter amber nodules and cause fractures in larger lumps. Miners bring the gray sandstone to the surface in sacks and spread it on the ground in the sunlight for sorting. The "ore" is picked over and then flushed with water to remove waste sandstone and sediment. The raw amber is then transported to workshops in the village where it is cut and polished (see Fig. 7.17a & b).

Most amber workshops are located in Santiago or Santo Domingo or are cottage industries in small villages. In a typical Dominican factory, amber is graded according to size; then the large pieces are cut to desired shapes using steel hand tools such as a hacksaw; and, finally, polished by using power-driven cotton buffing wheels and tripoli polish. The amber pieces range in size from an inch up to the size of a small grapefruit, and most pieces have a thick dull brown crust. Larger pieces are rare and, of course, much higher in price. Porous lumps may soften after a few days of exposure to a humid moist atmosphere.

When amber mining was first begun by the Dominican government, much rough material was sold to German amber industries. Unconfirmed reports indicate that this material was used in producing ambroid or reconstructed amber items. Since Dominican amber reacts like Baltic amber to heat treatment, it is likely that similar pressed amber products could be produced.

It soon became evident that more profits would be gained if local workers were trained in the art of polishing amber, then finished jewelry could be sold. With government support, in the early 1950s, Brouwer, then Minister of Social Security, organized the first school for polishing amber in the Dominican Republic called the Coindarte Foundation or Cooperative of Industrial Artists. Emilo Perez, a local amber worker from the hills of Tamboril noted for skill in carving amber pipes and cigar holders, became the first instructor for the school in Santiago. Brouwer and Perez studied the German amber industry to develop techniques and to secure polishes suited for working the beautiful gemstone. Brouwer recruited the first group of students—local mountain boys from the Palo Alto and Palo Quemado region. They were paid 30 pesos per month while attending school in Santiago to learn the skill of polishing amber and the art of making fine jewelry. To this day, the amber jewelry from the Dominican Republic has its own distinct style and is found in shops from Puerto Plata to Santo Domingo. The name of the school was changed to the National Crafts Center when Trujillo's government fell (Rice and Rice 1980, 40; see Figs. 7.18 and 7.19). Currently there are many home artisans who have technical experience and are experts in this work.

Jewelry designed by Dominican artists tends to reflect the Taino Indian culture of the past. In making jewelry, small nodules—called marifinga in the Dominican Republic—are sometimes tumbled just enough to remove the outer crust so the amber is left in a semipolished state. Also, nodules are drilled through and strung as bracelets and necklaces. Another common type—a signature of Dominican Republic jewelry—is made with gold-filled wire wrapped around each bead requiring no holes to be drilled. Transparent amber, which is most common from this area, is general-

ly constructed in this manner. The technique requires amber pieces to be shaped, polished on a felt wheel, with grooves cut around the periphery of each bead to take the wire. The luminous quality of the Dominican amber is uniquely displayed in the wire-wrapped jewelry employing beads of various shapes, including round, square, diamond, and oval.

Another characteristic style of jewelry construction uses pieces

Figure 7.17. a and b, Workers grinding and drilling amber into distinctive jewelry in an amber workshop in the Dominican Republic.

Figure 7.18. Dominican green amber carved in the figure of an Indian chief's head with feather headdress. Purchased from Jim Work Company in 1985. Courtesy Burgess-Jastak Collection.
Figure 7.19. Dominican green amber carved in the figure of a salamander. Purchased from D. McAuley Co. in 1985. Courtesy Burgess-Jastak Collection, University of Delaware.

of polished amber with small eye-pins inserted in each end, rather than drilled through as for beads. The loops on the eye-pins are then linked together to form a necklace (see Fig. 7.20a & b).

To the northeast of Santo Domingo, about 24 kilometers into the sugar cane growing areas of the Bayaguana basin, a lowland region composed of clay and mudstone also produces amber. However, this amber is of a different quality than that from the Palo Alto mountainous region. The amber is very aromatic when burned or heated. It does not take as good a polish as the Palo Alto amber and appears to be of a younger age. Bayaguana amber tends to be soft and has a greasy luster. The first geologic survey of this amber basin was completed by Salvador Brouwer in 1980 (Brouwer and Brouwer, 1980). According to Brouwer's preliminary report, the basin was bordered on the southwest and northeast by volcanoes (Brouwer, personal communication, 1980). North and west of the basin is a mountainous region with peaks composed of limestone formations called Los Haitos karst platform, similar to the karst topography in northwestern Yugoslavia where the limestone is continually being dissolved by reaction with water. South and east the basin is bounded by igneous rocks belonging to the Cordillera del Seibo. The Bayaguana amber-producing basin consists of sedimentary materials such as red clay, silt, and mudstone along with a course-grain white quartz sandstone (see Fig. 7.21 & 7.22).

A large number of amber excavations are carried on at Sierra de Agua simply by digging pits using a pick and shovel (see Fig. 7.24). The amber diggings are located along the streams of Sierra de Agua, where miners dig pits or wells about 5 meters deep through sedimentary layers until reaching a horizontal black lignite layer that contains small amounts of amber. The amount of lignite tends to decline westward. Under the dark layer is a grayish-white coarse-grained sandstone and carbonaceous mudstone that also contains amber (see Fig. 7.24). One pit may produce from 4 to 6 pounds of amber ore and takes approximately 2 weeks to dig using hand tools. Rains often fill the pit with water and mud, causing work to stop until the pits dry or are bailed out. When all amber appears to be extracted, the pit is abandoned and a new pit is dug. There is more controversy about the age of the amber from Bayaguana and Cotui mines than about other regions in the Dominican Republic. Some researchers (Lambert, Frye, and Poinar 1985, 43–51) date this softer fossil resin as ranging from 15 to 17 million years old or mid-Miocene, whereas Grimaldi (1995, 205) found pieces to be only 300 years old and classified it as copal.

Cotui and Miches (on the northeast coast) amber is causing confusion about what is Dominican amber and what is simply copal. This is because the material recovered from Cotui and Miches is much softer and considered to be much more recent. The resin from Cotui was

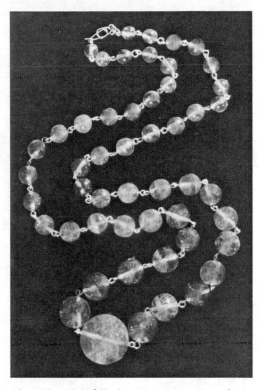

Figure 7.20. Dominican amber jewelry is usually constructed with gold-filled wire wrapping, but the amber pieces are polished in a variety of ways from round beads to flat disks in various geometric shapes: a, Round amber beads connected with gold-filled wire.; b, A variety of different styles of Dominican jewelry. Courtesy Pansy Tice Huff Collection.

271

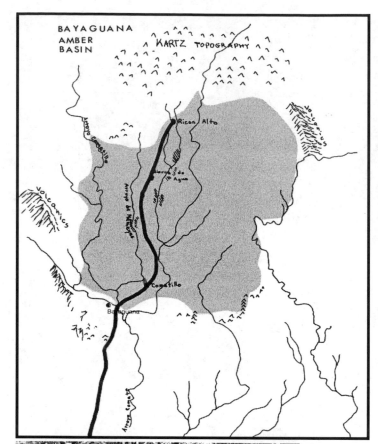

Figure 7.21. Bayaguana region amber deposits.

found in 1981. Grimaldi (1995), after having a piece carbon 14 dated, which indicated it to be only 300 years old, suggested this resin should be classified as copal. Miches amber was tested by laboratories in London and found to be a clear yellow, easily friable material that produced an IR spectra with sharp intense peaks typical of copal and similar to those from Sierra Leone, Congo, and Zanzibar. Fraquet (1987) also reports the coating of recent copal-like resins from Cotui and Miches with an epoxy compound to produce a more stable gem material that would be less likely to craze. This makes it difficult for gemologists to determine if jewelry is genuine amber or simply a recent resin. This type of process has also been used with turquoise to stabilize the lower-quality soft material (Fraquet 1987, 126).

Amber reportedly was found by the Greater Antilles Ltd, while core drilling near Porta Plata on the northern coastal plain. The first layer of amber-bearing earth was located at a depth of 40 to 43 feet below the surface with a second layer at approximately 60 feet. The company spokesperson dated the amber as coming mainly from a Lower Miocene formation with the lower layer as Upper Oligocene. The amber is associated with marine saltwater deposits and is dispersed in scattered pockets that are now mined using a simple open-pit technique. When the amber lumps are first removed from the earth, they have a chalky white powdery outer crust

Figure 7.22. Author inspecting the mines in the Bayaguana region on the way to Sierra de Agua. Notice the undercutting in the black lignite layer.

that must be polished off by hand to display the transparent golden-brown amber.

Specimens containing botanical inclusions suggest that Dominican amber is not an exudation from a coniferous tree, but is from an ancient leguminous plant or pod-producing tree, possibly a species of *Hymenaea*, similar to the African variety that produces copal today (Francis Hueber, phone conversation, 1980). Poinar (1992, 37) describes the tree as the now extinct species of *Hymenaea protera*, which grew during the Paleogene period (see Fig. 7.25). This tree most closely resembles the present-day *H. verrucosa Gaertner*, which occurs in East Africa. Today's related trees, such as Costa Rica's *guapinol* or literally "stinking toe," or the mimosa and poinciana, now occur in tropical areas of South America and Africa. They grow to large sizes and produce an abundance of resin. They are broadleafed evergreens that shed their leaves throughout the year. The large seed pods and leaves are studded with tiny pockets of resin, which chemically repel caterpillars, weevils, and other herbivorous insects. Remains of leaves and flowers are often found as botanical inclusions in Dominican amber.

Figure 7.23. Mining at Sierra de Agua. Workers climb down ladders to reach the bottom of a 5 meter deep pit digging for amber.

Dominican amber was placed in the retinite group—using the older classification. Using Anderson's classification, it is placed in Class Ic. Its specific gravity is 1.048 to 1.08, and it has a hardness of 1.5 to 2—somewhat softer than Baltic amber. There are different varieties of amber mined in the Dominican Republic, thus hardness varies according to the area from which it was mined. Some believe the degree of hardness depends on the environmental conditions under which the resin dried and on the geologic processes that took place during the polymerization stages. For example, Kirchner (1950, 70–5) reported that amber, associated with wet watery conditions while forming, tended to be softer than that associated with wood and drier conditions. This appears to be evident in the amber from the

"BUGGY AMBER JURASSIC PARK, all over again! On Friday, May 19, 1996, the Associated Press reported on the discovery of "live" bacteria discovered in a Bee trapped in Amber from 5,000,000 to 40,000,000 years ago! WOW-now you too can own a part of ANCIENT HISTORY with some of our inexpensive GENUINE AMBER from the same period. . .The typical amber specimen is 1/2" to 1" in length with small beetles, ants or termites visible. These exotic Amber pieces contain spiders, crickets, moths or particularly large insects. . . .

105.49 Gms	2 &1/4 x4 & 1/2"	Large Moth, Cricket, Millipede	golden-Orange	$5,000.00
204.70 Gms	3 x 4"	13 Winged Insects	golden-Orange	$3,000.00
13.51 Gms.	32 x 39 mm	Fly Swarm	brownish-Orange	$2,000.00
23.34	3/4 x 1"	3 Termites	Yellow	$1,500.00

Smaller pieces under one gram and about 3/8 to 1/2 inches with small visible insects could be purchased for $40.00 each.

Figure 7.24. Sample 1997 advertisement for "buggy" amber specimens. Courtesy The House of Onyx Company.

Bayaguana region in the Dominican Republic. However, other researchers believe this amber is in reality still copal and much younger in age than that from the Cordillera Septentrional.

In general, Dominican amber ranges from transparent to only a few opaque nodules. Its color varies from pale yellow to brown, with brownish-yellow being most common. Occasionally, deep reddish-browns and fluorescent blues occur. Some reports indicate that red pieces may bleach to yellow or brown shades after a few hours of exposure to sunlight, though most pieces retain their original surface reddish hues unless polished too deeply.

Dominican amber polishes beautifully, but, because of its softness, care must be taken to prevent scratching or chipping. Dominican amber jewelry is generally less expensive than Baltic. Beautiful transparent reddish-brown and fluorescent blue amber occur, but are rare and highly prized by collectors and local amber connoisseurs, thus demanding a higher price. Pieces of red and blue amber sell for 50 percent more than yellows and browns.

Fossiliferous amber with rare fossil inclusions can run from hundreds to thousands of dollars. For example, Conti (1996, 38–42) reported on his buying trip to the Dominican Republic specifically looking for fossils for scientific study. After much effort to find a dealer who specialized in fossiliferous amber for scientific study, he reported the retail value of a small fly or gnat to vary from $15 to $30.

However, a small pseudoscorpion, which is a small arachnid similar to a scorpion without a tail, typically ran from about $200 to $300. He found that a small lizard, which is extremely rare, cost about $25,000 or as much as $50,000! Thus the size of the fossil is not as important as the rarity of the specimen. In fact the prices quoted to Conti while in the Dominican Republic may be quite low. To illustrate, see The House of Onyx Company's winter 1997 wholesale newsletter advertisement (Fig. 7.24).

Dominican amber contains abundant inclusions, from unidentifiable botanical debris to leaves and a wide variety of insects and other arthropods. Though Dominican amber is considered one of the most fossiliferous of the world ambers, it is estimated that only about one out of a hundred pieces contains a fossil inclusion. In 1978, after examining thousands of fossils in Dominican amber, Woodruff, of the Florida Department of Agriculture, located representatives of 19 of the 29 orders of insects, plus spiders, millipedes, and centipedes. He found the ants, stingless bees, and ambrosia beetles to be the most abundant of the fauna enshrined in Dominican amber.

Figure 7.25. Example of leaves of the present day Hymenaea tree, called the algarrobo in the Dominican Republic.

Because island insects evolve more rapidly than continental species as a result of isolation, insects preserved in Dominican amber are closely related to the ancestral stock of the island and may show relationships between species in other areas to those of the island. Fossils may also indicate how the islands were colonized millions of years ago (see Fig. 7.26).

Sanderson and Farr (1969, 1313) identified several orders of insects, including Blattaria, Isoptera, Corrodentia, Heteropter, Hymenoptera, Hemiptera, Coleoptera, Lepidoptera, and Diptera. Also, observed were spiders, fragments of wood, roots, flowers, leaves, and air bubbles. Some of the bubbles were large enough to contain water that could easily be seen with a jeweler's loop (10x). Dominican amber fossil collections are found in the Smithsonian Institution, the American Museum of Natural History, and the Museum für Naturkunde in Stuttgart.

Grimaldi reconstructed the amber forest, as it probably looked in the Dominican Republic 20 million years ago, in his amber exhibition held at the American Museum of Natural History in 1996 and in the Smithsonian Institution in summer of 1997. This reconstruction was based on the study of numerous fossils in Dominican amber. The resin dripping in stalactites from the reconstructed trees with the models of insects and asso-

Figure 7.26. A new weevil (Velates dominicana gen. n., sp.n.) in Dominican amber and key to the genera of the Anchonini. Description published in Nouv. Revue Ent. 2003 19:373-381. Courtesy George Poinar. Photo by G. Poniar.

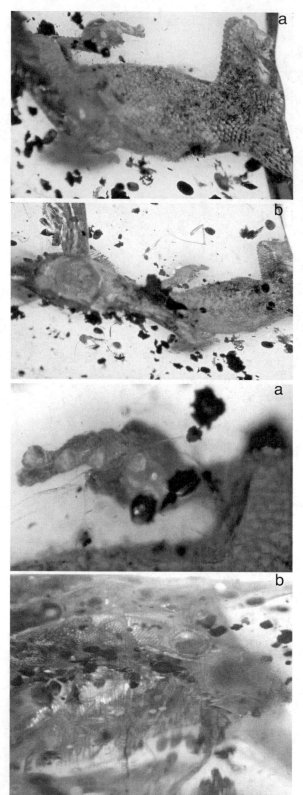

ciated plants and environs took the visitor back in time for a walk through the amber-bearing forest. He described this reconstruction in his book, *Amber: Window to the Past.*

Dominican amber received much attention with the increased interest in fossiliferous amber in the 1980s. Many papers have been written on the unusual fossils in Dominican amber such as the Poinar et al. reports on sun scorpions (1989), a mushroom (1990), snails (1991), as well as rare mammalian hair (1988), and even small frogs (1987). Of special interest were the finds of remains of anole and gecko lizards, small frogs, and bird feathers, because they represent the oldest fossils of land vertebrates located in the Caribbean (Poinar 1992, 310–1). The variety and size of the fossils found in Dominican amber is what attracts the attention of today's scientists and collectors. Unfortunately, many of these fossils are collected without the miner or dealer carefully noting the exact location from which the amber was produced, thus some of the scientific value is lost since fossils are used for studying biogeography and continental drift, evolutionary relationships, and even rates of DNA evolution (Grimaldi 1995, 204, 1996*a*, 203–7). An amazing discovery was announced in 1980, when Baroni-Urbani (213–23) from the Natural History Museum in Basel, obtained a rare well-preserved specimen of a lizard of the genus *Anolis* in amber from the La Toca mine in the northern mountain range (Rieppel 1980, 486–7). The rarest and most highly prized fossils found in Dominican amber are the small *Anolis* lizards and *Sphaerodactyl* geckos and small frogs of the genus *Elethrodactylys*. These genera still inhabit the Caribbean with hundreds of species of small lizards and frogs (see Figs. 7.27a, b. and 7.28a, b). X-ray

Figure 7.27. Dominican amber inclusions: a, Gecko in amber. b, other side Gecko (Geckkonidae, Sphaerodactylus). Figure 7.28. a, Gecko foot, b. Gecko eye. Courtesy Amber Fox, Richard Fox.

positives of the gecko lizards embalmed in amber show tiny bones in legs and feet and vertebrae. Occasionally the fossil shows signs of struggle such as broken vertebrae, perhaps occurring before being trapped in the resin or even while trying to free itself. Other evidence of vertebrates living in the ancient forest is found in feathers, hair and particles of reptilian skin embedded in amber. Grimaldi (1996*a*, 108) attested to one feather identified microscopically as originating from a woodpecker. Some hair found in amber is thought to be from a sloth. Thus, Dominican amber reveals that land vertebrates roamed the amber forests during the Oligo-Miocene and were very similar to today's varieties. Dominican amber also provides a glimpse at the numerous varieties of bloodsucking arthropods that indicate a diverse vertebrate population.

Up until 1996 evidence of the presence of vertebrates in Dominican amber was limited to small frogs and lizards whose skeletons had been verified. Other debris indicated the presence of mammals and birds in the environs of the ancient forest, but because of the size and strength of these creatures it was thought they would have been able to pull away from the sticky resin. The first mammal specimen, a tiny shrewlike creature in 18- to 20-million-year-old Dominican amber, was reported by workers at the American Museum of Natural History (Lewin 1996, 215).

Amber's exceptional quality for preservation not only has preserved the fossil themselves, but often embalms them in association with other fauna in their environment, presenting us with a glimpse of developmental stages of these insects, their prey, and plant hosts. Amber has even shown parasites leaving the body of their hosts, and the social behavior of insects.

One behavior that has been noted in both Dominican amber (Grimaldi 1996*a*, 79) and in Baltic amber (Kulicka 1996*a*, 11) is "commensalism." This behavior usually involves the method of transport employed by mites and others, in which they travel by attaching themselves to other stronger creatures. This process is known as phoresy to entomologists. For example, a mite may hitch a ride on a fly to travel from mushroom to mushroom, and phoretic mites still on a fly have been preserved in ancient amber. Both Grimaldi and Poinar report amber with sweat bees with dozens of tiny mites still attached as they went for their last ride.

Another interesting fossil reported by Grimaldi and by Poinar is the pseudoscorpions, tiny scorpions without a stinger, that live under bark, feeding on mites and tiny arthropods. These creatures have been found with wasps or flies or wood-boring ambrosia beetles. Some pseudoscorpions live in galleries made by wood-boring beetles, and to disperse, they attach themselves to a beetle to be taken to another tree. Sometimes amber that contains wood-boring platypodid, or ambrosia beetles, will also have numerous particles of sawdust plugs that the beetles pushed out of their tunnels in the wood. These beetles lived on and attacked the Dominican amber-producing trees. In 2002 Poinar described and extinct palm flower in Dominican amber (see Figure 7.29).

Ancient DNA in Fossiliferous Amber. In March 1982 Poinar and Hess (1241–2) published electron micrographs of the detailed internal structure of muscle preserved in 40-million-year-old Baltic amber. Since striated muscles were observed, the researchers questioned if they possibly still held the DNA of the organism. In order to preserve DNA, the organisms had to be trapped in amber at death and had to have undergone extraordinary preservation with dehydration of the tissue. This led Poinar and Hess to investigate what chemical constituents of amber resin achieved this remarkable preservation? As evidence of the extraordinary qualities of preservation amber possesses, Poinar and Singer (1990, 1099–1101) describe the delicate tissues of a gilled mushroom, *Coprinites domincana*, preserved in 35-million-year-old Dominican amber.

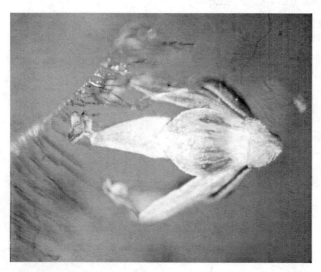

Figure 7.29. A flower of the extinct palm, Trithinix dominicana in Dominican amber. Published in "Poinar, Jr., G. O. 2002. Fossil palm flowers in Dominican and Baltic amber. Botanical Journal of the Linnean Society 139: 361-367". Courtesy G. Poinar. Photo by G. Poinar.

For many years resins have been used as preservatives in wine making, as local antibiotics, and in waterproofing, yet not one chemist had actually described the components of amber resin that provided this capability. It became important to understand which chemicals were responsible for dehydration, tissue fixation, and exclusion of bacteria effected by resin promoting preservation. It appears that Dominican amber is more effective than Baltic amber in preserving tissues so completely that DNA is actually preserved. Some chemists believe much of the resin's preservative power is derived from its ability to exclude oxygen and destructive bacterial agents and the fact that this process takes place quickly. This preservation of an organism's tissues takes place as soon as the organism dies. First internally, the tissue's own enzymes begin the process. In Dominican amber, no white coating is found around the fossil organism as is often encountered in Baltic amber. This white coating has been explained as putrefaction and moisture from the insects' decomposition. The next process takes place internally and externally, by bacteria with fungi in the environment, yet the organism must be quickly dehydrated and protected from further destructive bacteria. Thus, dehydration is followed by a process of polymerization as the simple terpenes become linked together to form long chains, as the amber becomes solid.

DNA Isolated in Dominican Amber. As previously mentioned, it was from Dominican amber that ancient DNA was first reported as being sequenced from a 25-to 40-million-year-old bee and then shortly after from the 25-million-year-old termite late in 1992. This was followed by the recovery of 130-million-year-old DNA from a weevil from Lebanese amber in 1993. These spectacular achievements from researchers at the American Museum of Natural History and the University of California were more than just finding the oldest DNA. Now entomologists could use this information to understand the development of the species of arthropods. This is exactly what the American Museum of Natural History team attempted (Grimaldi 1996b, 84–91). They compared the DNA sequences from a living cockroach and termite with that of the 25-million-year-old DNA of an Australian termite to show an evolutionary link between cockroaches and termites (Lewin 1996, 213; see Fig. 7.30).

The wide publicity received by the ancient DNA studies raised new questions, such as "Should the museum fossil collections be disturbed and even destroyed to isolate DNA when this process virtually obliterates the specimen and it will no longer be available for others to study?" This question is being discussed vigorously by the researchers themselves. As expressed by Lewin (1996, 213), there are two issues in ancient DNA research: (1) The value of museum collections and how best to use them; and (2) The dangers of ancient specimens becoming contaminated with modern DNA.

Figure 7.30. An extinct termite, Mastotermese electrodominicus, the basis of David Grimaldi's research in isolating 25-million-year-old DNA from Dominican amber. Courtesy Natural History Museum, New York.

Since bacteria, the simplest form of life, has also been found in amber, this led microbiologists to the next question. If we can isolate the bacteria and if we can isolate DNA, can bacteria that has been dormant for millions of years be revived? Cano and his associates took on this challenge (Cano 1994, 129–34). First, by partially dissolving amber in turpentine, and straining and centrifuging the resulting substance, the bacteria in the amber could be studied under a high-powered electron microscope. Thus, he located bacilli, cocci, and spirilla bacteria of different kinds. Cano then found a 25- to 30-million-year-old stingless bee preserved in amber from the Dominican Republic.

This bee could be compared to today's stingless bees to see if the same bacteria were present, since a strain of bacteria exists in symbiosis with the bee, helping in the digestion of pollen, other forms of nourishment, and protection against certain diseases. First, Cano had to retrieve bacteria from the stomach of the fossilized stingless bee. This in itself required fine precision microscopic instruments. The DNA analysis of the bacteria showed that they were related to the same species found inside recent bee's stomachs. He carefully plotted the bacteria from many fossil bees with that from recent bees and bacteria from surrounding areas to show they were of the same strain. This in itself was a tantalizing task and revealing for ancient bacteria strains, but his next step was to isolate the ancient bacteria obtained from a fossil bee's stomach to see if he could revive it from its dormant state to prove there was still life in the 40-million-year-old organisms! With the assistance of Monica Borucki, his former graduate student, Cano, using careful precautions to keep from contaminating the bacteria with modern-day bacteria in the environment, placed the ancient bacteria in a nutrient solution and the bacteria reproduced themselves! It was previously known that bacteria survived for a long time in a state of inertia without being affected by their environment through encasement in endospores, but 25 to 40 million years caused amazement in the scientific world. Scientists the world over questioned if proper precautions were taken to eliminate contamination. Others simply stated that Cano had only reproduced bacteria from the laboratory environment and that he could not prove his bacteria were the exact ones extracted from stomachs of ancient bees. However, on March 31, 1997, the *New York Times* reported that Cano had patented his technique of recovering ancient bacteria. The article read:

> Patenting the Jurassic Age?
> Two molecular biologists have been issued a patent that appears to cover the concept of recovering not just ancient bacteria, but an array of other organisms found in amber or other resins, including fungi, protozoa, viruses, microalgae and arthropods.
> Teresa Riordan: Patents

Cano's patent lawyer stated that "The breadth of the patent is a reflection of the pioneering nature of the discovery. This is the first time anyone has revived spore-related organisms that have been entombed in these resins for millions of years." Cano and Borucki received patent 5,593,883, which was assigned to the Ambergene Corporation. Cano founded the Ambergene Corporation, based in

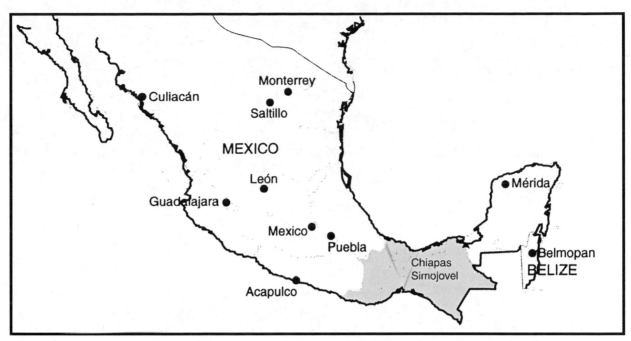

San Francisco, to commercialize his research. His corporation declares that it has been able to show that spore-forming organisms preserved in amber have the ability to be revived, and it is participating in collaborations in search of a variety of applications. These include new versions of antibiotic detergents, biological pesticides, cosmetics, and antibiotic medicine. Most interesting was the statement that Ambergene was developing a microbrewery beer made from an ancient yeast that was to be introduced when the sequel to *Jurassic Park* hit the movie theaters!

Figure 7.31. Mexican Amber Deposits in Chiapis.

Chiapas Amber from Mexico

Mexican amber is found near San Cristobal las Casas in Chiapas, in the form of lumps large enough to be polished (see Fig. 7.31; 7.32a & b). Although Chiapas amber has only been publicized since the early 1980s, it was well known to the area's ancient inhabitants. It is said Montezuma used a long amber ladle, carved and polished by skilled jewelers of Xochimilco, to stir his cocoa drinks (Buddhue 1935, 170–1).

Chiapas amber occurs in alluvium in the Simojovel district, where heavily forested mountains overlook the Huitapan River. After heavy rains, amber is washed out from the sides of gorges and carried downstream, where it is collected by the natives.

The Mexican amber is found about 50 kilometers from Tuxtla Gutierrez, the capital of the southernmost state, Chiapas. Fran Blom, archaeologist and anthropologist, spent several years studying the

Figure 7.32 a. & b.Chiapas amber with insect inclusions.

Figure 7.33. First puffball (Lycoperditius tertiarius gen. n.) in amber. Published by Poinar, Jr., G.O. 2001. Fossil puffballs (Gasteromycetes: Lycoperdales) in Mexican amber.<u>Historical Biology</u> 15: 41-43. Courtesy of G. O. Poinar. Photo by George Poinar.

Simojovel region where amber is obtained and also found amber near Yajalon, at a location called Hool Babuchil, or "Head of Amber," a deposit exposed by a steep fault (Ruzic 1973*a*, 1304–6). Amber is found along the surface in horizontal strata in the slope as well as at the foot of the cliff, where it has weathered from soft marine limestone often containing shell fossils. Often the amber is found where landslides have taken place, exposing the lignite and carboniferous veins that contain amber. The Hool Babuchil deposit was studied in 1954 by two scientists from the University of California, J. Wyatt Durham, paleontologist, and Dr. Paul Hurd, entomologist. This team collected many specimens that were later taken to Berkeley for further study (Sanderson and Farr 1960, 1313). One particularly unusual discovery was a small crab found in 25-million-year-old Mexican amber (Grimaldi 1996*b*, 84–91). The first puffball described in amber was found in a single piece of Mexican amber dated at about 22-26 Ma or Lower Miocene-Upper Oligocene. These puffballs, described by G. Poinar, Jr., probably grew on decaying hardwood during the mid Paleogene. (see Fig. 7.34)

Chiapas amber is similar to Baltic amber in refractive index and specific gravity, but it is not as hard. In other attributes it is more similar to Dominican amber. It occurs in various shades of red, yellow, and almost black. The Natural History Museum in Nuevo Chapultepec Park in Mexico City contains an exhibit of five pieces of Chiapas amber from Fran Blom, all are reddish-brown and range in size from a walnut to a small fist.

Chiapas amber was found to be a fossil resin formed during the Miocene to Oligocene epochs of the Paleogene. The source of this resin is thought to be a leguminous tree similar to the *Hymenaea courbaril*. The resin does not contain succinic acid, thus according to Anderson's classification it would be a Class 1c fossil resin.

Locally, natives collect amber for wear as amulets, or they make small trinkets, carvings, and pendants to sell to tourists. The amber is cut and shaped with saws and steel files and polished by rubbing with a mixture of water and ashes on a soft piece of balsalike wood (Ruzic 1973*a*, 1304–6, 1973*b*, 1400–6). However, many imitations are made from a recent gum, which is bright yellow in color. When held in the hand, such imitations become sticky to the touch. Being recent in origin, their surfaces do not keep a polish but become crazed and lose luster. When burned, such gums will drip, whereas amber burns and emits a fragrant resinous odor along with dark smoke.

Baja California has also been the source of a fossil resin that has received some study by scientists. Insect inclusions were found in the amber from this area. One piece included a whitish beelike insect and a grub.

Rumanite Amber from Romania

Figure 7.34 shows the locations of the Romanian amber deposits. Amber from Romania was named rumanite in 1891 by

Figure 7.34. Romanian amber deposits.

Otto Helm. It reached its height in popularity during the early 1900s, but is very rarely seen today. Besides being used for amber jewelry, rumanite was sent to Vienna to be manufactured into pipe stems and other smokers' supplies. Rumanite from Buzau was prized for its rich colors, which range from brownish-yellow to deep brown, with deep colors predominating. Colors such as dark garnet, smoky gray, and greenish to almost black are known. Other regions in Romania produce amber in a clear pale yellow. Many of these pieces have extensive internal crazing resulting from geologic pressure. The amber may have a mottled appearance, resulting from veins and cracks that reflect a metal-like sparkle. Romanian geologists refute some of Helm's findings because samples were not taken from all locations but reported on amber found from only one location. Thus, he gave the impression that all Romanian amber was dark in hue, when in reality a lighter shade of yellow is most common. In the early 1900s Murgoci (1903, 1924), of Romania, described 160 different shades of amber found in Romania. He described deposits at Le Lac that appear ruby red when retrieved from the summit of the hill, but those pieces retrieved from the base of the hill were transparent yellow. Murgoci found rumanite frequently associated with oil or tar deposits. Also he indicated that the amber located while boring for gas and oil near the North Sea was stained brown and black.

Rumanite does not include the opaque, bonelike varieties so commonly found among Baltic ambers. More often it is transparent, with an opaque crust darker in color than the interior of the mass.

A variety of rumanite referred to as "black amber" is actually

Table 7.3 Rumanite Solubility	
Solvent	Percent of Solubility*
Alcohol	6
Ether	16
Chloroform	10
Benzene	14
*Figures vary with different specimens.	

a deep ruby red, blue, or brown when held to a light source, allowing light to be transmitted through the resin. The term black amber is misleading, as there is no truly black amber. Most fossil resins that appear black have much lignite debris included. Kosmowska-Ceranowicz (1988, 9–10) reported on black fossil resins at the Sixth Annual Symposium on Amber, finding microscopically that the "resins are a structureless ground mass of dark gray color containing yellow orange and blood red inner reflections."

These black resins are found in the Ukraine; Bitterfeld, Germany; Sambia, Russia; and Poland. Kosmowska-Ceranowicz and Migaszewski (1990, 157) suggest black resins were formed by "thunderbolts" or strikes of lightning, causing intense heat and pressure that altered the physicochemical conditions. Stantienite is a black fossil resin, but it is not used in the jewelry industry. Black amber jewelry appears on the Polish market, but it is fabricated and tests to be a fine-grained pressed under high pressure and temperature (180–200°C). In Poland, jewelry is occasionally sold erroneously as black amber, but it is in reality jet. This material is sometimes confused with amber, but jet is a variety of lignite coal—a fossil wood similar to cannel coal (see Fig. 7.35a & B).

The deep colors of rumanite are thought to be influenced by sulfur deposits that occur in the same places where amber is found. The gases absorbed in the amber also tend to discolor it. In addition, inclusions of hydrocarbons, such as coal, or minerals, such as pyrite, may occur, and each contributes to the darkening. It is thought that the presence of such inclusions also contributes to the play of colors as seen in opal--caused by light refracting and diffusing in water and air--over the surface. Pieces of brownish-red amber occasionally emit a blue-green fluorescence, similar to that of the blue amber from the Dominican Republic. Rumanite is highly fluorescent under ultraviolet light.

A significant characteristic of rumanite is its sulfur content, which amounts to as much as 1.15 percent. When burned, the choking fumes of hydrogen sulfide are readily detected. Rumanite fuses at 300–310°C without swelling or altering shape. The melting point is 330°C as compared to the 287°C for Baltic amber.

Rumanite is similar to Baltic succinite in structure and breaks with a conchoidal fracture. Compared to the specific gravity of Baltic amber at 1.08, rumanite is 1.048. The refractive index of rumanite is 1.438, compared to that of Baltic amber at 1.54. Rumanite is also slightly harder than Baltic succinite.

A chemical analysis of rumanite indicates that it contains 1 to 5 percent, less than the amount found in Baltic amber. It is classed as a succinite and offers more resistance to solvents than does Baltic succinite (see Table 7.3).

The earliest record of Rumanian amber dates from the Paleolithic and was recovered from a cave at Cioclovina. Bronze Age amber was found in a necropolis at Buzau in the form of a necklace with seven amber beads together with bronze biconical beads, bronze wire spirals, and a bronze bracelet. Iron Age finds from 1200–600 B.C. from this area included 5000 amber beads. In Cetateni, Ages district, a necklace dated to 200 B.C. was found made of a large center amber bead, cowry shells, cockle shells, and large glass beads. In 1981 Carmen Coltos of Romania analyzed two beads from a Dacian tomb using IR spectrometry and found one to be rumanite similar to that from Buzau and the other from the Baltic (Fraquet 1987, 115).

Rumanite was known and used as a gem by the Romans, who established colonies in Romania and gave it its name. The Buzau area ruins yielding Roman artifacts also included rumanite objects, proving that amber was known at that time. Furthermore, Roman trade routes, which include Klaipeda, Lithuania, in the center of the Baltic amber area, pass through Buzau, one of the localities where rumanite is found.

The first historical account appears in the work by Raicevich in 1788, while the first geologic study mentioning the amber of Romania was published in 1822 by Zuicker, the German geologist who first investigated the area's geologic history and studied the resin from the banks of the river Buzau (Bauer 1904, 535–7; Webster 1975, 513). The Buzau deposits, the main source in the early 1900s, were located along the railroad lines between Bucharest and Braila. Though the Buzau deposits are the most important, rumanite is mined not only from sandstone near the Buzau River, but is also found in the Carpathian Mountain region. Recent Soviet reports on prospects of amber in Russia stated that the Carpathian area where rumanite and other fossil resins are found covered an area of Ciscarpathia in the Carpathians near the upper basins of the Prut and Dniester rivers as far as Khatin. The amber is said to be suitable for use in

Figure 7.35. a, Black amber carving of a bear, actually dark brown resin with botanical debris and lignite inclusions; b, Black amber sold in a jewelry store in Poland, but the material is really jet, a variety of lignite coal.

jewelry.

Other deposits are located near the Danube, in tar deposits of Putna, and along the banks of brooks near Valeng de Munty. Rumanite is reported elsewhere in Romania at Ramnicul-Serat and Prohova and along the shores of the Black Sea in the province of Dobrogea, also near Valzea in the province of Oltenia, and at Bacau, Meamt, and Sucava in Moldavia. A yellow amber is found near Sibiu and Alba in Transylvania.

A 1929 survey of the Dnieper deposits along the right bank from the village of Starye Pwtrivtsy to Vyshgorod was conducted to assess the quantity of amber for possible use in the jewelry and chemical industries. However, the amount of amber in the deposit did not exceed 7 grams per cubic meter and at that time was not considered economical to mine. Also between the wars, the Grigoresco Company began more systematic exploitation of amber at Colti, installing mining shafts and galleries. The walls were propped with fresh pine timbers that were moved to various locations, creating a problem of galleries collapsing when props were removed. In some areas amber was found in quantities of 1 kilogram per meter and on occasion up to 3 kilograms per meter. During this time, Fraquet (1987, 114–5) reports large lumps being given to the Geology Laboratory of Bucharest— one weighed as much as 3204 kilograms, while another weighed 2473 kilograms. It seems these were destroyed during bombing of the city in the Second World War.

In 1975 an amber lump weighing 1785 grams was found in Colti parish after a storm. This piece is exhibited in a local museum. The piece is clear with internal crazing or cracking, which is typical for Romanian amber. In the early 1980s, the Colti Collective Museum was constructed, which now exhibits Romanian amber (Fraquet 1987, 115).

Rumanite historically has been described as occurring in formations ranging from Cretaceous to Miocene deposits, but the report given by Ghiurca (1988, 15–6), from the University of Cluj-Napoca, Romania, indicates that the most frequent deposits occur in Oligocene flysch deposits of the Eastern Carpathians, from Bucovina in the north to the Ialomita Valley in the south. Also, research during this era established a main amber-bearing level within the Lower Kliwa Sandstone, which, based on locating nannoplankton assemblages, was assigned to the Lower Oligocene. The amber-bearing strata within the sandstone varied from 0.2 to 1.4 meters in thickness. The strata dipped at 60 degrees to 80 degrees with overlapping folds.

A study of the associated fossils by Protescu (1937, 1–46) indicated the source tree for rumanite as Pinaceae family resembling the

Abies genus. Ghiurca (1988, 15–6), studying the plant tissue in the amber-impregnated wood and fragments in the amber strata suggested the Taxodiaceae family, perhaps the *sequoioxylon gypsaceum Göppert* and *taxodioxylon gypsaceum Göppert*. Since fossil evidence of the *sequoioxylon gypsaeceum* was found at the Colti and Buzau districts, this species was considered to be the source tree for generating the amber-producing resin in the Lower Oligocene deposits at Colti. However, hydrocarbon studies, using IR spectrometry and X-ray diffraction patterns of crystalline constituents, indicate that the rumanite samples used in Langenheim's study may have been of a leguminous rather than coniferous origin (Langenheim 1969, 1157). Comparative chromatographic analyses conducted by Vavrà (1889, 3–14, 1993, 147–57), from Vienna, and Ghiurca (1988, 15–6), from Romania, revealed a distinction between Romanian and Baltic ambers based on their characteristic peak values with the average being 9.34 for Romanian amber and 8.53 for Baltic amber from Poland.

Many different flora and fauna fossils have been identified in Romanian amber. Ghiurca (1988, 15–6) reported eight different insects identified by either genus or species, spiders, ants, mosquitoes, wasps, moths, plant fleas, pseudoscorpions, different kinds of larvae, lichen remains, conifer needles, one cone, and pollen. From the occurrence of the variety of fossil inclusions these researchers assumed the paleoenvironment was a shallow sea or lagoon.

Thus, though the Carpathian and Dnieper basins produced commercial quality amber that was used in jewelry for beads, amber buttons, and in the smoking industry, it is rarely exported today. In the nineteenth century, Romanian amber was most popular for the smoking industry because it was thought to protect from germs and purify the smoke. Therefore, it was extensively used for pipe mouthpieces, cigar holders, cigarette holders, and Eastern hookah pipe mouthpieces. Romanian amber was often given as royal gifts to ruling classes all over the world by the royal family in the early 1900s. The amber was collected by the local workers and exported in the rough to Vienna and Constantinople. Romania had close connections with these two cities because the Austro-Hungarians and Ottomans (Turks) ruled the area until after World War I (see Fig. 7.36).

None of the amber is currently reaching the United States. During the late nineteenth and early twentieth centuries, the only

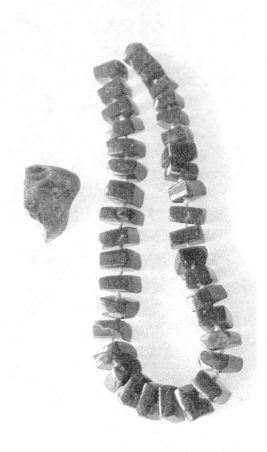

Figure 7.36. Romanian amber necklace, dark brown to blackish resinous material, with reddish appearance when light is transmitted. It is 19 inches long, but was originally 22 inches long as some small end pieces are missing. The beads are cut in squares drilled through the center. The largest beads are 25 centimeters across. Courtesy of Ann O. Fletcher collection.

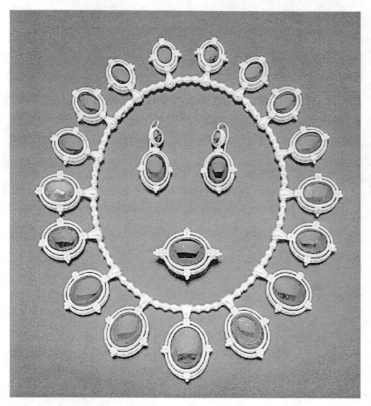

Figure 7.37. Necklace, brooch, and matching earrings, 1860–1870. Bequest of William Arnold Buffum. Courtesy, Museum of Fine Arts, Boston.

commercial rumanite came from the Buzau area. Although amber from the Carpathian Mountains is still of interest to the jewelry industry, it is not produced on a commercial scale. The Colti craftsmen work amber from the local area using motor-driven grinding wheels and powdered gypsum for a polish. Outside of Romania, the most likely source of rumanite jewelry today is from old collections, since rumanite is by far the least plentiful type of amber jewelry. In this connection, the only rumanite I was able to locate was in the amber exhibition organized by the American Museum of Natural History, New York, on permanent exhibition in the Field Museum of Natural History, Chicago, and in the Cranbrook Science and Mineral Museum, Bloomfield, Michigan. Several old faceted beaded necklaces, circa 1920–1930, sold by antique dealers as genuine rumanite, proved to be Bakelite, a synthetic resin.

Simetite Amber from Sicily

One of the most beautiful and highly prized varieties of amber used for jewelry in the past was found in Sicily. It was named simetite after the Simeto River on the eastern coast of Sicily near Catania. The mouth of the Simeto River was the major source, but amber was also found near brooks and streams in clay deposits in the central part of the island. The entire area around Etna and Syracuse is volcanic and amber found within this area possesses a special opalescence and luminescence. Some people think the volcanic activity is the cause of this unusually beautiful coloring. The early works of Ferrara (1805, 73) and Bombicci (1873) give complete descriptions of localities in Sicily where amber was found, such as at Giarretta south of Catania. Smaller amounts were found in the mouth of the Salso at Lilcata on the southern coast. Areas such as Asaro, Leonforte, S. Filippo d'Agiro, Centorbi, Terranuova, Girgenti, Ragusa, and Sciccli also were listed as possible locations. However, by the late 1980s small amounts of amber were found at Nicosia, which appeared to be the only location (see Fig. 7.38).

Simetite amber is rarely yellow; more often it is a transparent dark red, blue, or green. Red specimens are highly prized and were regarded as the most desirable variety in the late 1800s. Expounding on the radiant beauty of the "Ambra de Sicilia," Buffum (1896, xvii) wrote:

> The gems in her necklace flashed in the sunlight, showing color shades ranging from faint blue to deepest azure, from pale rose to intense, pigeon blood, ruby red. The varied and lustrous hues, here blended in lavish beauty, drew from me involuntary expressions of admiration.

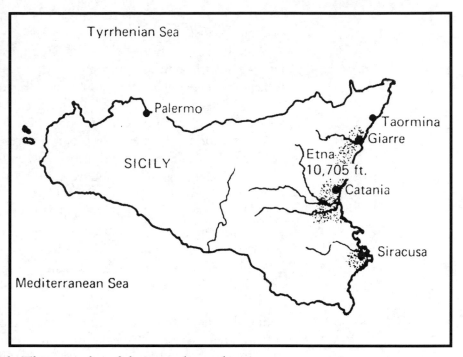

Figure 7.38. Sicilian amber deposits.

Buffum emphasized that he could not believe the hues were natural, but when assured that they were "pristine" hues, he exclaimed "They are hues, then of the primeval world, the imprisoned color shades of an earlier and more exuberant clime" (Buffum 1896, xix). Buffum was so impressed with this necklace that he included a picture of it in the frontispiece of his book, and had a duplicate made that was donated to the Museum of Fine Arts, Boston. The Sicilian amber cabochons were highly polished and set in gold in a classical design similar to the Etruscan granulated style. The necklace was made by the Castellani firm of Rome, which reflected a revivalist movement in classical art based on the excavations of Roman, Greek, and Etruscan ruins. The necklace was exhibited in the American Museum of Natural History amber exhibition in the spring of 1996 and at the Smithsonian amber exhibition in the summer of 1997. However, the various colors reported by Buffum were not evident to the extent described by him. In fact, the amber appeared to be a dark yellow (see Fig. 7.37).

By the turn of the century, amber in such beautiful colors was scarce in Sicily. The widespread, isolated, nestlike surface deposits were already greatly diminished. Buffum (1896, xix) cautioned: "It would not be surprising if these deposits were to cease and the amber of Sicily to disappear."

The refractive index and specific gravity of simetite are similar to those of Baltic succinite. Its hardness is 2.5 on Mohs scale, and it has the same conchoidal fracture and electrical quality as Baltic succinite (Webster 1976, 177–83). Simetite contains no succinic acid. When burned, the fumes are not irritating. However, along with the resinous odor, a sulfurous odor is emitted, resulting from the relatively high sulfur content, which varies between 0.67 and 2.46 percent.

Modern botanical investigations indicate that the simetite resin source tree is related to the *Burseraceae protium*, an angiosperm rather than a conifer (Langenheim 1969, 1161).

Simetite generally occurs in Paleogene lignite, dated to middle Miocene formation. Studies by Skalski and Veggiani (1988, 296) indicated the amber to be Oligocene age similar to that found in northern Italy. A black resin is sometimes associated with simetite, occasionally forming a coating over simetite nodules, but this is believed to be of different origin and has no commercial use.

Disagreements have developed over how long the existence of Sicilian amber has been known. Buffum (1896, 9–11) presented the argument that the Phoenicians were aware of it. Not only were they great traders, but, according to Buffum, they also knew that to sell their wares for the highest price they needed to guard their secret sources of supply. One report of a Phoenician ship master tells of his running his ship aground—wrecking not only his, but the Roman ship following him—thus preventing the Romans from learning the secret of his route. For this deed of valor and shrewdness, he was rewarded by his state. Buffum believed the Phoenicians traded with the inhabitants of "Trinacria," the Phoenician name for Sicily, perhaps receiving amber, but they took drastic steps to keep their trade secrets, telling tales of incredible dangers and monsters in that district.

Other authorities doubt that trade in simetite existed in this early period, since amber found in tombs excavated along Phoenician trade routes proved to be from the Baltic. Such tomb amber appeared red when found, but upon cutting, it displayed a yellow inner core typical of Baltic amber. The newer technological methods for identifying amber will more clearly differentiate which resins are alike, thus the old tomb amber pieces are being reevaluated and perhaps a new description of the trade routes will be developed when the Polish and Italian researchers complete their study.

Today ornaments or jewelry made of simetite are rare, and most examples can be found only in museum collections. Interestingly, Fraquet, a gemologist from London, indicates that many nineteenth century collections of Sicilian amber are now bright, clear, cherry red or a gradation between clear golden to orange-red, but fluorescent colors no longer are present. Local craftsmen of Taormina occasionally offer articles carved of Sicilian amber, and antique collectors are constantly alert for red varieties when viewing old estate jewelry collections, since this amber was more prevalent in the past (Fraquet 1987, 105).

Unfortunately, Sicilian red amber was often expertly imitated, especially with the synthetic Bakelite plastic, which provides a beautiful imitation with many properties similar to those of amber, including some of its electrical powers. Bakelite is often passed off as "cherry red amber." However, since Bakelite is much heavier, more uniform in texture, clearer, and harder than amber, it is not difficult to identify. For these reasons, antique red amber should be tested for refractive index and specific gravity.

In recent years local hobbyists have found Sicilian amber around Catania and other smaller towns, but amber pieces sold in the marketplace more often are not from this area. Schlee and Glöckner (1978) reported that they had been advised by Beiner of the Dominican Republic that red Dominican amber was being exported to Sicily. Since the two ambers, red and that with a natural blue fluorescence, were found in both locations, it would be easy to confuse Sicilian and Dominican amber.

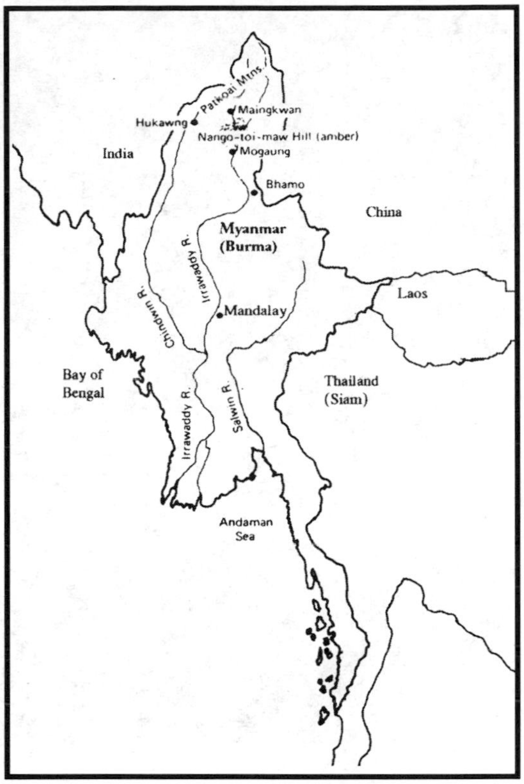

Figure 7.39. Burmese amber deposits in Myanmar (formerly Burma).

Burmite Amber and Other Asian Amber

Burmese amber, or burmite, has been known to Chinese craftsmen from as early as the Han Dynasty (206 B.C. to A.D. 220). Amber was transported from Myanmar, formerly Burma, through the Yunnan Province. It is possible that small amber deposits may also have been found there (see Fig. 7.39). During a later period, Baltic amber was imported via the Mediterranean and India. Up until World War II, most burmite went to China (Bauer 1968, 335–7).

Amber was highly valued in China because it was thought to have many symbolistic powers. One of the favorite figures carved in amber by Chinese craftsmen was the fish, because it held several symbolic values. Legend has it that messages were placed in the bellies of the fish. Therefore, the fish was symbolic of communication with a distant friend or loved one. The fish also symbolized strength and

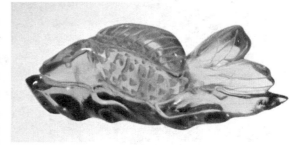

Figure 7.40. Oriental carving of a fish made out of amber to represent strength and perseverance and communication with a distant friend.

Figure 7.41. Chang E, the Moon goddess, a symbol of immortality, carved in China from transparent yellow Dominican amber.

Figure 7.42. Burmese amber pendant with a mottled appearance called "tiger amber."

perseverance because its scales reminded the Chinese of the dragon (see Fig. 7.40). Yet even in the Han Dynasty, Chinese writers recognized amber as a product of the pine, a tree held in particular reverence by the Chinese. Because the pine tree remains green year-round, it represents endurance and longevity. Although the pine trees growing on the mountainside are buffeted by winds and storms, they thrive year after year, only becoming gnarled as they age. Thus, the pine represents constancy in adverse conditions, therefore, amber a product of the pine, became symbolic of longevity.

Not only was Chinese amber and Burmese amber used but the significance of the material was so important that Chinese carvers imported fine large lumps of amber from the Dominican Republic for elaborate carvings such as the pure transparent amber carving of the Goddess Chang E, the Moon goddess, a symbol of immortality. In the Chinese legend, Chang E stole the drug of immortality from her husband, who then ordered her death. Quickly Chang E drank the elixir and flew to the Moon, where she lives in her palace to this day (see Fig. 7.41).

As early as the fifth century, T'ao Hung-ching stated: There is an old saying that the resin of fir-trees sinks into the earth, and transforms itself [into amber] after a thousand years. When it is then burned, it still has the odor of fir-trees. There is also amber, in the midst of which there is a single bee, in shape and color like a living one. . . . It may happen that bees are moistened by the fir-resin, and thus, as it falls down to the ground, are completely entrapped (Laufer 1907, 218).

Amber was known to the Chinese as "hu-p'o," meaning "soul of the tiger," as explained by Li Shih-chen: When a tiger dies, its soul (spirit) penetrates into the earth, and is a stone. This object resembles amber, and is therefore called "hu-p'o" [tiger's soul]. The ordinary character is combined with the radical yu [jewel], since it belongs to the class of jewels (Stevens 1976, 185).

Being the "soul of the tiger," amber was regarded by the Chinese as symbolic of courage and was supposed to possess the many strong qualities of the tiger.

Burmese amber is highly prized for its deep colors and, in European and American markets, for its rarity. Usually, Burmese amber is dark brown and largely somewhat turbid. Darker specimens verge on reddish shades, appearing deep red when intense light is transmitted through the mass. Peasants recognize 14 varieties of amber based on colors and shades. Cloudy varieties, typical of Baltic, do not occur in Burmese amber, but an opaque variety mixing white osseous and brown streaks, called "root amber," is known. The various mixtures of brown shades characterizing root amber result from penetration of calcite into pores and openings in the amber, with cracks often filled with calcite in thin layers, the latter commonly found intersecting one another. This combination of amber and calcite gives burmite a curious mottled appearance (Chhibber 1934, 90–1; see Fig. 7.42).

Figure 7.43. Largest lump of Burmese amber ever found, now located in The Natural History Museum, London. The specimen measures 50 cms left to right. Photo used with permission ©The Natural History Museum, London.

In this connection, a variety of Baltic amber, also imported into China, is similar to Burmese root amber because of having become impregnated with salts from the soil and losing its color and transparency. When first found, Burmese root amber has a hard dark crust resembling an ordinary pebble, but upon polishing, it assumes a rich brown mottled appearance similar to marble. Both kinds of root amber are used for carving ornaments with designs that cleverly utilize the natural swirls, color variations, and actual flaws in the material to create remarkable effects. In China, this type of carving is called *ch'iao-tiao*, meaning "clever or ingenious carving," and is, of course, also used in fashioning jade and hard stone ornaments (Stevens 1976, 185).

In 1970 Savkevitch, a Russian scientist who specialized in amber research, reported that Burmese amber contained up to 2 percent succinic acid. However, this information was not supported by the analysis conducted by Shedrinsky, an American chemist and conservationist, who found no succinic acid in Burmese amber (Shedrinsky 1996, 5–26). The chemical composition of burmite is approximately as follows: carbon 80.5 percent, hydrogen 11.5 percent, oxygen 8.43 percent, and sulfur 0.02 percent. It is the hardest of all amber (2.5 to 3 Mohs), and, though brittle, it is not difficult to work. If the piece does not contain too much calcite, it easily takes a beautiful polish. Its specific gravity of 1.034 to 1.095 and refractive index of 1.54 are similar to those of Baltic amber. Burmese amber is also highly fluorescent, with most varieties emitting a bluish or greenish color, while root amber pieces may fluoresce yellow or orange (Williamson 1932, 219–20).

Both botanical and insect inclusions are found in burmite. Botanical studies conducted by Langenheim (1969, 1160) stated that amber specimens from the Hukawng Valley in Burma was from the Eocene and were related botanically to angiosperms rather than coniferous trees. She lists the source tree from the family Burseraceae.

The rough is generally found as oval or elliptical lumps and occasionally in rounded pieces, but not in irregular or angular pieces, indicating that the amber has been transported. Most pieces are small, flat, and smooth. Large lumps were rarely obtained from shallow pits, as these tended to yield inferior amber. Large pieces were sometimes found in a finely laminated blue sandstone or dark blue shale. The largest lump ever found was the size of a man's head (Bauer 1968, 555). Today this piece, recovered from Myanmar, Burma, can be found in the Natural History Museum of London. The piece is a transparent deep red hue, and weighs 33.5 pounds (15.2 kg) and is 19.5 inches (49.5 cm) long (see Fig. 7.43).

The amber occurs in shales and sandstones that are alternately interbedded with layers of limestone and conglomerate. Carbonaceous impressions and, infrequently, very thin coal seams containing concretions of amber, are embedded in the sandstones and shales. Since both are soft, surface exposures are rare. In sections along streams, where erosion has taken place, pits and cuttings expose the Earth's structure and reveal the true nature of its geology.

The first geologist to visit Burma amber mines was Noetling in 1892 (130–5). He concluded that the deposits occurred in clay and wrongly ascribed a Miocene age to them. However, in 1922 Stuart (1–12) visited the mines and found occurrences of *Nummalites biarritzensis*, an index fossil for the Eocene epoch, among shales embedding amber. An Eocene age was also assigned in the work of Bather and Cockerell (1922, 713–4), based on studies of insect inclusions. Cockerell described about 42 arthropods from Burmese amber and, since many of these were primitive types, he also suggested the amber could be considered late Cretaceous and that it had been redeposited during the Eocene. In 1934, Ch'iao, geologist and author of *Mineral Resources of Burma*, visited the amber mines and conducted an extensive survey that confirmed that the Burmese deposits belonged to the Eocene and were about 45 million years old (Chhibber 1934, 85). However, Chhibber provided a complete description of primitive amber mine workings in the Nangtoimow Hills in northern Burma (today called Myanmar), the most famous mines being near the jadeite mining area. Approximately 200 pits at that time were located about 3 miles (4.8 km) south of the village of Shingban near a small stream between Shingban and Noije Bam. Other open pits were operating nearby in Pangmamaw at the time of Chhibber's visit (1934, 86).

The largest amber mining center, which supported about 150 miners, was located at Khanjamaw, about 3 miles (4.8 km) southwest of Singban. Workings consisted of pits lined with thin bamboos supported by heavy wooden posts. The deepest pit had been dug about 45 feet (14 m) deep, to a point just above the water table level, which appeared at a depth of 49 feet (15 m). At a depth of about 40 feet (12 m) below the surface, coal seams containing good amber were encountered.

Another small working, called Ningkundup, was located about 200 yards (183 m) northwest of Khanjamaw. In 1934 there were 6 pits worked by 15 Shan-Chinese. The strata here consisted of blue sandstones and dark blue shales. At Wayutmaw, 20 miners, working in small teams with primitive tools and bamboo hoists to raise baskets filled with soil, were engaged in rediggng old pits. This mine was next in importance to Khanjamaw. A little northwest of Wayutmaw, where amber-bearing strata consisted of bluish sandstones and finely bedded shales covered with a reddish overburden, the amber was washed out by sluicing during the rainy season (Chhibber (1934, 89).

In 1904 not more than 300 pounds (136 kg) of amber per year were produced, as the mine was operated by peasants with very primitive tools. Little improvement in mining methods had taken place by the time of the geologic survey conducted by Chhibber in the 1930s. Production reached its peak in 1927 and declined each year thereafter (see Table 7.4).

Some amber was polished in the mining area using simple cutting tools such as a saw, a *Kachin dah* or large knife, a file, and sandpaper. First, the crude amber was cut into small pieces with a saw made by fixing a piece of tin from a kerosene can into a bamboo handle and cutting teeth in the blade with a dah. Using this same knife, rough pieces were cut into shape, and shaping was completed with small flat files. Each piece was then sanded by rubbing with course sandpaper. The final polish was obtained by using a tree leaf with a prickly backed surface (Chhibber 1934, 92).

Table 7.4	Output of Burmese Amber 1926-1930	
	Quantity	
Year	Pounds	Kilograms
1926	3950.0	179.32
1927	7050.0	3197.80
1928	2940.0	1333.55
1929	1958.5	888.35
1930	207.3	94.03

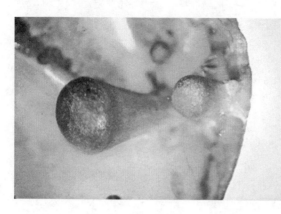

Fig. 7.44. The earliest known hymenomycete fungus in burmese amber. Description published in "Poinar, Jr., G. O. & Brown, AS. E. 2003. A non-gilled hymenomhycete in Cretaceous amber. Mycological Research 107: 763-768". Courtesy of George Poinar, Jr.

Figure 7.45. Antique Chinese carved amber platelet with four holes to sew the amber decoration to a headdress or other ornamentation. Made of Baltic amber, carved in China, the surface shows darkening by oxidation and some crazing from age. Size: 4 cm x 2.7 cm x 0.8 cm

The major portion of the production was sent to Mandalay and Mongaung. Again, very simple tools were used to fashion long round-beaded necklaces for Mandarins. Amber was also cut into beads for necklaces, earrings, bars for brooches, buttons, cufflinks and other trinkets. However, so many other articles were made that such adornments represented only a small fraction of the amount of amber used in the Kachin Hills, Hukawng, Naga, and Chin Hills. In Maingkwan, large amounts of amber were used for making nadaungs (Chinese style earrings). At this time, the important centers for amber trade were Maingkwan, Mogaung, Mandalay, and the Naga country of Assam. By the late 1930s, mining areas near Maingkwan and the Hukawng Valley were becoming abandoned pits and were quickly grown over by the jungle.

However, the current Poinar amber collection maintained at Oregon State University includes a specimen of Burmese amber from the Hukawng Valley in Myanmar that came from a mine in the side of a mountain apparently from the lignitic seam deposits that extend down to the valley were the earlier collections by Chhibber were found. Poinar reported that the nuclear magnetic resonance (NMR) spectra of the amber indicated an araucarian (possibly Agathis) source of the amber. A rare find identified by Poinar was the earliest known hymenomycete fungus in this burmese amber specimen (see Fig. 7.44). Poinar concurred with Cockerell (1922) and Zherikhin & Ross (2000) that the amber was Cretaceous based on primitive insect types found in the amber. This indicates that burmite is much older than previously thought making it between 80 to 110 million years old.

The symbolic value placed on amber led to much being imported from Baltic sources as well as from other Asian countries where amber was found. A particularly luminous, clear, pale yellow variety was imported from Thailand and was remarkable for its light sherry or golden color, similar to that of Baltic amber, and its usual uniformity in color. This amber was called *Ching Pah*, meaning golden amber, and was often worked into round or baroque shapes.

In the traditional necklaces of the Mandarins, all beads were required to be uniform in size, to be round, and to number exactly 108, because each bead had its own religious significance. Another shape of

Chinese amber beads is the flat, tabular type. Also among antique Chinese jewelry one can find amber platelets with surface carvings of flowers and holes to sew the platelets to headdresses and hair ornaments (see Fig. 7.45).

Another amber article highly prized by collectors today is the antique snuff bottle. Some of these are exquisitely carved, but others are plain, allowing the beauty of the amber to speak for itself (see Figure 7.46).

Since Burmese amber often was transparent brown with turbid shading or a striped appearance, it was called *hu peh*, which means "tiger amber" in Chinese. Two other varieties of amber received special names from the Chinese. A translucent yellow amber was called *mi la*, or "honey amber," whereas a darker yellow, opaque variety was called *chio-naio*, or "bird's brain."

Because of the high value placed on amber, it was often counterfeited in China. As early as the sixth century, T'ao Hungching (A.D. 452–536) warned "only that kind which, when rubbed with the palm of the hand, and thus made warm, attracts mustard-seed, is genuine" (Cammann 1962, 256).

Very little actual Burmese amber reaches markets outside China today. However, Chinese dealers often have carvings made of amber, much of it imported from both the Baltic and the Dominican Republic. You can find dark brown amber snuff bottles and small unique carvings from Chinese amber, but dealers often do not know the exact location where the amber was found. Ruyi, a dealer from Taiwan, attempted to open the Burmese mines for commercial purposes in about 1988 and placed small pieces of this newly mined Burmese amber on the market (see Fig. 7.47). However, the cost of this amber in the United States was much higher than amber

Figure 7.46. Chinese amber snuff bottles carved with oriental designs and Buddhist symbols. The bottles range from about 2.5 to 3 inches tall (5.7 cm to 7.6 cm) and have jade and coral caps. The amber pieces are a dark brown semitransparent and mottled amber. Burgess-Jastak Collections.

b

c

a

Figure 7.47. Amber nodules and beads from the Ruyi Burmese amber mine, ca. 1988: a, "Root" amber polished pendant-shaped piece, mottled with dark brown and caramel color opaque amber, 5.5 cm x 2.6 cm x 1.2 cm; b, Polished round beads, 15 mm, of opaque dark brown amber; c, Dark Burmese amber tassel with one bead and pendant, semitransparent, with botanical inclusions.

that could be purchased from other locations and very little contained fossil inclusions. One problem encountered was, since the amber was fragile, it would tend to crack or crumble as it dried or when being worked. The amber pieces were small, dark brown, and some were translucent with cream-colored mottling. For the collector, these pieces were unique and rare; for the Chinese, they were believed to have more mystical warmth and value than amber from elsewhere. A Canadian company, under a license, recently mined amber from the Hukawng Valley in the northern Kachin State, but the amber was extracted from the surrounding hills by removing about 1.5 meters of overburden to reveal the amber. Here the amber usually occurs in small pieces ranging in color from sherry to burnt orange with some pieces even a clear cherry red. The swirls in the Burmite appear under magnification to be minute dots of color which seem to change color depending on the direction of light transmitted through the piece. Burmite also has a great variety of insect inclusions.

Amber from Sarawak, Brunei, Malaysia

The largest piece of amber ever recovered was discovered in Sarawak, Malaysia (Borneo), on December 3, 1991. This amazing piece weighed over 150 pounds (68 kg). Unfortunately, to transport it had to be cut into sections (see Fig. 7.48). The enormous fossil resin is similar to dense coal, impregnated with fossil resin. The color of the amber after polishing was various shades of pink, orange, green, white, and violet. Under the microscope reddish-brown droplets were discovered, which give the pinkish opaque appearance. The amber was located in a Lower to Middle Miocene formation. Large quantities of these Paleogene fossil resins, called Bornean amber, have been found during recent years; perhaps the deposits are greater than the Sambian Baltic deposit according to Dr. Kosmowska-Ceranowicz. Yet the mine near the town of Kait and the Rajang River was reported by dealers as being nearly exhausted and now closed (personal conversation 2003) The fossil resin is similar to glessite (a fossil resin found along with Baltic) even though it is thought to originate from a differ-

ent source tree. The region where the amber was found is a coal-pro-
ducing area with amber being found in some coal seams. Usually the
amber pieces are from 1 to 40 centimeters in diameter and have a
bright yellowish color. Some of these pieces contain spiders, beetles,
wasps, ants, and centipedes. In general the fossil resin is reported to
be difficult to work and must be handled carefully to cut and polish it
into jewelry (Kosmowska-Ceranowicz 1994, 575–8). Studies reveal
the source tree for this amber is the Dipterocarpaceae family, a
straight, wide-girth timber tree with resinous wood. The
Dipterocarpaceae are represented by over 500 species that comprise a
significant portion of the vegetation of the rain forests in tropical Asia
and Malaysia. Langenheim (1996, 6–15) reports that all trees in this
family produce resin, with some, such as the *Shorea, Hopea, Vatica*, and
Dipterocarpus, producing it more copiously. Using IR spectra and
C13NMR, Langenhelm (1996) found characteristic sesqui- and
triterpenoids, indicating a relationship between the fossil resin found
at Sarawak, Malaysia, and the *Shorea* recent resin and, interestingly
enough, also with the fossil resin from Arkansas found in Eocene lig-
nites and those found in Utah coals.

Figure 7.48. The largest piece of amber in the world—weighing over 150 pounds—found on December 3, 1991, in Sarawak, Malaysia, on the Island of Borneo. With kind permission of the Natural History Museum, Stuttgart.

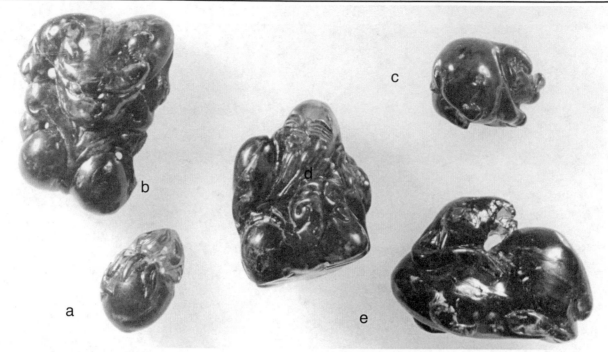

Figure 7.49. Carvings from Chinese amber: a, Miniature figures, that is, peach of mortality, dark brown amber, about 1 inch x 1.5 inches or 2.5 cm x 3.5 cm; b, Beast, dark brown amber about 2.5 inches x 2 inches x 2 inches or 6.4 cm x 5 cm x 5 cm; c, Pig, dark brown amber, 1 inch x 1.75 inches or 2.5 cm x 4.5 cm; d, Old beggar sitting, dark brown amber, about 2 inches x 1.75 inches or 5 cn x 4.5 cm; e, Old style Chinese horse with head looking back, dark brown amber, as 2.5 inches x 2 inches or 6.4 cm x 5 cm. Courtesy Burgess-Jastak Collection.

Figure 7.50. Amber carving of two Chinese goddesses with vases and flowers done in a Chinese workshop, commissioned by Ed Tripp, who sent a large lump of blue amber from the Dominican Republic to be carved. Carving is 6 inches across and 5.5 inches tall or 15 cm x 14 cm.

Chinese Amber from Fu Shun

A fossiliferous amber occurs in coal beds near the city of Fu Shun in Liaoning Province in China (Hong 1981, 166). This amber is from the Guchenzgi Formation of Fu Shun dated as Eocene, about 40 to 53 million years old. It is associated with the remains of a variety of plant species of both angiosperm and gymnosperm origin. The amber strata extend from slightly beneath the surface to about 200 meters underground. The coal seams that contain the amber are interspersed with beds of shale and sandstone (Hong et al. 1974, 113–49). Some samples of this amber are displayed at the Peking Geological Museum, though this museum is difficult for tourists to visit and requires a long, winding trolley ride from the center of Beijing. The fossils found in the amber nodules were studied and on the basis of the inclusions, it was concluded that during the Eocene the climate for the area must have been subtropical (Poinar 1992, 46).

Dealers from China often have small miniatures of Buddha, wisemen, animals, and snuff bottles carved from a dark brown to blackish transparent amber with some lignite or carbonaceous inclusions represented to be Chinese amber. These are very interesting carvings made in typical Chinese styles but one cannot be certain where the amber itself originated because much amber is imported into China from various locations (see Figs. 7.49, 7.50, 7.51, and 7.52).

Li Po, the great Chinese poet depicted in Figure 7.52a & b, wrote 20,000 poems that survive today. Legend has it that the drunken poet leaned out of his boat not only to embrace his own reflection, but also that of the Moon. While vainly admiring his reflection, he fell in and drowned in his "happy illusion."

Figure 7.51. Transparent light yellow amber carving of a water dragon on the side of a small snuff dish.
Size: 3 in x 2.5 in or 7.6 cm x 6.4 cm.
Courtesy Jimmy Chin, Hong Kong.

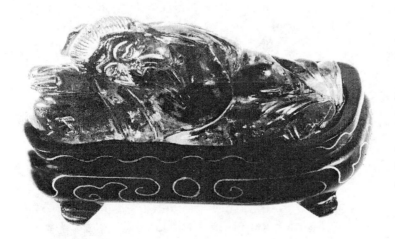

Left, Figure 7.52a & b. Chinese miniature carving of Li Po (701–762), one of China's greatest poets, and perhaps the most appealing T'ang figure. Figure a, Size: 2.5 in x 2 in or 6.4 cm x 5 cm. b, Size:3 in x 2.5 in or 7.6 x 6.4 cm.
Author's collection.

Li Po wrote:
Amid the flowers with a jug of wine
The world is like a great empty dream
Why should one toil away one's life
That is why I spend my days drinking
Lustily singing, I wait for the bright moon.

Paleogene Amber and Fossil Resins from the United States

In addition to the Cretaceous amber found in the United States, several locations produce small quantities of Paleogene amber usually found in lignite beds or coals. Most of these deposits are small and do not produce amber on a commercial scale. Some locations have received considerable study by geologists, petrologists, and botanists, mainly for their scientific value.

Arkansas. The first to report amber from Arkansas was D. D. Owen in 1858, who found it in the lignite of Poinsett County. In 1860 Owen reported small pieces of fossil resin in lignite near Camden, Ouachita County. The pieces found at Hot Springs County near Gifford were described as up to 3 inches by 3 inches by 2 inch (7.6 cm x 7.6 cm x 2.54 cm) but the deposits were poor. In 1971 amber was found near Malvern, Hot Springs County, Arkansas. Mapes and Mapes (Saunders et al. 1974, 979) recorded collecting 300 grams of amber containing several insect inclusions, while others found more than 8 kilograms of amber in the lignite coal of the Eocene strata. There were more than 900 organisms, of which there were 700 arthropods, including 8 insect orders. The amber pieces were generally small—from 3 to 8 centimeters—and ranged in color from clear, pale yellow to dark brown-orange. Numerous impurities, marked by dark opaque bands within translucent areas, were present. The material was too fragile to be of gem quality (see Fig. 7.53).

Figure 7.53. Samples of amber from Malvern, Arkansas. Nodules range from opaque yellow to brown, with barklike surface on some pieces resembling amber-impregnated wood. Pieces are very fragile and crumble easily. Author's Collection.

A sampling of Arkansas amber is preserved in the Museum of Comparative Zoology, Harvard University. IR spectra of the Arkansas amber produces a similar pattern to present-day *Shorea* resin, which suggests the botanical source to be Dipterocarpaceae. This surprised investiga-

tors since no trees in this family are now found growing in North America (Saunders et al. 1974, 979–80; see Fig. 7.54).

New Mexico. In New Mexico fossil resins occur in lenslike sheets filling fractures in several coal deposits, but the material is small in size and none is of gem quality. However, jet is found and polished by the local Indians.

Washington. In the state of Washington fossil resins are found in the Tiger Mountain Formation near Issaquah, a part of the Puget Group. The resin is determined to be Middle Eocene or about 45 million years old (Mustoe 1985, 1530–6). This amber is dark red and transparent. Being brittle, it fractures easily as it is extracted from the amber-bearing clay strata. Much plant debris that appears to be similar to bark on cedars in the family Cypressaceae are found, suggesting this to be the original source. No insects have as yet been found.

Figure 7.54. Shorea tree, a relative of the tree that produced the amber in Arkansas (United States) and in Sarawak (Malaysia). An extinct species of Shorea or other dipterocarp tree was the resin-producing tree related to the amber found in these two distant locations.

Kauri pine trees, *Agathis australica, Waipoua Forest on Northland's west coast,* **New Zealand**

NATURAL RESINS THAT
RESEMBLE AMBER

The oldest known substitute for amber is copal, a resin very similar in appearance to amber. Before synthetic resins were produced, copal was the principle substitute for amber. In earlier periods, copal was worked into ornaments in those cultures having ready access to supplies of the rough material. Ancient Egyptian graves commonly contained lumps of copal, providing evidence of the earliest known use of copal in jewelry for royalty. Articles made of semifossilized resin, including a double ring within the resin engraved in the form of the royal seal, were found in the tomb of King Tutankhamen. Two large scarabs—one carved on the surface with a relief design of a bird—and two necklaces—one with graduated resin beads and one alternating resin and lapis lazuli beads—were also found, as well as resin earrings, hair rings, and a knob for a box. The resin in all of these objects is dark red, perhaps from oxidation, and very brittle. Tests showed that the resin was readily soluble in alcohol and acetone, both of which scarcely affect true amber (Lucus 1934, 337).

The Chinese also used resins other than amber for carving and ornaments. Early authors described Chinese articles made of copal, shellac, and colophony, which were a clear golden color, and even contained insect inclusions. Sometimes the copal ornaments were coated with a varnish or shellac to prevent the volatile and essential oils from escaping, which would result in extensive crazing. The crazing over the surface of copal objects is somewhat different from the fine crazing on amber. Copal crazes with a white appearance of fine, shallow lines on the surface with flaking that results in resin crumbling off where the material has dried. Coating the surface with a lacquer was a technique used by the Chinese that has protected some of the old nineteenth-century copal pieces. But when exhibited in illuminated show cases, the drying continues. Even old Victorian necklaces of clear resin with a champagne color have proven to be

Figure 8.1. Antique Chinese toggle button made of copal with several insect inclusions. Note the heavily crazed surface.
Size: 2.5 in. x 1.75 in. or 6.4 cm x 4.5 cm. Patty Rice Collection.

copal and can be easily detected today because the surface has a dull, dried, crusty effect. Figure 8.1 shows an old Chinese toggle button made of copal containing insect inclusions. Since toggle buttons were not only used to fasten outer garments, but were also decorative and often had symbolic meanings, the symbolism attached to the resins and inclusions attributed to the desirability of such buttons. As can be seen in Figure 8.1, the copal objects, even after having been polished, develop a finely checked or crazed surface, often in just a few years, as a result of the escape of volatile substances (Cammann 1962, 78–9). The word copal was derived from the Spanish word *copalli*, meaning incense. Some believe the origin of the word reverts back to the Sanskrit word *chandaras*, meaning moon juice, a word that was also used in relation to the resin sandarac. Copal refers to a group of resins exuded from various tropical trees and used principally in varnishes. Such resins are soluble in oils and organic liquids, but are insoluble in water. In making varnish, copal is melted in a copper pan, and at 212°F water contained within the resin evaporates, with a considerable amount of gases evolved as the material decomposes. The copal preserves its yellow color but becomes more soluble (Rebaux 1880; see Fig. 8.2).

During the time when natural varnish resins were in great demand, several regions of the world exported various types of resins. Raw copal resin was obtained by directly tapping the trees, and was collected for shipment to varnish manufacturers. However, synthetic resins have now replaced most natural resins, and only those varieties commonly confused with amber will be discussed further.

Copals occur both in a semifossilized form called true copal and in freshly obtained gum called raw copal. The semifossil resins range from over 1000 to as little as 100 years old. They are either transparent or translucent and typically are yellow or brown in color. Copals are softer than amber, but are very brittle and break in conchoidal fractures. When heated, they emit a distinctive resinous odor and burn with a smoky flame. The refractive index (1.54) and specific gravity (1.06 to 1.08) are similar to those of amber, but the lower hardness and easy solubility can be used to identify copal when it is masquerading as amber. A drop of ether on copal causes it to become sticky, whereas on true amber no reaction is noted. Extensive crazing also suggests the material is copal rather than true amber. Copals fluoresce white under shortwave ultraviolet light.

Insects, leaves, and other botanical debris are found in both semifossil copals and recent copals. These enclosures are always of recent, currently living varieties rather than the primitive, extinct

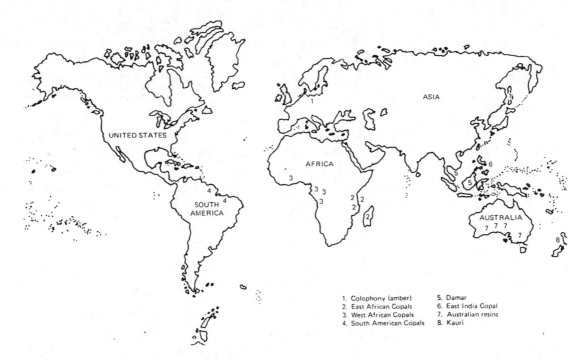

1. Colophony (amber)
2. East African Copals
3. West African Copals
4. South American Copals
5. Damar
6. East India Copal
7. Australian resins
8. Kauri

forms found in amber. Enclosures may be artificially placed in either copal or amber by inserting the insect between two layers of the substance and fusing them together. The layered material may be detected by careful examination under a microscope or with a high power jeweler's eyepiece.

Figure 8.2. Copal producing regions.

KAURI COPAL

The New Zealand kauri copal was and is exuded by the kauri pine, *Agathis australis*, which belongs to the family of Araucariaceae and flourishes in the northern region of New Zealand, the main area where the kauri has grown for many thousands of years (see Fig. 8.3). The forests grow in the Province of Auckland, from the lower Waikato and Coromandel Peninsula northward. Kauri trees often live over 1500 years and grow to giant size, with trunks that range from 5 to 12 feet (1.8–4 m) in diameter and heights of 120 to 160 feet (40–50 m). The timber is close-grained, white, and so durable it was used for the masts of sailing vessels (see Fig. 8.4).

When a tree developed a rupture in the bark or a fracture, the resin congealed forming lumps of gum that would fall to the ground and then become buried in the forest floor. The resin would also collect in the fork of the branches or trunk. During violent storms a tree trunk could suffer a fracture of the trunk and then resin would collect internally. The internal pocket would be filled with a chalky white gum, sometimes collecting in large volumes around the roots.

In certain areas where a number of forests flourished, two or more layers of copal resin may be found in the ground. The best quality gum was usually found buried in shallow depth on hills and was bright and hard. It was usually found no deeper than 1 meter in the hill country. However, in peat bogs and wetlands some pieces were buried as deep as 5 meters covered with sediment. Interestingly, it is reported that dull, opaque lumps could be transformed into higher-quality transparent material by exposure to sunlight or bush fires. In sand dune areas, such as Aupouri Peninsula, the resin, and even logs and stumps of the kauri tree, occurred in layers up to 4 meters below the surface. In some instances, kauri copal has been found under a layer of sandstone and limestone as deep as 300 feet (100 m) below the surface. Kauri resin taken from the ground, though extremely old, is still soluble and therefore is considered to be in a semifossilized state. Its color ranges from yellow to reddish-yellow to brown. It sometimes contains many insect inclusions. When lumps are taken from the ground, they are covered with soil and a crust that must be scraped off.

Kauri does not contain succinic acid. Its specific gravity is 1.05 and it melts between 187°C and 232°C (360–450°F). As a result of this low melting point, the friction produced during polishing commonly causes the surface to become sticky. Also because of its softness, it does not polish well; yet, Penrose (1912, 43–4) reported that clear, transparent varieties were used as substitutes for amber in mouthpieces for pipes and cigar and cigarette holders, and that a certain amount was carved into small ornaments (Reed and Collins 1967; Hornaday 1915, 181–2). New Zealand is also the source of another true fossil resin called ambrite, which is transparent and yellow in color. It has been known for many years but does not appear to have been used to any great extent.

Before 1850 the kauri gum was collected from the surface. Later, to increase the yield, gum was extracted by digging. In the 1850s manufacturers of varnish began using kauri resin to produce one of the better quality varnishes, with the industry reaching its peak about the turn of the century. One report indicates that in 1856 (Hornaday 1915, 181–5), close to 1440 tons of

Figure 8.3. New Zealand's kauri-producing area. Waipoua Kauri Forest has been under the protection of the Department of Conservation since 1987.

kauri resin were exported to varnish manufacturing companies in the United States and Great Britain. Though many trees were tapped for the purpose of collecting the raw gum, much buried material was mined in a simple process. A long, pointed, steel rod was used to probe the earth, and was thrust into various spots until a lump of kauri resin was felt, at which time spades were used to dig up the lump. In contrast, when prospecting in peat bogs or swampy areas, a hook was inserted into the soft mud and the resin lumps snagged to bring them to the surface. Exceptionally large and clear specimens were prized by the "gum diggers," who often polished such pieces in natural form or carved them into ornaments. Large lumps of kauri copal were not uncommon, with some pieces weighing between 10 and 12 pounds (4.5–5.4 kg). The largest lump of kauri copal on record weighed almost 100 pounds (45 kg).

The Matakohe Kauri and Pioneer Museum in New Zealand exhibits many large cobblestones and massive pieces of kauri and presents a history of kauri gum digging. In the past, the museum published the *Gum Diggers Gazette*, which gave a history of the kauri industry. These records provide the best-documented history of the removal, treatment, exportation, and use of a natural resin since this industry became an integral part of New Zealand's history (Poinar 1996*b*, 52–6; see Fig. 8.5).

The Maori landed on the island of New Zealand about 1000 years ago and were the first to discover that the gum or resin that runs down the tree could be used as chewing gum. Gum that had been buried and was hard could be softened by soaking in water and mixing with milk of the Puwha thistle. In fact, kauri chewing became part of Maori social culture and at family gatherings, they would mix a

Figure 8.4. Te Matua Ngahere, oldest living kauri tree, is estimated to exceed 2000 years of age. Reprinted with permission from Matakohe Kauri Museum.

Figure 8.5. a, Anthony Petrie (l) and his gumdiggers displaying large pieces dug from Tomorato Swamp near Wellsford (ca. 1930s); b, Very old gumdigger scraping gum until the day he died. Reprinted with permission from Matakohe Kauri Museum.

large lump of resin with the juice of the thistle, then pass the piece from person to person and each one would take turns chewing. The Maori also used the gum to start fires and on their torches to produce light used by fishermen at night to attract fish. At the same time, smoke of the burning kauri worked as an insecticide. Tattooing, a prized art of the Maori, also benefitted from kauri gum, whether from its effect on the skin or from its ability to seal away bacteria is unknown. The Maori would burn the gum, then powder it. This would be mixed with animal fat or oil, then poured into tattoo cuts (Fraquet 1987, 134–5).

The Maori would simply wade into swamps and when they felt the semihard copal lumps, they would pick up the pieces with their feet. The Maori called kauri gum *kapia*.

The large lumps of hard kauri copal were often saved to use in making ornaments. The gum digger would shape the pieces into a figurine of the Maori warrior heads. The figurines were painted with oil paint showing the tattoo marks covering the face. One such piece is preserved in the Museum of Natural History in Cleveland, Ohio. These pieces did not craze quickly like natural kauri lumps do after exposure to the atmosphere because the oil paint covering the figure assisted in preserving it (see Figs. 8.6 and 8.7).

As the industry developed, Maori from Northland, then Waikato, King Country, and Bay of Plenty worked in the fields. Later immigrants from Russia, Germany, France, Malaya, Finland, and Croatia joined the kauri industry in digging for kauri gum. Since gum diggers spent nights far out in the bush, they would often entertain themselves by making ornaments from the large pieces. Some even would carve out the resin to insert lizards or frogs in the hollow, and then reseal it with melted resin. Many of these fakes have found their way into museums.

When the Europeans arrived, they soon discovered that the kauri copal could be melted to produce varnish, adhesives, and, later, linoleum. Soon the industry for retrieving and selling copal became more organized. Immigrants arriving from the Dalmatian coast and Austria became the workers who specialized in retrieving the copal, or gum digging, thus they were given the label of "gum diggers."

By the 1880s it became easier to recover the kauri gum from around the base of large trees. This area provided an accumulation of resin from hundreds of years and was called "bush gum." By the 1890s over 20,000 people were employed as gum diggers in Auckland. In 1900 kauri copal was Auckland's most valuable export, with over 10,000 tons being exported.

Figure 8.6. Carving of a Maori head, 3 inches or 7.6 cm tall plus the base. Two lumps of kauri copal. These figures were usually made by the Dalmatian gum diggers. Courtesy Museum of Natural History, Cleveland.

Fresh resin could also be collected from forks of the branches, high in the tree tops. Soon the workers began bleeding trees by cutting V-shaped taps horizontally spaced about 40 centimeters apart around the trunk. The workers would return in about six months and collect the exuded gum. They would climb these enormous trees with spikes and hatchets. This was very dangerous. Occasionally, workers would drop their spikes stranding themselves in the tops of the trees. Skeletons of unfortunate workers have been found at the base of tall kauri trees. It also became apparent that this "gum bleeding" was killing many of the large old kauri trees and would eventually damage the whole forest. In 1905 legislation was enacted banning the bleeding of trees in state-owned forests. However, bush gum collected from around the base of the tree or from the soil was allowed.

In 1906 more than 275,319 tons of kauri copal were exported

Figure 8.7. "Paddle" made of kauri gum, given to Cleveland's Museum of Natural History, Cleveland, Ohio, in 1922 by the McGlidden Varnish Company. In 2002 the curator of the Matakohe Kauri Museum, New Zealand, identified this as a kauri replica of New Zealand Maori weapon called a "patu" or "mere" that was typically made of wood, bone or stone and were used in short thrusting jabs in combat. Size: about 10.5 inches or 27 cm long with its widest area being 3.25 inches or 83 cm. Courtesy Cleveland Museum of Natural History.

to varnish factories. This trade peaked in 1900, with 70 percent of the export being sold for the production of varnish in the United States and Great Britain; the remaining 30 percent was used for a variety of manufacturing purposes. By 1910 the best gum had been extracted, and spear-found gum could very seldom be recovered. At this time large pieces were exhausted but kauri was being used in the linoleum industry and inferior quality resin in the form of smaller chips and dust could be used. It became necessary to work the swamps and alluvial flats to find gum. This was accomplished by draining the land and using saline floats or vacuum tanks to extract the kauri. Between 1853 and 1971, about 460,000 tons of kauri copal were exported. As the twentieth century progressed, synthetic varnishes and other products took over the market for kauri gum. Some special varnishes were made for use in the violin industry and in the manufacture of dentures (Fraquet 1987, 136). Poinar (1996b, 52–6) reports that in 1973 a company called Kauri Deposit Surveys Ltd, later called Kaurex Corporation Ltd, began extracting remaining pieces of copal from peat bogs. They dredged up peat and copal mixed together, then extracted the resin using hot solvents and recovered the resin by a "fraction crystallization process." Poinar states that "It is calculated that by working 1,450 hectares of peat for 30 years, Kaurex would recover some 600,000 tons of resins and waxes, valued in excess of NZ\$1 billion on the world market"(Poinar 1996b, 52–6). In spite of this, the company went out of business in 1988.

Visitors to Auckland, New Zealand, will find a trip through the kauri forest and a stop at the Matakohe Kauri and Pioneer Museum a very interesting part of their trip. The museum demonstrates the importance of the copal to the lives of the people. Another museum exhibiting kauri gum history and actual specimens of kauri copal in different forms is the Auckland Museum, which overlooks the Waitemata Harbor. This museum provides a natural history of New Zealand and collections of Maori artifacts.

COLOMBIAN COPAL

In the current fossil marketplaces such as the large gem and mineral shows of Tucson, Arizona, and Denver, Colorado, dealers are displaying large quantities of copal from Colombia, South America. Interest in the sequel to *Jurassic Park* has brought a demand for fossil insects imbedded in amber, thus, signs will state "Insects in Colombian Amber." This is deceiving to uninformed purchasers. Poinar (1992, 6), defines copal as "a recently deposited resin that can be distinguished from amber . . . by its physical characteristics . . . including hardness, specific gravity, melting point and solubility." He states that deposits younger than Tertiary period (or less than 2 million years old) fall into the copal category. However, scientists still have not agreed on the exact time when copal changes into amber. The Colombian copal material is readily available since it is of recent origin (up to about 1000 years old). The insect inclusions are of recent origin. Radiocarbon testing of a sample of copal from Colombia indicated that it is less than 250 years old.

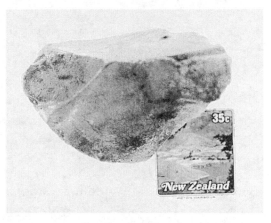

Figure 8.8. Kauri gum specimen purchased in New Zealand in 2002. Size about 3 inches x 2 inches or 7.6 cm x 5 cm. Patty Rice Collection.

The tropical tree, *Hymenaea courbaril*, as well as other species, produces copal resin in Colombia and Brazil. The resin of *Hymenaea c.* is typically greenish-yellow in hue. Berkeley researchers, studying Colombian copal, related it to the *Hymenaea protea*. It is reported to come from Santander Province, Pena Blanca, and about nine different localities. Copal is not a single substance, but a mixture of resinous products from many different kinds of tropical trees (see Fig. 8.9).

According to Poinar (1996*b*, 52–6) as early as 1860, there were reports of large quantities of Colombian resin being shipped to Hamburg, Germany, as Colombian amber. The material was from the Magdalena River Valley in Colombia and a commercial venture between Colombia, Germany, and England had started to profit by it. Hearing of this, Dr. Schiefferdecker, of northern Germany, who was well familiar with amber, obtained a couple of pounds of the material to test. He found that the resin was fairly soft, burnt easily, producing a sweet-smelling odor usually associated with copal. He then gave some samples to Dr. Spirgatis, who tested it for solubility, finding it soluble in alcohol, ether, and other organic solvents. Therefore, Schiefferdecker concluded that the resin was copal, which had been mentioned by earlier Portuguese and Spanish travelers. Even these early reports described the Jotaba tree and the Algarroba tree, both of the genus *Hymenaea*, as producing quantities of resin that collected around the roots and between the bark and wood. The early reports also indicated that the resin was generally found in river valleys in areas where the trees grew in Brazil and Colombia. The rivers carried the copal from the mountainous areas down to the valleys, depositing it in the alluvial soil.

With this information known for so long, it is surprising that we find so much Colombian copal represented as "Colombian amber" on the market today. The pieces are obviously pale, clear yellow with a whitish powdery crust covering the rough material. Copal is softer than amber and the surface will move when polished on a wheel caused by the heat melting the surface. It easily melts with a hot needle. It is reported to be soluble in heavy hydrocarbons and becomes sticky in alcohol or acetone.

In an attempt to investigate the ubiquitous material called "Colombian amber" that dealers were offering at the Tucson Mineral Show, Poinar (1996*b*, 55) determined that most came from Santander, Boyaca, and Bolivar, specifically near the towns of Bucaramanga, Bonda, Pena Blanca, Giron, Medellin, Valle du Jesus, and Mariquita. Often the material was discovered by farmers as they burned their fields during the slash-and-burn agriculture of the region, because they recognized the sweet odor of the resin. Since none of the dealers had been to the region where the resin was gathered, it was impossible for Poinar to describe the strata where the material was deposited. Not only were exact locations kept secret, but, more importantly, danger from terrorist groups and drug establishments made the region dangerous. Poinar recalled two Americans going to search the sites several years ago and never returning! Thus, instead of visiting the sites, Poinar sent the material to an independent lab for carbon dating to verify his findings. He purchased samples from three separate commercial dealers and sent the samples to Beta Analytic in Miami, Florida, who provided radiocarbon dating. The results varied unexpectedly. The first sample ranged from 380 to 500 years old based on carbon dating reports. The second sample ranged from 210 to 310 years old. More surprising was the third sample, which was dated only between 10 and 80 years old.

Schlee (1984*b*, 299–337, 1986, 1–15) also performed carbon dating on a piece of Colombian material from Pena Blanca in the late 1980s, reporting it to be under 250 years old. But, for further verification, Poinar sent samples of Colombian resin to Joseph Lambert at Northwestern University for nuclear magnetic spectrometry analysis. The results from this study indicated that the resin was similar to the modern *Hymenaea courbaril* tree. Thus, verifying the early 1860 work of Schiefferdecker, who also suggested this genus of tree as the source for the Colombian resin from Magdalena Valley.

Colombian semifossilized resin contains a variety of insects in large quantities. Both plant and animal inclusions are found. Many of the plant materials described by Poinar can be found growing in Central and South America today. The Algarroba trees are widely distributed throughout this region and produce an amber-colored resin that accumulates around the roots of the trees. Farmers today burn the copal for incense, melt it down for making varnishes, and even use it medicinally as home remedies. Because the wood is highly prized and valuable, many older trees are being cut for timber and the species is becoming somewhat scarce. Some plant inclusions in the copal were also from a different Hymenaea tree, such as the *H. oblongifolia*. Poinar describes three species of Hymenaea that occur in Colombia and all produce resins. Therefore, they all could be contributing to producing the Colombian copal.

Figure 8.9 Colombian copal, purchased from a dealer who sold it as Colombian amber, produced by the Hymenaea courbaril tree, which grows in Colombia and other countries of South America. Size: about 5.5 in x 3 in x 2 in or 14 cm x 7.6 cm x 5 cm.

To comfort all those who have purchased Colombian copal as a fossil specimen, Poinar (1996*b*, 56) states that Colombian copal does have an important value, since it provides evidence of organisms that existed several hundred years ago, giving biologists an opportunity to study habitat change over a short period of time. In this manner, ecologists can determine changes in the flora and fauna of the localized area caused by human intervention.

Colombian copal from different regions may be different ages, but all of the material is not very old. There may be fossilized resin in Colombia, but it would be found in an earth strata where it had been deposited eons ago. As yet the pieces appear to be from recent trees, not over a 1000 years old. Grimaldi (1996*a*, 19) also indicates that there are many particularly large pieces of Colombian copal with swarms of termites that are sold to amateur collectors (and even some museums) as "Pliocene amber," which would make it about 2 million years old. Yet carbon14 dating indicates that it is only several hundred years old, "as are all the other Hymenaea copal deposits."

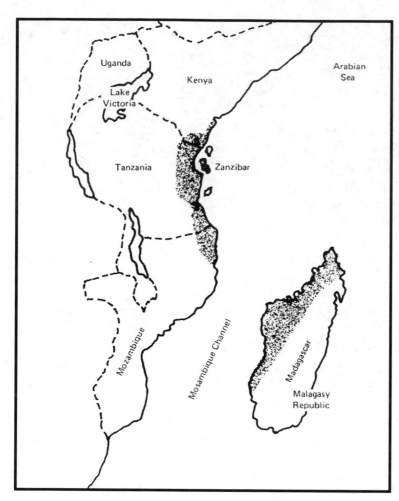

Figure 8.10. East African copal areas.

EAST AFRICAN COPALS — ZANZIBAR

A major industry producing copal for varnish flourished in Zanzibar, now part of Tanzania. Semifossilized copal from Zanzibar, sometimes termed "jackass" copal (*chakasi*) by the natives, was also known as Zanzibar animè. The copal is found embedded about 2 to 4 feet (0.6–1.2 m) underground over a wide area of East Africa on the coast opposite the island of Zanzibar (see Fig. 8.10). Shallow mining during the rainy season when the ground is soft produces pieces from pebble size to masses weighing several ounces. The pieces of fresh copal are very pale yellow and include many insects. Occasionally, lumps up to 4 to 5 pounds (1.8–2.3 kg) may be found. Zanzibar raw copal is also collected from the living *Hymenaea Caesalpinioidea*, a leguminous tree. During the peak of the industry these trees were called *Trachylobium verrucasum*, but are now called *Hymenaea verrucosa*. The trees occur naturally from Somalia to Tanzania, Mozambique, Madagascar Island, Zanzibar Island, and islands farther from the East African coast such as Mauritius and Seychelles. The copal was obtained near the roots and on the ground. In some areas where copal is found, the trees that produced the resin no longer grow there.

East African copal was commercially used for manufacturing varnish in China and India. In 1898 Germany imported over 512,600 pounds of East African copal to manufacture high-grade varnishes. This is the hardest form of copal and since the region was formerly German East Africa, this copal was easily available for the German amber industry for making substitutes combining amber and copal. Several patents were issued in Germany, as described earlier, indicating copal and amber were mixed to produce an imitation amber or pressed amber during the nineteenth century. German patents indicate that copal was even blended with harder fossil resins to make amber substitutes. For example:

German Patent 207 744 (1907) Spiller, F. Amber substitutes. Copal heated with water at pressures of 16–20 atmospheres. Acid, alkaline, or neutral additions may be made to the water.

German Patent 247 734 (1911) Spiller, F. Converting copal into a mass resembling natural amber. After freeing copal from its crust, the material is softened with carbon disulfide and heated uniformly in air-tight compressed forms.

German Patent (1913) Bahket, J. Artificial amber. Copal is dissolved in acetone with or without the addition of dyes, kneaded until the acetone is volatilized, then melted at 300°C and poured into forms where it is allowed to cool under pressure (as translated by Fraquet 1987).

About 1976 a find of so-called young amber was made somewhere in Tanzania and became available on the U.S. market. It was reported by Geological Enterprises to be from a Pleistocene formation, and would more appropriately be termed a semifossilized resin. Pleistocene age would make it older than most copal resins, but much younger than Baltic or Dominican amber, which are considered to be true fossilized resins. Some current researchers suggest lowering the age placement used for determining when a resin changes from copal to amber. However, the significance is in reality; to the fossil collector, the question is, Are the fossils rare and ancient?; and to the jewelry industry, the question is, Will the substance hold up for years as beautiful ornaments? This material is transparent, a pale champagne color, and softer and more brittle than Baltic amber. Most very pale yellow resins are suspect as "young" and are still losing volatiles, thus will craze extensively over time. The resin has a similar refractive index and specific gravity as Baltic amber. Small insect inclusions are present, but these are varieties that can be identified by genus and species of recent insects. The material polishes well, but because of its softness, it will not take much heat during polishing. The surface tends to move as if it were melting, when too much pressure is applied at high speed on a buffing wheel. It will not take much wear because it is soft and not as resistant as amber.

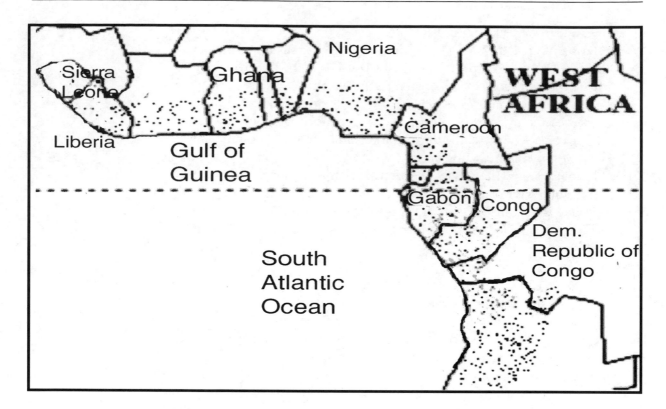

Figure 8.11. West African copal-producing regions.

WEST AFRICAN COPALS — SIERRA LEONE, DEMOCRATIC REPUBLIC OF CONGO (ZAIRE)

West African copals come from the West African coast from Sierra Leone to Angola, including Ghana, Guinea, the Congo region, and Niger (see Fig. 8.11). Raw copal from Sierra Leone is produced by a leguminous tree of the subfamily Caesalpinioideae, *Copaifera guibourthiana*, which is the same tree that produces much of the Congo copals. However, Sierra Leone has a variety of trees that produce copal resins. The resin is often found in large peat swamps, and, in the past, was cleaned by washing in a rotating drum. Following this treatment, the copal was sorted and sent to various varnish manufacturing factories.

Congo copals, or those from the regions of the Democratic Republic of Congo (Zaire) , are similar to those from Sierra Leone. They are usually found in the hardened semifossilized form. The source trees for the Congo copals and Angola copals are the *Copaifera* and the *Colophospermum mapane*.

In Ghana and Niger, the resin-producing tree is generally the *Cyanothyrsus*, while there are also various species of *Daniellia* and *Colophospermum mapane*. These trees are not cultivated species, but occur naturally depending on climactic conditions.

MANILA COPAL — DAMMAR RESIN

Both copals and dammar resins are produced in Manila. Because *damar* is the Malay word for all resins it is sometimes used for resins from different Burseraceae and Dipterocarpaceae trees. Also, contributing to the dilemma, several species of trees growing in Malaysian forests were previously called *Dammara*. These trees, which produce copal, were renamed *Agathis dammara*, because of errors in identification in the past. Other trees were also renamed. For example, two trees previously placed in the genus *Dammara* are now renamed *Protium* (Burseraceae) and *Shorea* (Dipterocarpaceae). Therefore, by the time resins are collected and they pass through dealers, one cannot be positive of what the original source was.

a

Manila copals differ from West African copals in that they are produced by trees of the genus *Agathis* in Indonesia and the Philippines. Copal, a clear, pale yellow copal-like resin, melts at about 140°C and is soluble in aromatic hydrocarbons (such as benzene) but is insoluble in alcohol. It is one of the harder resins used in making lacquers and clear varnishes. Though copal is collected from a variety of different species of trees, the main sources are from *Agathis dammara* (previously named *Dammara orientalis*; see Fig. 8.12).

In these principal regions, the copal is collected from ground deposits or from living trees intentionally tapped to produce a resin flow. Various names are used to denote the resin in different stages of hardness, with the hard, semifossilized forms being called *boea* (which, despite its hardness, is alcohol-soluble). The native Philippine name for resin that is somewhat hardened by delaying its collection for one to three months after tapping the tree is *loba*. A still softer kind gathered about two weeks after trees are tapped is called *melengkete*. Another form of hard, semifossilized Manila copal is found in Borneo and is called *pontianak*.

Dammar is chiefly obtained from Dipterocarpaceous trees of southern Asia, especially in Malaya and Sumatra in the Pacific Islands. The dammar trees grow in mountain regions and can become quite, large attaining gigantic heights and girths. The resin may be either semifossilized or recent and is softer than African copals.

Even though most varnishes today are made from synthetic epoxy resins, from 1979 to 1982 the Department of Mineral and

Figure 8.12. Manila copal from Philippines donated to the Cleveland Museum of Natural History, Cleveland, Ohio, by the McGlidden Varnish Company in 1922. The specimen is labeled, "Resin from a tall tree, Agathis dammara, Philippines, Manila." Size about 4 inches x 3 inches x 3.5 inches or 10 cm x 7.6 cm x 8.9 cm. Courtesy Museum of Natural History, Cleveland, Ohio.

b

Energy of Papua, New Guinea, included a development plan for dammar gum extraction. Extension officers were trained in the processes of collecting the resin, cleaning and packing it for shipping with a hope of reviving the industry. It was predicted that they would provide an income for 20 full-time gatherers, since the dammar is used in Asia for domestic purposes and in a somewhat inferior varnish.

OTHER RESINS ASSOCIATED WITH THE AMBER AND VARNISH INDUSTRIES

Infopedia (1995) defines resins as a term "applied to a group of sticky, liquid, organic substances that usually harden, upon exposure to air, into brittle, amorphous, solid substances." Many natural resins secreted from plants are classified into three main groups based on hardness and chemical constituents: (1) hard resins, (2) oleoresins, and (3) gum resins. Hard resins are generally thought of as amber, copal, mastic, and sandarac. These generally exhibit a glasslike conchoidal fracture. Hard resins can be obtained as a distillation product of an oleoresin. Oleoresins are sticky, amorphous semisolids containing essential oils. This groups includes balsam, turpentine, and others. The crude oleoresins were often used for caulking and waterproofing sailing vessels during the nineteenth century. The group generally called gum resins are frankincense, myrrh, benzoin, and asafetida because they contain gums. The resins more closely associated with the amber industry are discussed here.

Oleoresins and "true balsams" have not yet been shown to produce amber, even though there are many varieties of Pinaceae and Dipterocarpaceae that produce oleoresins. However, the resin from the *Pinus*, is distilled to make turpentine and rosin, also called "colophony," that was used in the varnish industry (Langenheim 1996, 5, 10–17; see Table 8.1).

Sandarac

Sandarac, a resin from North Africa, is brittle, faintly aromatic, and transparent yellow. It originated from the bark of a northern African tree of the cypress family, the *Tetraclinis articulata*, and from any of several Australian trees (genus *Callitris*) of the same family. It is also called "juniper gum" because the common juniper also produces sandarac. The ancient Egyptians used sandarac as a base for paint pigments on jars and walls of tombs. Sandarac was also used as incense and as a hardener added to softer resins. In the past, it was used as a spirit varnish for bookbinding, wood preservation, labels, and photographic negatives.

Mastic

Mastic is a resin from Greece that is used as a varnish, either by itself or blended with other resins. Mastic resin is produced by the *Pistacia entiscus* tree in pale yellow tear-shaped droplets. It is collected directly from the tree to prevent it from becoming contaminated with dirt. This resin was sometimes mixed with sandarac for use as paint pigments. Artists used it to harden their oil paints, then liquid dammar was used to coat the paintings.

Frankincense

Frankincense is a resin familiar to all at least by name. The Bible mentions it 22 times by virtue of its resinous aroma when continuously burned in the Hebrew temples. Most recognize it as one of the gifts offered to the Christ child by the Magi, but little attention has been paid to exactly what the material was. During this period, A.D. 2, about 3000 tons per year are recorded as being transported to Roman and Greek cultures. Frankincense is a pinkish resin from the Boswellia trees, perhaps from three different species. The resin, which is white, yellow, and tan, is collect after making cuts on the trunk and branches of living trees. The trees grow to be very large and have a star-shaped, greenish flower. (Poinar 2001, 80). It originally was harvested in southern Arabia and Turkey. The Arabian merchants brought frankincense in camel caravans, following the trade routes to Palestine and Egypt. Here it was traded, transferring hands, to be taken on to Greece and Rome. It was highly prized as the finest incense because of its resinous aroma.

Myrrh

Myrrh is a resin that comes from Arabia, Abyssinia, Turkey, and Somali coast of eastern Africa. The Myrrh tree are small or low thorny shrubs that grow in rocky terrain. Their stems and branches exude a tan to whitish resin. Myrrh also is mentioned in the Bible in connection with royal gifts. Myrrh was used particularly in cremation and embalming the dead, as a base for perfumes, and as an incense. The Biblical Hebrews used it for an anointing oil. The resin exudes from the Commiphora tree in thick yellow drops. As it hardens, with exposure to the atmosphere, it darkens to a reddish-brown hue. It is waxy, brittle, and has an aromatic balsamic odor.

Rosin and Labdanum

All violin players are familiar with rosin since it is used extensively on the bows of stringed instruments. Rosin is produced by chemical means from the oleoresin of a variety of living pines. The oleoresin is distilled into turpentine and colophony, which is also called rosin. The resulting material is a translucent amber-colored to almost black, brittle, friable resin used in the varnish industry for making a fine varnish for musical instruments. Acrobats used rosin on their shoes to keep them from slipping. Also, rosin was used in making paper size, soap, and soldering flux. Another oleoresin still collected today in Lebanon is a produced by the rock rose plant and is called labdanum. It is often used in the manufacturing of perfume. Several species of *Cistus.* a shrubby tree, that grows along the cost and mountains of Lebanon produce Labdanum (Poinar 200, 81).

Varnish

Varnish is a manufactured product long connected with the amber industry. The waste amber was sent to the smelter to be made into colophony used to make amber varnish. As early as A.D. 1200, distillation of amber for making varnish was introduced by Arnoldus de Villa Nova. Flemish artist Van Eyck, in the 1600s, mixed amber varnish with paints for his masterpieces to give the paint a flowing texture. The Mayerne manuscript of the seventeenth century includes a treatise by Van Eyck, along with various notes from other Flemish painters, on the preparation of amber varnish. In Italy amber varnish was used on musical instruments from Cremona, and is reported to give the wood a more resonant tone because it allowed the wood to breathe better than today's synthetic varnishes. Being weather resistant, amber varnish was also chosen as the best for carriages.

Amber varnish was slow-drying, hard, and dark in tone. There seems to be three methods for making amber varnish, the very earliest involving beeswax and resin. True amber varnish is a combination of linseed oil, oil of turpentine, and colophony of amber. First, waste low-grade amber pieces must be dissolved in an oleoresin such as turpentine, or amber was fused then dissolved in oil (this gives a dark slow-drying varnish), or the amber was added to boiling oil without fusing (in which case it does not dissolve completely). In the 1850s as the copal industry developed, it became the main base for producing varnish because it was cheaper and because copal dissolved more easily than amber. When synthetic varnishes were introduced, the copal industry declined.

Table 8.1 Common Resins Used Commercially with Names for the Resins from Various Trees

Name		Plant Family
I. OLEO-RESINS (relatively fluid—high proportion of volatile terpenes. Rosin (colophony) nonvolatile fraction after distillation of resin from Pinus		Many species of Pinaceae, Dipterocarpaceae
		Pinus spp.
Labdanum	Rock rose tree	*Cistus labdaniferus* (Cistaceae)
II. BALSAMS (relatively soft; malleable initially)		
A. Balsam	Balm of Gilead	*Commiphora* sp. (Burseraceae)
Canadian balsam		*Abies balsamea* (pinaceae)
Tolu and Peru balsam		*Myroxylon balsamum* (Leguminosae; Papilionoideae)
Malaysian balsam		*Canarium* spp.; *Dacyrodes* spp. (Burseraceae)
B. Elemi (highly scented, semisolid initially)		*Protium* spp., *Canarium* spp., *Dacryodes* spp. (Burseraceae)
		Amyris spp. (Rutaceae)
		Calophyllum, Symponia (Guttiferae)
C. IIncense (highly scented)		
Frankincense	Frankincense tree	*Bosewelia* spp. (Burseraceae)
Myrrh	Myrrh tree	*Commiphora* spp. (Burseraceae)
Mexican incense		*Bursera* spp. (Burseraceae)
D. Storax (comprised of aromatic phenolics)	Storax tree	*Liquidambar* spp. (Hammamelidaceae)
		Styrax officinalis (Styracaceae)
III. DAMMARS (Malay word for all resins)		Many genera Dipterocarpaceae; sometimes Burseraceae, et al.
		Agathis dammara
IV. SANDARACS	Sweet wood	*Callitris, Tetraclinus* (Cupressaceae)
V. MASTIC	Mastic tree	*Pistacia* spp. (Anacardiaceae)
VI. COPALS (extremely hard with high melting point)		
Brazilian, Colombian Copal		*Hymenaea* spp. (Leguminosae; Caesalpinioideae)
African copal		*Copaifera* spp., *Daniellia* spp., **H. verrucosa**
Manila copal		*Agathis alba* (Anacardiaceae)
Kauri copal		*Agathis australis* (Araucariaceae)

Source: Adapted from J. H. Langenheim (1995) Biology of Amber-Producing Trees: Focus on Case Studies of Hymenaea and Agathis. In Anderson & Crelling eds. p. 10. and from Poinar & Milki (2001) p. 81).

Figure 8.13. Genuine Baltic amber beads purchased in Morocco. The dark coloring results from staining by the oils used while wearing. The beads are strung with Barbary silver and carnelian dangle beads. Size: 18 in. or 45 cm long, with largest amber beads 1 in. or 2.5 cm. Patty Rice Collection.

AFRICAN AMBER — IS IT AMBER, IS IT COPAL, OR IS IT SYNTHETIC RESIN?

During the past 20 years great interest has developed in ethnic jewelry and decorations along with cultural interests. This has spilled over into the gemstones often associated with various cultures. Amber has had a long association with many ethnic groups such as African, Middle Eastern, and Asian, thus amber jewelry and ornamentation from these areas are popular. However, these regions also had access to copal resins and developed methods of using these semifossil resins for ornamentation. Related to ethnographic interest these ornaments have value, but the problem is encountered when they are willfully represented as true amber, and therefore sold at a higher price. The recent resins that are confused with amber are semifossilized and are generally known as copal, dammar, and kauri.

Since the 1970s a lively interest in old African art and ornament resulted in the appearance of much "African amber" and many African glass trading beads. "African amber" or "copal amber" are popular names attached to oblate or barrel-shaped beads supposedly made of copal resin found along the east coast of Africa and elsewhere on the continent, and used to some extent by native groups for jewelry. A traveler can buy postcards of Moroccan ladies with large African amber beads around their necks. Some such beads were made of true amber imported from the Baltic area during the period of slave-trading on the coast of East Africa. Erikson (1969, 58) states: "Beads, of great size, brown, yellow, blue and colored, are reported as trade items for African gold, ivory and slaves during the 16th century." African chiefs during this period adorned themselves, their many wives, and their children in great quantities of beads. The slaves in Zanzibar were bedecked with beads to enhance their appearance and to attract buyers. This trade continued until 1899, when the last recorded slave ship was wrecked off Wasin Island, not far from Mombassa, East Africa.

In this connection, the *Dispatch*, a ship leaving Bristol, England, on September 30, 1725, records a variety of unusual beads in its hold for the purpose of trading, such as "1378 lb. bugles [Venetian beads]; a sort of bugle called Pazant; Maccaton, that is, beads of two sorts; Christal pipe beads; and yellow amber." Today, we cannot be sure of the meaning of all the names for the beads listed, but we do know amber was included in this early trade with Africa (Erikson 1969, 60).

Most amber sent to Africa was opaque, yellow, oblate-shaped

large pieces considered by Europeans to be of an inferior type. Amber beads were strung along with colorful glass beads by the natives. Sometimes only one large amber bead was placed in the center of the necklace. Occasionally, a strand of African amber currently on the market will contain a few of these true amber beads (see Fig. 8.13).

However, since about 1971, long heavy strands of amberlike beads have appeared on the U.S. market that were imported from Africa. These beads are generally cloudy or opaque, ranging from yellow to brown in color. Some have swirls of transparent material within the otherwise cloudy mass. Shortly after the first influx of large beads, a smaller, barrel-shaped bead was introduced, very even in size and color, as compared to earlier strands, which were quite diverse in respect to the color and quality of the beads on a single strand (Allen 1976c, 26; see Fig. 8.14).

Because of the revived interest in African heritage in local bead shops, the strands are romantically represented as "African amber," or copal, from the depths of the jungle, used by African natives as ornaments in the past. Fascinating tales are told of primitive methods of digging for amber and of savage barter as members of tribes attempt to obtain their "African amber." Yet, when tested, such beads generally prove to be neither amber nor copal, but a resole resin. or phenol-formaldehyde resin (Kosmowska-Ceranowicz 2003) One can only assume that, because of increased demand for "African amber," synthetics are being used to produce "pseudocopal," or an imitation of a substitute (see Fig. 8.15).

The hot point test, which is accomplished by touching a hot needle or wire to an inconspicuous spot near the perforation of a bead, aids in identifying synthetic imitations. An unpleasant acrid odor will be emitted from synthetic materials, rather than the resinous scent from either copal or amber. If a fine thread of the material adheres to the hot point as it is pulled away, the substance is most likely a thermoplastic. Such beads are often lighter than amber and will float in a solution of salt water.

Around 1974–1975, African amber beads in a pale yellow shade, as well as a turbid red, began appearing on the market and were fashioned in a variety of unusual shapes. The origins of these beads were given as several African countries, including Mali, Morocco, Nigeria, and Ethiopia, which are not sources of genuine copal. When tested, these beads usually were found to be either thermoplastics or thermosetting plastics.

The dark reddish African amber beads, referred to as "heat-reddened copal," again, more often than not, proved to be imitations, mostly made from thermosetting plastic. Some cloudy yellow ther-

Figure 8.14. Moroccan lady wearing large amberlike beads made of synthetic material. Photo by Bill Daugherty.

mosetting plastics turn red when heated in an oven at a temperature of 350°F for 10 to 20 minutes, the color penetrating completely throughout the bead. I tried this and it was easy to change the color of the beads to a darker shade. Close attention should be paid to remove the beads when the desired color is achieved. If similar treatment were given to true copal, the copal would soften and begin to decompose, giving off a resinous sweet-smelling vapor.

Erikson (1969, 122) stresses the possibility of finding plastic reproductions of older bead forms: "In the markets of faraway Pakistan, plastic beads are combined with brass or gilt reproductions of older bead forms and long strands of plastic amber beads may be seen hanging in the Arabian market stalls in Acre, Israel."

Originating in the Middle East and Afghanistan, in about 1979, small barrel-shaped beads called "prayer beads," were imported and distributed widely. Most were orange to orange-brown in color, and semitranslucent, with a grainy appearance under magnification. They were usually uniform in appearance and shape. Occasionally, flat ends of the beads were stained a darker color on the surface. Small orange beads of this material were marketed as either Egyptian or Afghanistan amber rather than copal. Once again, the majority proved to be either thermoplastic or thermosetting plastic imitations—not true amber. Sometimes the beads were described as "amber from Russia, cut into beads in Afghanistan" (Allen 1976c, 26).

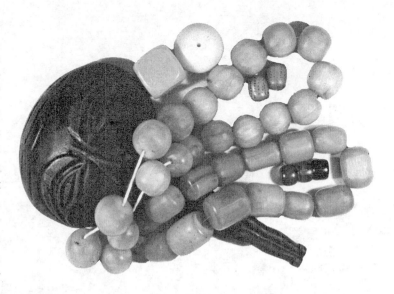

Such barrel-shaped beads are commonly seen in conjunction with old Middle East silver beads. Prayer beads are often strung with an unusual central bead designed to be used in forming a tassel that hangs from the center of the strand. Authentic Moslem prayer beads of the past were required to be made of amber and contained 99 beads.

Figure 8.15. Synthetic amberlike beads sold in the marketplace as "African amber."

Allen (1976c, 22), historian of ornamentation and authority on African amber, relates an interesting experience associated with such beads:

> In East Africa, in market places and elsewhere, were many strands of amber-like beads. They had a very even, manufactured look to them, were barrel shaped, yellow or brown in color, and were called "Somali amber." After making a few inquiries [a friend] was informed by the Nairobi Museum in Kenya that the natives cut these beads from long, broom-handle-sized rods. These rods have been imported from Europe and are made of synthetic resin, or plastic. They are represented to the natives as amber, and not knowing what amber truly is, or what the composition of the rods is, the natives repeat the story. My friend brought back a strand of Somali amber, as well as a strand of "real amber" which is probably from Zanzibar, and may be authentic copal. The plastic "Somali amber" strongly resembles most African and Middle Eastern "copal" in appearance.

Allen (1976c, 22) advises that when examining copal-type beads, a fairly uniform grain suggests the beads were formed or pressed, because "a uniform grain running parallel to the axis of many similar, large-sized beads indicates they were originally one long piece."

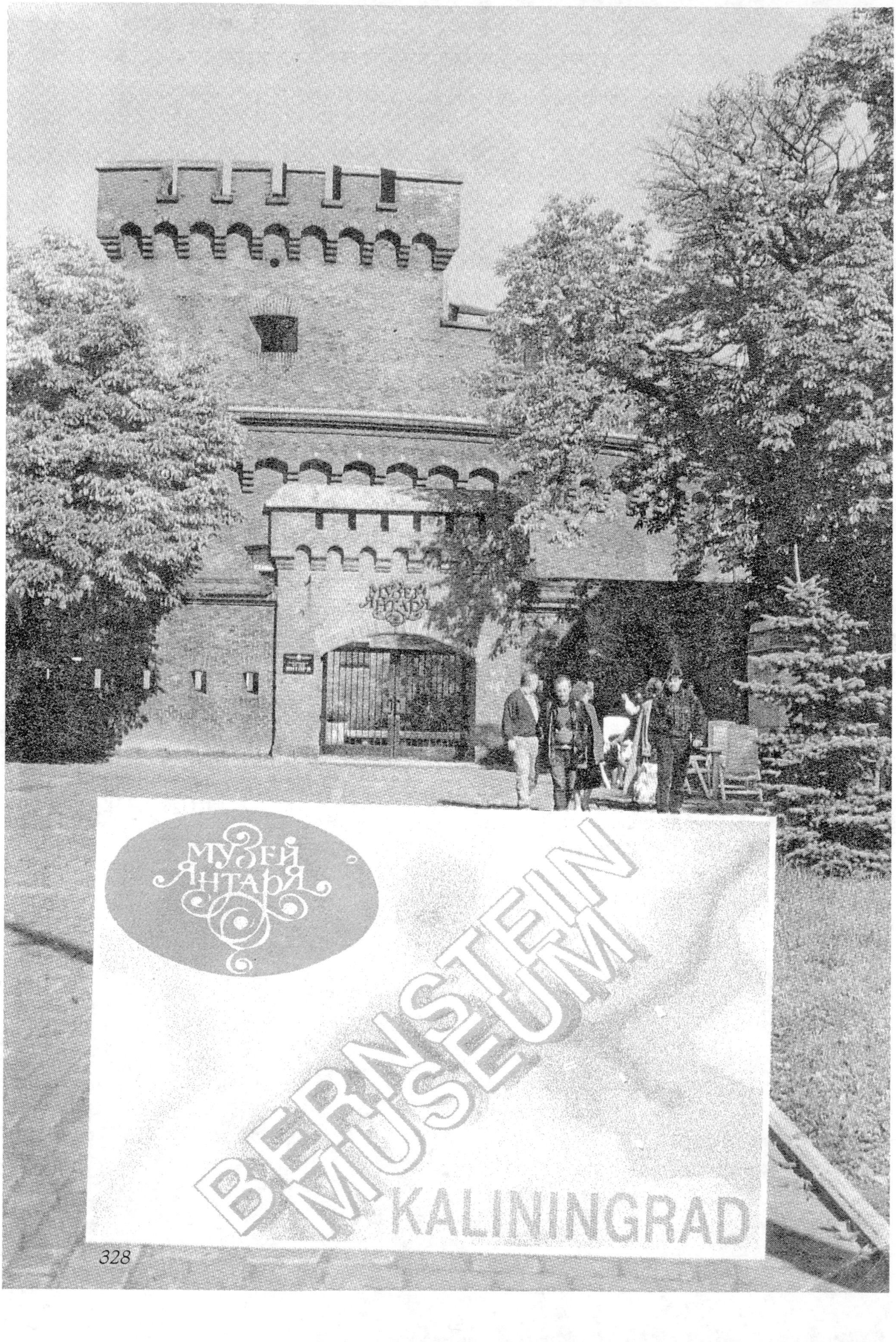
МУЗЕЙ ЯНТАРЯ
МУЗЕЙ ЯНТАРЯ
BERNSTEIN MUSEUM
KALININGRAD

PART IV

COMMERCIAL ASPECTS
OF AMBER

Vase made with large chips of amber embedded in brown synthetic resin. The vase has silver handles. Purchased in Poland.

IMITATIONS OF AMBER AND THEIR DISCRIMINATION

Without question, there are far more substitutes and make-believe gemstones in circulation throughout the world than there are natural, properly identified stones. This is an ancient state of affairs persisting since the beginning of man's experience with gems. Pliny's "Natural History" summed up the situation beautifully for all time. "Truthfully," he wrote, "there is no fraud or deceit in the world which produces larger gain and profit than that of counterfeiting gems." Very likely there are more profitable endeavors and very likely the statement is a little strong and uncharitable toward those engaged in this trade. Most frequently gem counterfeiting is not intended as fraud. When fraud occurs it is not the fault of the materials or their producers necessarily, but rather the motives of the buyer or seller. Gem counterfeiting caters to a very large market composed of those thousands of people who feel they cannot afford the price of an attractive natural gem. Also, there are many who appreciate the decorative value of gems, but prefer to make there investments elsewhere.
—Paul Desautels, *The Gem Kingdom*

THE NATURE OF PLASTICS

Since ancient times attempts have been made to imitate amber. Ancient Chinese literature describes an imitation amber made by boiling chicken eggs with "the fat of the dark fish" and suggests using amber's electrical properties to distinguish this imitation from the genuine material. As time passed, imitations of amber were continually improved, and today synthetic products have many characteristics similar to those of genuine amber, including external appearance and even some electrical properties. Before 1920 products imitating amber generally consisted of pitch, bitumen, rosin, copal, shellac, even casein, celluloid, and early phenolic resins, and these were referred to as compositions. As the processes for making synthetic materials developed, products that when hot and pliable could be

Figure 9.1. Old Victorian jewelry purchased at estate sales and antique shops as genuine amber. All the items are made of synthetic material from the early 1900s and are the most common imitations of amber. The old faceted red and red-brown beads are Bakelite disguised as Sicilian or Rumanian amber. The bangles are typical of the 1920s and 1930s. The box is also a Catalin or Bakelite. Courtesy A. Calomeni Collection and Patty Rice Collection.

shaped into molds or articles became called plastics (see Fig. 9.1).

Amber and other natural resins are composed of carbon, hydrogen, oxygen, and a few other elements, the same as those used in the manufacture of plastics or synthetic organic compounds of high molecular weight called polymers. The word *polymer* is of Greek origin, meaning "many parts." Polymers are built up from monomer units, or one-part molecules, in the process known as polymerization, where such molecules are combined to form long chains. Though plastics, or polymers, are produced in hundreds of varieties, there are basically two groups: the thermoplastics and thermosetting plastics. In polymers the long chains of molecules make up giant molecules that may be linear, branched, or cross-linked, depending on the plastic. Linear and branched molecules are thermoplastic or the type that softens when heated; whereas the cross-linked molecules are thermosetting or the type that hardens when heated.

Thermoplastics require heat to soften and mold them, and they can generally be reheated and remolded if necessary. No chemical changes occur during the essentially physical process of softening and solidifying. In contrast, thermosetting plastics harden or *cure* when heat is applied, and they generally cannot be remolded by reapplication of heat, since such irreversible hardening is brought about by a chemical reaction and is not merely a physical process. Both types have been used as imitations for amber.

CELLULOIDS

The first synthetic plastic widely used as an amber imitation was celluloid, or cellulose nitrate, invented in 1867 by John Westly Hyatt, and originally made to imitate ivory. The high cost of natural ivory after the American Civil War led manufacturers to search for a substitute. A $10,000 prize was offered to anyone who could invent an appropriate imitation. Hyatt, an inventor from Albany, New York, began experimenting with cotton and acids, and by 1868 he had produced a good imitation especially suitable for billiard balls. He called his invention celluloid, and this was the first plastic made in the United States and was granted a patent in 1870. The patent was disputed by the British inventor of xylonite, a similar product.

Celluloid is thermoplastic and thus is repeatedly fusible, softening when heated and hardening when cooled, but decomposing at high temperatures. Cellulose nitrate is made by treating cellulose, usually cotton, with nitric acid and sulfuric acid. The resulting pyroxylin is mixed with pigments and fillers in a solution of camphor in alcohol to make the product less brittle and easier to mold. When heated, the material is pliable or plastic and can be molded into a variety of shapes. Upon drying and cooling, celluloid hardens.

Today celluloid is one of three major categories of early plastic jewelry collectibles. Celluloid quickly became an important material for making many articles, including handles, combs, buttons, buckles, false teeth, and photographic film. It was most often used for collar clips and cuff links, vanity sets, and toilet articles, but some jewelry items were made. Some advantages of celluloid were that it was inexpensive and durable, it took a high polish, it did not warp or discolor, and it was not affected by moisture. Imitation amber beads made of celluloid were called "amber antique" (*Encyclopaedia Britannica* 1975, s.v. "plastics").

Amber antique, or celluloid imitations of amber, can be detected by examining the swirl lines commonly appearing as sharp delineations between clear and cloudy areas. Mold lines sometimes may be detected. Parings can be shaved from celluloid with a knife, but genuine amber splinters or crumbles when cut (*Encyclopaedia Britannica* 1975, s.v. "plastics"; see Fig. 9.2).

When touched with a hot point, celluloid adheres to the point and may flare into a bright flame with emission of a camphor odor. Because genuine amber has a higher melting point, it does not adhere to the hot point and burns slowly, giving off a piney odor. Celluloid is

Figure 9.2. Celluloid imitations of amber jewelry are difficult to find because they were flammable and were only made for a short period of time. This yellow celluloid necklace looks like tumbled amber, is lightweight but heavier than amber, is readily sectile, and gives off the odor of camphor when touched with a hot point.

readily soluble in sulfuric ether, though amber can lie in this liquid for 15 minutes without serious damage (Webster 1976, 182; see Fig. 9.3).

Safe Celluloid

Because of the dangerous inflammability of celluloid, it could not be used in smoker's articles. Manufacturers attempted to eliminate this major disadvantage and finally devised a *safe* celluloid, made by treating cellulose with a mixture of glacial acetic acid and sulfuric acid. This produced cellulose acetate, which is still extensively produced and used under a variety of trade names. According to Fraquet (1987, 74), Cross and Bevan invented the process for making safe celluloid as early as 1894. But the future rapid development of plastic material was the result of the work in 1920 of Hermann Staudinger (1881-1965), a German chemist who theorized that plastics were giant molecules. His experiment led to a major breakthrough in the chemistry of plastic. Thus, in the 1920s and 1930s, many new products were introduced such as the cellulose acetate that could be used in molding resins (Infopedia 1995, s.v. "plastics"). As an example of cellulose acetate being used as an amber substitute, in 1920, a U.S. patent (1 319 229) was issued to W. G. Lindsay, for an amber substitute. He described it as an imitation made from acetyl cellulose (cellulose acetate). However, improvements were made and in about 1930 Eichengrun produced a celluloid acetate molding material that could be used for injection-molded cellulose acetate, which began to be used extensively. Cellulose acetate was used for toys and eyeglass frames.

Figure 9.3. Celluloid imitation of amber in the form of yellow oval beads from the early 1900s. Courtesy A. Calomeni Collection.

Though celluloid is not used today as an amber imitation, older amber beads dating from the late 1800s should be tested for specific gravity to determine whether they are celluloid or genuine amber. A saturated solution of salt water with a density of about 1.13 floats amber, but the heavier celluloid with a density of about 1.38, sinks in this solution (Anderson 1973, 210-1). The refractive index of celluloid is 1.50, compared to amber at 1.54, but the substitute may be just as easily identified by placing a drop of 5 percent solution of diphenylamine in sulfuric acid on the surface of the specimen; if the material is celluloid, a blue color is produced; if it is amber, no color change takes place (Webster 1975, 514). Under ultraviolet light, celluloid fluoresces yellowish-white.

CASEINS

Casein is another one of the three major categories of collectible plastic jewelry in today's market. The material is also called galalith or erinoid. Caseins were promising because they were wonderfully decorative with an outstanding range of colors and luminous pastels. However, casein was not strong or moisture resistant and would warp occasionally. Caseins were very popular in Europe (Davidov and Dawes 1988, 17).

Casein, a hardened milk protein, was occasionally used as an amber imitation. It was first made in 1890 by German chemists Adolph Spitteler and W. Krische by condensing milk protein with formaldehyde to produce an insoluble, tough material. Galalith, or milkstone, was patented by Adolph Spitteler of Hamburg, Germany, in 1897. It was a plastic of glossy beauty, harder than celluloid, and it could also be molded by heating (Anderson 1973, 212). By 1914 England, France, and Germany were still making the plastic under the names of galalith, erinoid, karolith, and aladdinite. Because of its nonflammability, casein promised to be a substitute for the dangerously flammable cellulose nitrate. It could also be produced in lighter shades of color than phenolic resins. In 1919 it was introduced into the United States, but was found to be unsuitable for most uses other than making buttons. Cheap, rather crude beads imitating amber were made by adding fillers to the plastic (Miller 1988, 78-9).

Caseins are denser than amber, with a specific gravity of 1.33, and the refractive index is 1.55. Under ultraviolet light, the material fluoresces white. A drop of nitric acid on the surface results in a bright spot, but because the body color of the imitation is often yellow, the change in color may be difficult to see. Caseins are sectile but tough (Webster 1976, 141). A touch of a hot point to casein produces a typical burned milk aroma. No frictional electricity is developed on caseins (Webster 1975, 514).

PHENOL FORMALDEHYDE RESINS OR BAKELITE AND CATALIN

According to Mumford (1942):

It is all "over, over there," but wherever shells whirr, wherever women preen themselves in the glitter of electric lights, wherever a ship plows the sea or an airplane floats in the blue— wherever people are living, in the Twentieth Century sense of the word, there Bakelite will be found rendering its enduring service.

Bakelite, the first plastic superior to the previously mentioned products, completely revolutionized the entire industry of plastic jewelry making. It was tougher, more adaptable, harder, warm to the touch, tasteless, odorless, and stainless. The colors would not fade. It could be polished to a gem quality luster. Chunks and rods of the material could be carved, lathed, and polished. It soon became the most widely used synthetic resin for imitating amber. Unfortunately this success causes problems when distinguishing genuine amber from the collectable Bakelite, and the buyer needs to know what the material is that is being purchased (see Fig. 9.4).

The general process for producing thermosetting plastics is by reacting phenol or its derivatives with formaldehyde in a condensation reaction, through three stages, each producing a product with different physical and chemical properties. The first stage, polycondensation, is carried out in water and leads to water-soluble or water-dispersible intermediates. The second stage obtains a resinous, insoluble, and difficult-to-fuse mass that can be converted into fine, dry powder; this is sold to manufacturers of synthetic amber and molded and shaped into desired forms. Application of heat and pressure in the third stage converts the product to a resin of maximum hardness and permanent infusibility. If this material that now has cross-linked molecules is burned, it does not soften or melt, nor does it burn by itself, but simply carbonizes into a black cindery material (*Encyclopedia Americana*, s.v. "plastics"). Thermosetting plastics are generally dense—much more so than genuine amber -- with their specific gravity ranging from about 1.11 to as much as 1.55 when fillers or pigments are added.

Bakelite was formed in two general categories—cast and molded. Though several varieties of phenoformaldehyde resins were made, the cast phenolic resin with a high formaldehyde content was most often used for imitation amber. It was made in transparent colors, mostly yellows and reds, for use in beads and bracelets. Opaque shades were made in bright colors such as yellows, reds, oranges, and greens, and were used for drawer knobs, napkin rings, rings, bangles, beads, and even pan handles. Amber mouthpieces for pipes and smoking industry uses were replaced by this new material, since it did not melt or burn but became harder with heat. Soon customers began reporting irritation of the mucous membranes of the mouth by the new mouthpieces. This was attributed to the freeing of formaldehyde from the composition of the material. In 1921 in Austria workers in factories manufacturing Bakelite reported dermatitis from exposure to the fumes of phenol, formaldehyde, and ammonia in the plants making the new Bakelite material. Proper therapeutic measures and protection for workers was described.

Figure 9.4. Bakelite could be beautifully faceted into long Victorian, red, transparent beads that are extremely common, dating from the 1920s to early 1930s. Beads retain their sharp facets and show little wear at the edges of the perforations. No crazing is found on the transparent beads. Courtesy Pansy Tice Huff collection.

Bakelite

In 1906 a Belgian-American chemist, Leo Handrik Baekeland, while searching for a substitute for shellac, invented the first and most versatile of all thermosetting plastics, a phenoformaldehyde condensation resin that was given the name Bakelite. Baekeland produced a dark syrup that hardened when heated, could be molded into any shape, was not dissolved by common solvents, was nonflammable, and was inexpensive. It seems the first formaldehyde polymers were described by Butlerow in 1859, but it wasn't until 1907 that variations of actual phenoformaldehyde resinous products started being patented. In 1909 Baekeland received a medal from the American Chemical Society for being the founder of modern plastics, and between 1920 and1930 a large number of new products were introduced. However, it was about 1926 that Bakelite was applied to its greatest extent to making hard gemlike resin for imitation amber both in the United States and Europe *(Encyclopedia Americana*, s.v. "plastics").

It must be remembered that synthetic resins have been in use for over a hundred years, and it is important, when examining antique beads, to test for their presence. In 1920 when amber reached a new high in popularity, long imitation amber necklaces were produced in a variety of faceted bead shapes and colors, the most popular being a red imitation of highly prized Sicilian amber. Deep black and dark ruby-red beads were also produced to resemble rumanite amber. The majority of deep red amber beads found in old estates dating from the early 1900s are made of Bakelite, and their external appearance may closely resemble the genuine article, making it very easy to mistake an imitation for true amber.

Another useful clue to imitations depends on the fact that amber is softer than most plastics. Therefore, amber, with constant rubbing of the stringing material, in time wears away around the hole in the bead. Sometimes the edges chip as the beads rub together. Since Bakelite is tougher than amber, even Bakelite beads that have been strung for 60 or 70 years or more still have clean smooth holes. Furthermore, old amber beads commonly develop shallow crazing over the surface, resulting from temperature changes over the years, whereas plastic beads still retain a smooth polish.

Other physical properties of phenolic resins can be used to identify them. Although the different types of plastics also differ in their specific gravities, the denser types most commonly used for imitating amber average 1.26 in specific gravity, much higher than amber at 1.05 to 1.09. On the other hand, the newer thermoplastics are generally lighter than amber, and the simple flotation test in saturated salt solution will not separate them from amber. As an added distinction, the refractive index of 1.60 to 1.66 for Bakelite is well above amber at 1.54, but testing carved pieces or mounted specimens without flat or easily accessible surfaces creates difficulties in determining either refractive index or specific gravity. If these tests are impossible, the hot point test may be used on some inconspicuous spot to detect the acrid odor and resistance to burning of Bakelite (see Fig. 9.5).

Bakelite jewelry of the 1930s is collectable in its own right today. Many collectors pay high prices for authentic Bakelite or other cast phenolic jewelry. Originally the material was planned as a substitute for ivory, coral, and amber, but it began to take on a style of its own. However, today, because it was made in yellows, browns, and reds in both opaque and transparent forms, it is confused with amber and sold for genuine amber. The transparent yellow forms were often carved on the reverse side similar to the old masterpieces described as *verre églomisé* from the late Middle Ages.

Catalin

In 1928 the Catalin Corporation in the United States also began producing a phenolic resin of high formaldehyde ratio that enabled a resin syrup to be cast in molds and cured in ovens. It was this process that first produced the transparent material that was extremely attractive in a variety of new colors. And, most of the pre-Second World War imitation amber was produced by using the dark transparent material. Catalin specialized in clear cherry red (sometimes called "antique amber"), opaque rust red (sometimes called "Russian amber"), opaque golden or yellow, and even opaque green, which can still be found today. It seems the height of amber color was between 1930 and 1935, at which time numerous items beside jewelry were made of this new resin (see Fig. 9.6).

Articles, such as pipe stems, cigar and cigarette holders, and beads, were machined from large blocks of the resin, giving the ornaments a hand-polished look. Raw materials from the Bakelite and Catalin factories came in tubes, rods, and sheets of varying lengths and sizes. These were cut with saws, ground, sanded, drilled, and worked like amber. The material could be carved into intricate shapes and polished in big rolling tumblers. Rods were used by the makers of buttons. Tubular stock was machined into bracelets and then carved into artful designs by machinists using drills and carving tools. Even the carved items were given a final polish applied in a large tumbler filled with wooden pegs and pumice. After the polish, the metal trim or other embellishments were drilled or cemented into place.

Figure 9.5. Estate beads purchased as true amber, but tested to be heavy thermosetting plastic. When touched with a hot point an acrid odor was emitted, being a thermosetting plastic they do not melt. Courtesy A. Calomeni Collection.

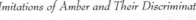

POLYSTYRENES

Polystyrene, a thermoplastic or a linear polymer, has gained in popularity for costume jewelry because of its suitability for injection molding. It was first produced commercially in about 1937 and is characterized by high resistance to chemical alteration at low temperatures and by very low absorption of water. Thus, it was used as a good electrical insulation and as a component of radar during the Second World War. Plastics based on styrene were made available shortly before the war from Dow Chemical (United States), IG Farben (Germany), and, after the war, from British Resin Products and Mobil Chemicals. Styrene plastics were produced with the trade names of Distrine and Erinoid Polystyrene after the war in England when plastic plants stopped manufacturing war products.

Polystyrene is less dense than amber with a specific gravity of 1.05 and has a refractive index of 1.59-1.67. Its hardness is 2.25 to 2.5 on Mohs scale. The material softens at a temperature of 70-90°C. The material dissolves in organic hydrocarbon liquids such as bromoform, methylene iodide, and benzene. Its low specific gravity poses some problems in identification when using the usual flotation test. Allen (1976c, 22) and others suggest the following procedure for separating imitation substances from genuine amber when using the flotation in salt water test: Fill three 10-ounce glasses with water. One tablespoon of salt is added to the first glass, two tablespoons to the second, and three tablespoons to the third. The lighter thermoplastics float in all three salt solutions, and some may even float in water without salt. Amber generally floats in only the second or third solution, depending on the specimen, whereas Bakelite and the heavier synthetic resins sink in even the most saturated salt solution.

Modern plastics such as polystyrene have a specific gravity lower than natural fossil resins, thus the flotation test used for discriminating Bakelite from amber is not as useful. However, it can be used to determine if the object is the collectable plastic Bakelite or a modern plastic. In addition to flotation, another reliable test for identification of polystyrene is its ready solubility in ketones, benzene, toluene, and other aromatic hydrocarbon liquids (Fraquet 1987, 73-83; see Fig.9.7).

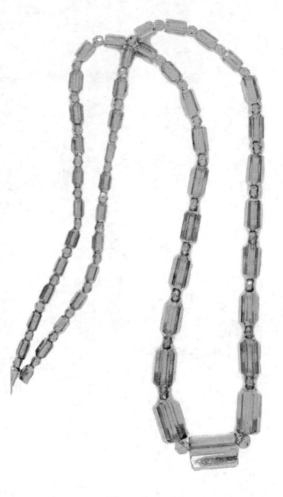

Figure 9.6. Transparent red Victorian style beads were often called "Cherry amber" and were produced in a variety of shapes strung into long necklaces. The beads will sink in a solution of salt water. Courtesy A. Calomeni Collection.

Figure 9.7. Thermoplastic imitation of antique amber beads. When touched with a hot point, the bead melts easily and a fine thread of plastic adheres to the point.

ACRYLIC-TYPE RESINS

The most important plastics in this category are methyl methacrylate polymers that are produced under the names of Perspex, Diakon, Plexiglas, and so forth. The methyl methacrylate polymers are hard and can be used in jewelry. They burn with a fruity or floral odor. The refractive index is 1.50 with a specific gravity of 1.18-1.19 (see Fig. 9.8).

Slocum Imitation Amber

In the 1970s, a U.S. firm, J. L. Slocum Laboratories of Royal Oak, Michigan, introduced an imitation amber material that can be purchased in block form by lapidaries to be polished and shaped into ornaments. It is a transparent synthetic resin produced in both pale orange and red shades and is available with either insect inclusions or with discoidal sun-spangle fissures. The inclusions are, of course, of recent origin and are much more numerous than would be found in either Baltic or Dominican natural resins. The circular fissures can be distinguished in most specimens from those in natural amber because of their frosted appearance (see Fig. 9.9).

The Gemological Institute of America (GIA) reports the refractive index of the Slocum imitation to be 1.50 to 1.55 or very similar to that of natural amber. It has a specific gravity of 1.17, heavier than true amber, and, with a hardness of 3, it is harder than amber. When touched with a hot point, it emits the odor of burned fruit, in contrast to the piney odor of amber, indicating it may be made from an acrylic, styrene, or polyester resin.

Figure 9.8. Plastic imitation amber beads in an orange-brown barrel shape sold as Afghanistan amber. Emits a fruity odor when touched with a hot point.

THERMOSETTING ETHENOID PLASTICS

Polyesters were first made in 1847 by Berzelins (Fraquet 1987, 81). They were used commercially in the United States during the Second World War and were originally used to reinforce glass. They are produced by complex ester compounds that polymerize at room temperature and are thermosetting. Because the resin can be hot- or cold-cured, it is used heavily in the model and craft market. Polyester materials are frequently used to embed specimens in natural history museums or classrooms. Museums today are using this as a method to preserve rare amber specimens with genuine inclusions, since amber does deteriorate. The resin can be in an epoxy form and used for molding or injecting in the liquid form then set aside to harden.

Figure 9.9. Imitation amber produced by Slocum Laboratories contains insect inclusions and discoidal petal-like inclusions. It is made in red and orange hues. Courtesy Miner's Den, Royal Oak, Michigan.

Bernit (Bernat)

Gebhardt Wilhelm, a jewelry manufacturer in Germany, produces imitation amber jewelry called *Bernit* or *Bernat*. This material closely resembles amber because it contains stress-spangle inclusions like those in genuine amber and proper coloration. It is also similar in refractive index to amber, but is heavier, with a density of about 1.23. The discoidal stress spangles mentioned are circular, but lack the radiating lines within each fissure typical of those found in genuine amber. Under magnification, the circular fissures appear too clear, as they are much more transparent than those found in genuine amber. In addition, some of the fissures in the imitation are unnaturally curved or bent, unlike those in genuine amber, which are almost invariably flat. When a bernit imitation is compared to a lump of genuine amber possessing the same sort of spangling, the difference between true amber and this impostor can be detected. In spite of these differences, bernit imitations are extremely realistic and must be examined carefully to identify them.

Bernit is also available in pieces containing plant fragments and insects. Although these pieces closely simulate natural amber with inclusions, the imitation often can be detected because the plastic surrounding the insect is too clear and gives no signs of the insect having been embedded alive. In true amber, small air bubbles emitted from the insect's respiratory system are found close by; in some cases, the insect's struggles may have reformed swirls in the resin. Large insects are particularly suspect and should be examined with care.

Figure 9.10. Necklace of polybern beads with amber chips embedded in synthetic resin purchased in Poland. Courtesy H. Evens Collection.

Figure 9.11. Close-up of polybern bead showing the appearance of amber chips within the synthetic resin.

Polybern

Polybern is an amber imitation made by combining polyester resin and small pieces of real amber that was developed in the 1960s in Germany. It was produced in Poland and Lithuania for making sculptures, souvenirs, and inexpensive jewelry. It is easily confused with genuine amber because it does contain embedded natural amber chips. The name *polybern* was derived by combining the words *polyester* and *Bernstein* (the German word for amber), the two substances of which the material is composed. Polybern is manufactured into oval beads for necklaces, long oval cabochon shapes for bracelets, and a variety of rectangular shapes for brooches (see Fig. 9.10).

A similar imitation, but containing less genuine amber within the synthetic matrix, began flooding the market in the 1970s. This material, manufactured in Poland, produced bulky, cubic beads formed in graduated molds. The molds were filled half-full with a synthetic resin, followed by a layer of small amber chips, and finally filled with additional synthetic resin. The resulting cubes were perforated and strung in graduated sizes, with the center bead often as large as 1 inch by 1 inch by 0.5 inch (25 mm x 25 mm x 12.5 mm). Careful examination of the large beads readily reveals the layered structure (see Fig. 9.11).

Another imitation amber product manufactured in Germany and introduced into the market in the 1970s combined polyester resin with amber dust. It was attractive, closely resembling amber, and was practically impossible to distinguish from the natural material without careful examination and testing. When touched with a hot point, the polyester-amber dust product even emits the piney odor of true amber because of the abundance of amber dust, and, depending on the matrix resin used, it may float in a saturated solution of salt water (see Fig. 9.12).

German laboratories became so skilled in producing amber substitutes that physically and optically closely resemble genuine amber that only by using sophisticated tests and instrumentation beyond the resources of the ordinary gemological laboratory could the imitations be identified as such. Therefore, in 1968 courts in Germany established consumer protection laws requiring that customers be given facts regarding the type of stone purchased and tests to establish its identity. Furthermore, such articles must be marked as natural or synthetic. When purchasing amber articles from German manufacturers, the buyer should look for such labels (see Fig. 9.13).

Sun-spangle Imitations

Many manufacturers of imitation amber are attempting to duplicate the popular sun-spangle amber. How the sun-spangle inclusions in imitation amber are created has not been divulged; one manufacturer simply stated that he "cross-linked the molecules in a thermosetting plastic." In about 1950 natural amber began being treated to be transparent and have the discoidal sun-spangle inclusions. In the natural fossilized resin, these are produced by heating and by a sudden release of pressure in an autoclave, causing the air bubbles within to suddenly have more pressure than the atmosphere outside the amber piece. At this point, a minute internal explosion occurs and the structure of long cross-linked molecules within force explosion fissures into a discoidal pattern (see Fig. 9.14).

Figure 9.12. Vase molded with amber chips set in plastic. Purchased in Poland. Courtesy A. Fletcher Collection.

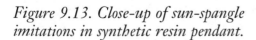

Figure 9.13. Close-up of sun-spangle imitations in synthetic resin pendant.

Figure 9.14. Amber shop in St. Petersburg, Russia, in 2003. Many of the beautiful amber items were made from pressed amber and were very attractive. They had the same high price as pieces made of natural amber because there was more workmanship and factory expenses involved in the production process.

GENERAL RECOMMENDATIONS FOR DETECTING AMBER IMITATIONS

When purchasing a supposedly amber article, where it is not feasible to perform more reliable tests such as refractive index and specific gravity, careful examination of each bead can often detect imitations. Modern synthetics are similar to natural material, making it necessary to compare all articles purchased as amber in flea markets, art fairs, gem shows, and antique shops with a known piece of genuine amber (see Fig. 9.14). Carrying a piece of genuine amber along as a comparison guide when shopping for amber articles is very useful. For fine details of inclusions, bubbles, crazing, and so forth, examine the piece with a 5x or 10x magnifier.

Swirl lines in synthetic materials differ in appearance from swirls within true amber. This is a result of the different ways the substances were formed. Synthetic resins are poured into a mold, causing turbulent swirls. Amber resins slowly oozed from trees to accumulate in layers over a long period of time. Because of this, swirl lines in the imitations do not show the same flow, or "movement," as in true amber. When a piece of genuine amber, which possesses swirls within the mass, is placed beside a synthetic piece containing swirls, the difference is obvious.

Examination of the back of bracelet sections or other flat cabochon shapes is useful in detecting pieces formed in molds. An irregular surface on the back signifies synthetic resins solidified while exposed to the atmosphere. Careful examination may also reveal other mold marks, indicating that the specimen was formed in a mold, rather than cut from a piece of rough (see Table 9.1 for a comparison of the characteristics of amber and some of its common imitators.)

Glass Imitations

Glass imitations are usually easily detected because of the greater hardness of glass and its much greater weight. Glass beads are also cold to the touch, whereas amber feels warm. Differences in the luster of glass and amber are just as easily distinguished. Glass beads are often faceted in a manner similar to antique faceted amber, and beads range in hue from yellow to orange to brown. Yellow glass colored by uranium oxide will show a brilliant yellow-green fluorescence under ultraviolet light. Manganese is also used to produce yellow-colored glass, and this gives a dull green fluorescence quite different from the fluorescence of amber.

Figure 9.15. Antique claw watch fob carved in China from horn, sold in an estate sale as amber. Courtesy A. Calomeni Collection.

Amateur collectors may test glass beads by simply rubbing them together. The hardness can be felt as the beads scratch against the hard surfaces of one another.

Another method used by seasoned collectors to detect an impostor of genuine amber is to gently *bite* the beads. Collectors swear their teeth can feel the difference between the hardness of amber, glass, and even ambroid or plastic. If the beads are amber, the material will be soft and brittle. Since ambroid is only slightly harder than amber, it may be difficult to detect using this unsophisticated method. Some say plastic and glass beads also have a different taste on the tongue. In any case, beginners using this bite test may want to practice on their own beads to get the feel of it.

Horn Imitations

Another organic substance used to imitate amber is, surprisingly, horn. Horn beads are usually grayish-tan in color, but they may be dyed various colors. Small, opaque to translucent, yellowish, barrel-shaped horn beads are imported from Ireland and occasionally passed off as amber. Horn beads are sometimes fashioned into a necklace with amber and synthetic amber. Often these beads are found strung as a rosary.

Horn usually has a higher specific gravity than amber and sinks in a saturated solution of salt water (though an occasional bead may float). The hot point provides a useful test. Horn emits an unpleasant odor, like that of burning hair, when touched with the red hot point.

The carved, claw-shaped watch fob in Figure 9.15 resembles pale

yellow pressed amber and was identified as such in several examinations. Had it been tested with the hot point, the obvious odor of burning hair would have been detected. A gemologist's hot point tester is more reliable and easier to use than heating a needle. The hand-held needle usually does not retain the heat long enough to do a successful test.

Figure 9.16. "Reconstructed Burmese amber" carvings purchased in Hong Kong in 1983 at a cost of $10 to $20. "Reconstructed Burmese amber" may at one time have had Burmese amber mixed with copal and other synthetic additives, but today most are simply a block of red synthetic resin carved or molded into figurines.

Reconstructed Burmese Amber

In about the mid-1970s, large carvings made from a transparent red synthetic resin began appearing on the American market and were claimed to be reconstructed Burmese amber, but obviously they were much too large to be true amber. However, to produce blocks large enough to be carved, pressed amber has been used in China for many years. During the manufacturing process, amber chips are mixed with linseed oil and red coloring added. Nevertheless, the majority of such large carved red art objects are manufactured from a synthetic resin similar to Bakelite plastic, such as styresol: polyster. The only way to be sure any piece of this type is genuine is to test it as explained in Chapter 5. I purchased a pair of beautiful, obviously hand carved, red elephants from an antique dealer who claimed they were reconstructed Burmese amber at a price of $400 in 1979. Then, in 1983, while in China, I found more of these carvings ranging from figurines of Chinese wisemen to water buffalo, but at a price of $10 to $20 dollars each! The Chinese dealer still called them reconstructed Burmese amber, but was more aware of the value of the art objects (see Fig. 9.16.).

Table 9.1   Characteristics of Amber and Common Imitations

Material	Specific Gravity	Refractive Index	Under Knife	Hot Point Odor
Amber	1.08	1.54	Splinters readily	Resinous
Copal	1.06	1.53	Splinters readily; softer than amber	Resinous
Phenol Bakelites (Bakelite, Catalin, etc.)	1.26-1.28	1.64-1.66	Sectile, tough	Acrid
Urea Bakelites (beetle, etc.)	1.48-1.55	1.55-1.62	Sectile, tough	Acrid
Caseins (galalith, lactoid)	1.32-1.43	1.49-1.51	Sectile, rather tough	Burned milk
Acrylate Resins (perspex, diakon)	1.18-1.19	1.49-1.50	Sectile	Burned fruit
Polystyrenes	1.05	1.59-1.67	Sectile	
Slocum Imitation	1.17	1.50-1.55	Sectile	Sweet acrid, burned fruit
Cellulose Nitrate (celluloid)	1.37-1.43	1.49-1.51	Readily sectile	Camphorous
Cellulose Acetate (safety celluloid)	1.29-1.35	1.49-1.51	Readily sectile	
Bernat, Bernit	1.23	1.54	Readily sectile	

*Russian pressed amber
black and clear faceted
beads and bracelet.*

348

PREPARATION AND
WORKING OF AMBER

CLARIFICATION AND HEAT
TREATMENT OF NATURAL AMBER

That cloudy amber could be clarified apparently was known as early as the first century, for Pliny, in his *Natural History*, mentioned that pieces of amber were "dressed by being boiled in the fat of a suckling pig by Archelaus, King of Cappadocia." This was perhaps the first attempt at clarifying amber. Andreas Aurifaber made observations about clarifying amber in 1572, as did Johannes Wigand in 1590. It was the latter who discovered that other oils could be used rather than the "fat of suckling pigs" (Ley 1951, 38), and, by the end of the seventeenth century, methods for ridding amber of internal obscurations had been well established. In 1897 Dahms described the clarification of succinite using rapeseed oil and sandbath methods (Dahms 1901, 201–11).

According to Bauer (1904, 538), the method followed at the turn of the last century was as follows:

> The rough material is completely immersed in rapeseed oil in an iron vessel, and then very slowly heated to about the temperature at which the oil boils and begins to decompose. It is then allowed to cool and this must take place just as slowly and gradually as the preliminary heating otherwise the clarified amber will become cracked and possibly fractured. The smaller the fragments of amber operated upon the quicker is the process completed; the heating of large pieces must be continued for the considerable period, and not infrequently needs to be several times repeated. The time required for the operation depends also upon the character of the material, for different pieces of amber of the same size will not require the same length of time to complete the operation. The clarifying process begins on the surface and spreads gradually inward.

To understand how this process works, it must be remem-

Figure 10.1. Natural Baltic amber with sun spangles—discoidal explosion fissures, resulting from heating, which causes increased pressure in minute internal bubbles.

Figure 10.2a. Antiqued amber, a result of slowly heating the amber to increase the rate of oxidation, and reconstituted amber bracelets. These types of jewelry and ornaments are produced by the Kaliningrad Russian Amber Factory. Courtesy Gifts International, Chicago, Illinois.

bered that minute air bubbles give amber its turbid appearance. When these air spaces are filled with oil of the same (or nearly the same) refractive index as amber, the material appears transparent, since light is then allowed to pass through the material without interference. The refractive index of rapeseed oil, a light yellow oil obtained from pressing rapeseeds, is 1.475, which is close to that of amber, so this is the oil most often used. Coloring can be mixed with the oil as desired. The discoidal fissures, or sun spangles, in amber were believed to be caused by droplets of trapped water that flattened to circular forms. However, with the knowledge gained from studying the chemistry of polymers, it was found that molecular formation (cross-linking into long chains) forced internal fissures to follow the structure when pressure in minute air bubbles within the amber became greater than the pressure on the surface of the amber piece. This is usually accomplished when amber is subjected to heat causing pressure in the air bubbles to rise. Thus, when amber with sun spangles was found in old tombs, the researcher could assume it had been submitted to heating at some time, whether this was by nature or by humans (see Fig. 10.1).

As mentioned previously, sun-spangle features in amber jewelry became very popular with manufacturers and almost all large transparent pieces were heated to force sun spangles to develop. When amber was clarified, great care was taken to make the fissures conspicuous rather than to produce a completely transparent piece. If properly done, the fissures assume an elegant appearance and display characteristic radiating circular forms within the transparent mass. These fissures are considered very attractive by amber connoisseurs and enhance the sparkling beauty of the jewelry piece. The presence of sun spangles with brown edges distinguishes clarified from natural transparent amber. The brown edges can be controlled by the amount of heat used, as higher heat darkens amber causing it to oxidize faster.

To produce a rich brown antique color, amber in the past was imbedded in pure sand in an iron pot, then increasingly heated for 30 to 40 hours (Williamson 1932, 233). The resulting rich brown color closely resembles the oxidized amber hues produced by aging. Such treated amber is often sold under the name "antiqued amber" (see Fig. 10.2).

COMMERCIAL METHODS OF TREATMENT

According to the 1978 catalog of the Soviet firm, Almazjuvelirexport (1978, 1):

> One kind of yellow cloudy amber is not as desirable as other varieties. Therefore, a hardening process has been developed. The color is removed in an autoclave, then the amber is hardened in electric ovens. The amber becomes beautiful in color and markings.

The current catalog for the Russian export company (Almazjuvelirexport 1988, 2) describes the relatively easy way of cutting amber and the beautiful variation of colors and luster that can be achieved through proper polishing. The catalog states:

> The discs of inclusions in transparent type of amber that are so attractive in jewelry may be achieved, as well as certain changes in coloration, through heating amber stones in sand, oil or even water. As they appear, they are basically the invisible structure of a very fragile stone that becomes noticeable through heating process.

These meager descriptions are about the only explanations obtainable from current amber wholesale firms relating to improving color and clarity of natural amber. The jewelry trade, for obvious reasons, does not provide details of the processes used, but they are generally along the lines of the treatments described below.

An autoclave or pressure cooker is used to force oil, such as rapeseed (Cannola) oil, other vegetable oils, or dry distillate amber oil, into air pockets of the

Figure 10.2b. Russian pressed amber bracelet produced in 2002. Purchased from Russian amber dealer at the Tucson Gem show Arizona.

Figure 10.3. Russian amber necklace made from clear pressed amber then treating the beads to create black edges. The Russian pressed amber factory has perfected pressing of amber that was first ground into a fine powder. These pieces of jewelry may be priced higher than tumble amber necklaces because they require more workmanship to produce. Purchased in Russia in 2003. Mary Lion's Collection.

Figure 10.4a. Souvenir of amber mushrooms using natural yellow opaque amber and brown heat-treated natural amber. Typical construction of Kaliningrad Russian Amber Factory—heat-treated dark brown caps, yellow amber polished and left with natural crust. Courtesy M. Kizis Collection.

porous material to produce clarification. Next, the amber is placed in an electric oven and gradually heated to soften it. During this stage, any air bubbles left within the mass of amber expand to form discoidal fractures. At the same time, heating hastens the process of oxidation and causes the amber to deepen in color. Oxidation occurs from the outside in, causing the outer layers to be generally darker than the cores. In some pieces the darkened outer surface may be only 2 millimeters thick with a clear inner core. Some researchers question the use of oil, however, amber craftsmen indicate they have used oil to clarify amber.

Today nitrogen is used instead of oil in the autoclave treatment, because it is better for the environment and atmosphere ozone, according to a Lithuanian manufacturer. Also Stephien Christopher of Vessel International indicates that oxygen would allow amber to oxidize or even burn at the high temperatures needed for processing. Burning does not occur if an inert gas such as argon or nitrogen is used in the autoclave. The amber is polished to remove the crust and coarsely formed into desired shapes before being placed in the autoclave. The rough polishing facilitates the clarification process of amber.

The pressure in the autoclave is raised to 50 to 70 atmospheres while the amber remains in the autoclave from 12 to 36 hours. This prevents oxidizing as the temperature is raised, and when the amber is heated it softens. At the point when the amber is soft and pliable, the pressure is increased, which presses out the bubbles. This allows the light to pass through the amber making it appear clear. Sudden release of the pressure causes the discoidal explosion fissures to appear. When the amber is removed from the autoclave, the color has been removed. The amber then appears the color of weak tea or even slightly greenish. Next, the amber is placed in an oven, or hot autoclave, and heated to about 150°C (238°F). Depending on the size of the piece and the hardness of the amber, the material remains in the hot air autoclave from 2 to 48 hours. The heat accelerates the oxidation process, which can also take place over time in nature. The darker pieces of amber have been heated for a longer period of time. The pressure changes cause the discoidal explosion fissures to appear.

Pale yellow opaque amber can be clarified to a transparent

pale champagne-colored amber. Pale clear transparent clarified amber can have sun spangles induced after nitrogen treatment by releasing the pressure in the autoclave; then heat is introduced for about 20 seconds to *toast* the edges of the sun spangles. Darker hued opaque amber will result in a darker transparent amber, but again higher heat exposure is most important to darken amber. A greenish-hued amber is produced by special heating techniques. The amber is treated with nitrogen in the autoclave, then heated and allowed to cool quickly. It is then polished and reheated. Many of these techniques were experiments made in cottage industries in small shops throughout Poland and other Baltic countries.

Many souvenirs from Poland, Lithuania, and Russia are made of natural amber, with portions darkened by heating to present contrasts for the art object (see Fig. 10.4a,b).

The longer the amber is heated, the darker it becomes. After porous pieces have been subjected to this treatment, they tend to be harder and more durable than they were previously. The process produces a beautiful amber, in great demand by amber connoisseurs, and one that is still considered to be genuine amber. It is most commonly found in the finished pieces imported into the United States from manufacturers in Poland, Russia, Lithuania, and Germany.

These methods of processing amber into jewelry are using natural amber and not adding or subtracting from the natural material. Poland, Germany, and Denmark generally export natural amber articles, using very modern methods for enhancing the beauty of the gem, just as a lapidary polishes rough ore into beautiful gemstones.

Figure 10.4b. Pressed amber is made in the factory in Kaliningrad, Russia. These articles were made by Amber Room Restoration Project Tsarskoye Selo, Russia. Boris Igdalov, Director of the project, exhibited amber at the Tucson Amber and Organic Gem Show in 2004.

MANUFACTURE OF ADULTERATED AMBERS

As early as 1879 Germany issued a patent to Schrader and Dumke for a method and an apparatus for melting amber. The technique described not only amber being melted but also pine resin being added while treated with superheated steam. In 1886 or 1887 Lehmann, Burger, and Seifert, of Germany, received a patent to improve the melting of copal and amber using conical vessels that allowed superheated steam to surround the containers of copal and amber. During this same period England imported much copal from her African colonies that was primarily used in the production of varnish. However, about this same time a company in England obtained a patent to melt amber in pressure molds at temperatures about 300°C, which improved the previously described process for "concreting amber shavings" or making pressed amber.

The most interesting was a German patent, which was obtained by P. Haller in 1892, issued not for adulterating amber, but for substituting other material and changing its appearance to look like amber. Haller developed a manufacturing method for imitating amber by carving objects in bone and boiling them in oil, with potassium dichromate added. The objects were boiled until the oil foamed and the resulting carving acquired the appearance of amber.

Many of the adulterated ambers were developed for use in the smoking industry. It was thought that amber had a hygienic value and helped in purifying the smoke. Thus, manufacturers were looking for a harder substance and less expensive methods to obtain the material. In 1907 Spiller began experimenting with converting copal into a mass that resembled natural amber and then carving pipe stems and mouthpieces from this material. The copal was heated with water at pressures of 16 to 20 atmospheres. Sometimes dye additions were added to the water (Fraquet 19887, 67).

Another method of using the copal was to soften it with carbon disulfide. The resulting softened material was heated uniformly in airtight compression chambers. In 1913 Bahket, of Germany, developed an artificial amber by dissolving copal in acetone with addition of dyes, then kneading it until the acetone was volatilized. Next the material was melted at 300°C and poured into forms to cool under pressure. These materials have the same refractive index and same specific gravity as amber. Therefore, it was also necessary to describe methods to detect these new materials and to differentiate them from natural amber. Thus, in 1889 Weiss and Erckmann published an article describing the "double refraction which natural amber shows under crossed Nicol prisms, while pressed amber shows this, but is marked by a lack of uniformity" (Fraquet 19887, 65–7). Conwentz (1890*b*)

described the artificial coloring of ambroid in colors such as red, blue, light green, and heliotrope (pinkish-purple).

Ambroid or Pressed Amber

One method known as "Spiller" was used in the past to form pressed amber, and was occasionally used up until recently. Around 1880 in Vienna, Spiller discovered that small pieces of amber could be fused together by the use of heat and pressure, producing large blocks of so-called "pressed amber" or "ambroid." By 1881 techniques used in the German amber industry had been refined to make the process commercially feasible. Amber chips, after the external weathered layer had been removed, were evenly heated in a cubical steel mold with an airtight, sealed cover. The mold was submerged in hot paraffin or glycerin to gradually heat amber within the airtight chamber. Then hydraulic pressure of 400-500 kg per cm^2 was applied to the cover. With no means of escape, the amber welded together into a solid mass typically in the shape of plates or a flat block (Gierlowski 2005, 6).. However, when such material is examined under magnification, hazy outlines of fused chips can still be seen. Pressed amber formed in this manner is generally a transparent dark brown color because of the darkening caused by heat, but some pieces retain their original cloudy yellow appearance and produce an uneven coloration in the block (Fraquet 1987, 65).

In 1892 Egge, of Germany, developed a method to manufacture large articles from small pieces of amber by pressing amber chips at 240–250°C for 12 to 30 minutes in steel forms. Irregular and small pieces of amber, otherwise unusable in the manufacture of jewelry or art ornaments, as well as fragments left over from working amber articles, are the raw materials for pressed amber (Fraquet 1987, 65).

A more complex technique, called the Trebitsch method, made it possible to obtain cylinders or various shapes. Natural amber pieces are carefully selected according to size, and the weathered crusts are scraped and cleaned to remove all impurities. When exposed to a temperature of 170–190°C the amber softens to a rubber like consistency without disintegrating. To prevent decomposition during the process, the amber is placed in a deep steel tray with a perforated partition and heated to about 200–250°C with air being excluded. Hydraulic pressure of about 50,000 pounds per square inch (or no less than 3000 kg per cm^2) is applied to force the soft amber through the perforations into a second compartment, where it cools into a solid block. A variation of this method forces the softened amber through perforated plates into molds of desired shapes and sizes. According to Gierlewski (2005, 8), the first industrial use of this method was in 1881 in Vienna

Table 10.1 Ambroid Characteristics	
Refractive index	1.54
Specific gravity	1.04 to 1.08
Solubility	Softens under ether
Polariscope, dark position	Even light appearance in all positions
Microscopic examination	(a) Roiled appearance, undulating lines
	(b) Elongated air bubbles
	(c) Texture varies from uniform mistiness to hazy outlines of fused pieces

for production of smoking accessories.

In a 1913 issue of the *Mining Journal*, Bellmann (1913*b*, 122) defined compressed amber or ambroid as:

Amber produced from small pieces without admixture of foreign substances, fused by heat under high pressure in a vacuum. According to the size and color of the material used, the strength and duration of the pressure and the height of the temperature, the color shades including the characteristic cloudy spots of raw amber can be reproduced in the compressed article. Other characteristic properties of the natural "stone" suffer no alteration providing the crude material used does not contain any impurities whatever, such as dirt, dust, weathered cortex, or foreign substances inside the lumps. To free the lumps thoroughly from the cortex is very tedious and slow work, in which about 150 girls in the factory and 500 female home-workers are constantly engaged, partly in Palmnicken and partly in Königsberg. The amber is pressed into plates, bars, cones, and of various shapes and measurements.

The Staatliche Bernstein-Manufaktur Königsberg did "honestly" mark the products created by pressing amber as Pressbernstein (Gierlowski 2005). Passing the softened amber through perforated plates causes fragments to lose their shape and intermix, with the resulting finished product tending to be less clear and brilliant than nontreated amber. It has a hazy, roiled, or "cirrus cloudy" appearance. This was a major drawback because the early pressed amber products not only did not retain their original color and transparency, but also became cloudy with age. From an article entitled "Amber" (1933, 21), we learn that:

The cirrus clouds, which appeal to some in amber and appear in the process of manufacturing ambroid, became, after a few months, unpleasant in appearance, while articles made from the transparent ambroid lack the luster and warmth of these made from a natural mass because the structure reveals undulating lines.

The flow lines referred to give the appearance of water and glycerin flowing together. Such may be seen under magnification in transmitted light, and this roiled appearance indicates pressed amber. When a pressed piece is placed beside natural amber these appearances become readily apparent. Bubbles in ambroid are more elongated and somewhat flattened, as compared to those in natural amber, which retain round or spherical shapes. Also, in contrast to natural amber, pressed material tends to soften in ether, but the refractive index and specific gravity are similar to those of natural amber. Polariscopic examination reveals an even light appearance when rotated between darkened polaroids (Liddicoat 1969, 327; Anderson 1973, 323; see Table 10.1).

Because amber naturally darkens, becoming a pleasing reddish color after being polished and exposed to the atmosphere, pigments are sometimes added to pressed amber to provide deeper shades suggestive of age. Green hues have also been imparted either by dyes added, fillers or inert gas such as nitrogen. The bracelet in Figure 10.5 is made of pressed amber beads with a green hue, the result of green pigments being added during the pressing process. The photo clearly contrasts the pale olive-green beads with the natural yellow amber slab on which they are resting.

Except for the addition of minute quantities of color, pressed amber contains no foreign substances. Yellow pressed amber had fewer impurities in the amber chips and dust when the pressing process occurred. It follows that rough amber pieces with impurities would result in darker pressed amber material. When the process was first used, ambroid was more expensive than natural amber, but this is no longer the case. From the time the Royal Amber Works in Königsberg purchased the ambroid industry, around the beginning of the twentieth century, pressed amber was considered especially valuable for smoker's articles because it was more hygienic than wood, horn, bone, or celluloid and was also reputed to possess curative powers of its own. Ambroid was excellent, too, for making large art objects. Königsberg remained the center for manufacturing pressed amber products until World War II. After the amber area of Samland became incorporated into the former Soviet Union, the factory was renamed Kaliningrad Amber Conglomerate in Yantarny and work continued to improve the amber pressing technology. Poland

Figure 10.5. Pressed amber bracelet, ca. 1900-1920. Green dye was often added during the manufacturing process of pressed amber. The yellow lump is natural Baltic amber. Courtesy A. Calomeni Collection.

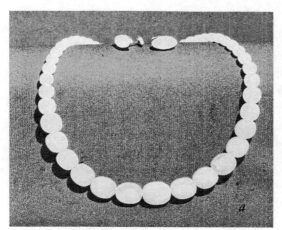

Figure 10.6a. Yellow pressed amber from Soviet company. The beads are beautifully polished, and uniform in color. Purchased in 1977 in Zagreb, Yugoslavia. b. Transparent brown mottled "reconstructed amber" from Almazjuvelirexport, Moscow, Russia. Note undulating lines in the amber's appearance. Courtesy H. Evens Collection.

b.

and Germany also began to manufacture pressed amber in mostly small firms (Gierlowski 2005, 6). Experimentation continued using varying degrees of amber granulation, different fills and dyes, atmospheres of inert gasses, changing temperatures and heating time and pressure. Gierlowski states that today amber grains are heated to as much as 220° C with a pressure from 2400 to 2700 kg per cm2. These newer techniques made it possible for artists reconstructing the Amber Room at Tsarskoye Selo to use pressed amber with a darker pigment added to it for the rococo frames in the panels. The engraving and intaglios at the bottom of the 156 cm long frames are semi-cylinders giving it a perfect look.

Many large companies selling amber articles have cards printed stating "Natural Amber," "Pressed Amber," and "Reconstructed Amber." Each of these categories means something different. Today an article marked "Natural Amber" will be made entirely of the fossilized resin or genuine amber. However, in the process of polishing to form jewelry, the amber itself may have been enhanced by heating, polishing, clarifying, or may have had nitrogen introduced in the processing, thus reducing air bubbles and allowing light to be transmitted. "Pressed amber" is made from natural chips without adding artificial materials, but scraps and cuttings of amber are pressed together as described. However, the label "Reconstructed Amber" indicates a different process and the material is 90 percent synthetic with about 10 percent resin added.

Before the breakup of the Soviet Union, the Soviet catalog for amber exports designated the articles that appeared as a yellow, uniformly colored, cloudy (roiled), hazy, material as "pressed amber" (see Fig. 10.6a). The transparent dark brown material with swirls of darker streaks throughout the mass was designated as "reconstructed amber" (see Fig. 10.6b).

In 2004, the Kaliningrad Amber Factory in the Russian region near Yantarnyy was the largest producer of pressed amber in the world. When asked about the addition of synthetic material into the reconstructed amber, their response was that this was done in the past but it is no longer produced in this manner. Today, instead of using the very cheap small pieces of amber, they select the choicest pieces, but small in size, of milky amber or "bastard" amber without inclusions or crust, thus using very clean, cloudy amber pieces. These are ground up like flour. This powdered amber is subjected to heat in the complete absence of oxygen and placed under enormous pressure of about 1000 atmospheres. The amber reaches the malleable stage and is pressed together to produce the highest grade of absolutely transparent amber. This pressed amber costs more than natural amber because there is less waste when working the amber into carvings. Almost 90 percent of the amber jewelry, framed in gold, produced by the Russian factory is made of this pressed amber (see Fig. 10.2b; 10.3; 10.7) (Shedrinsky 1996, 5–26; Kostiaszowa 1997*a*, 44–45).

Problematically, pressed amber products made for the internal market of the Russian Federation are designated identically to those made of natural nuggets. Because of the extremely high quality of some of the pressed products, it is very difficult to distinguish them from those made of natural amber pieces. Perfectly transparent, honey tinted and cognac beads are made and used in very desirable and highly sought after necklaces (Gierlowski 2005).

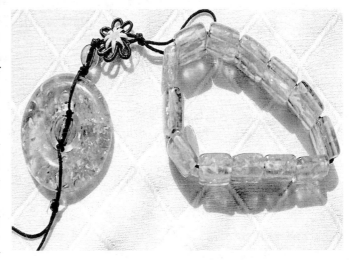

Modern techniques in processing ambroid have been developed in which sun-spangle inclusions can also be expertly induced in pressed amber. Upon close examination of the fissures, using a jeweler's eye loop, they appear unnatural, without the distinct circular or discoidal outlines so characteristic in natural amber.

In summary, the hallmarks of pressed amber are a roiled appearance throughout or along junctions between the separate amber pieces, sometimes only visible under magnification, and a tendency to soften under ether. Insect inclusions have been introduced into pressed amber, but may be distinguished by their flattened appearance, rather than the fullness that characterizes natural insect inclusions. Such inserted insects are whole, for the most part, and lack the dismemberment that is so often seen in natural amber, where insects struggled to free themselves, losing legs or wings in the process.

Figure 10.7. Pressed amber can be produced with the discoidal fissure inclusions and closely resembles natural amber that has been treated in an autoclave. Close examination with a microscope is needed to determine whether the piece is natural amber or pressed amber.

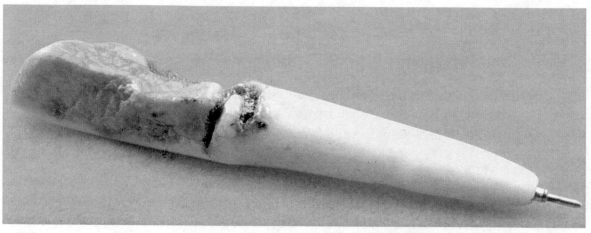

Figure 10.8. Amber folk art from Poland: a, Amber with natural crust polished into a cylinder for the pen; b, Ring constructed of two pieces of amber.

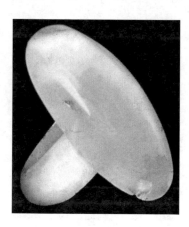

EARLY METHODS OF WORKING AMBER

Amber artifacts from Stone Age graves in the Baltic region show that early inhabitants were familiar with methods for crudely shaping and finishing many kinds of small objects. Prior to the advent of metals, such shaping was accomplished by use of sandstones and other gritty stones that served as saws and files. Wood, supplied with fine sand, could be used for smoothing surfaces. Holes probably were drilled, as in other early cultures, by simply twirling a pointed stick pressed against the amber. The point of the stick was supplied with slurry of water and fine sand as an abrasive. Wood was also used for polishing and, when supplied with wood ashes or some other very fine abrasive, could readily impart a polish to amber.

With the appearance of bronze (and later, iron) in the Baltic region, more sophisticated work was achieved, but methods for shaping, smoothing, and polishing remained little changed over the centuries. As late as the seventeenth century, natives of the region worked amber, especially during long winter evenings when there was little else to do, with the most primitive yet effective implements. The stages through which amber was processed consisted of removal of the outer soft, cracked, and discolored crust, generally by scraping; then sawing larger pieces into smaller ones as required; shaping pieces by use of scrapers and files; and, lastly, smoothing and polishing. When metal files were unavailable, gritty stones of various types were used, lubricated with water to keep friction heat from generating and to take away amber particles and dust. Smearing a paste of wood ashes or chalk on pieces of wood and rubbing the wood briskly over the surfaces of the amber obtained polishes. By such simple means, peasants were able to produce large round or oval beads, bars for brooches, finger rings, medallions, and even crucifixes. A popular carved object was the heart, which, in addition to its obvious ornamental value, was believed to confer protection upon its wearer. Amber buttons were made in a variety of shapes—round, square, triangular, and so forth. Some even were shaped into representations of flowers. Finger rings, shank and all, were carved from single pieces. At other times, rings were made in two pieces—the ring itself with a "stone" attached to the top. Products of such cottage industries reached their highest development in Poland during the last century in the districts of Kaszuby and Kurpie, in the Bory Tucholskie Forest in Pomerania, and in the Puszcza Myzsyniecka Forest in the Narew River basin. Amber pits were worked in these regions until the middle of the last century, providing a local source of raw material (see Fig. 10.8).

Though handmade objects were produced from amber for centuries, it was not until the eighteenth century that a kind of mechanization was introduced; in this case, the adaptation of a spinning wheel into a crude lathe for turning round and cylindrical objects from amber. The amber lump was attached to the axle and the latter turned with a foot treadle, while shaping was accomplished by placing a piece of broken glass against the amber. As may be expected, this crude lathe was soon superseded by more sophisticated devices, and the so-called "spinning wheel lathe" is now seen in use only in a few isolated areas or museums in the Kurpie district of Poland (Zalewska 1974, 98; see Fig. 10.9).

Amber carvings are still produced in quantities in Germany, Poland, Lithuania, and in the Kaliningrad District of Russia. In Lithuania pictures are made in mosaic fashion. These objects are, however, few in number with exports oftenlimited to cottage-industry pieces brought to the United States by relatives visiting from Lithuania. In regard to the mosaics and related inlay, thin pieces of amber are cut to fit together like pieces of a jigsaw puzzle. By clever choice of colors and textures, geometric designs or even pictorial

Figure 10.9. Spinning wheel lathe from Kurpie region of Poland. Courtesy Muzeum Ziemi, Warsaw, Poland.

Figure 10.10. Folk art mosaic from Lithuania. Thomaslunis Collection.

designs or scenes are created. Mosaic pictures from Lithuania usually are mounted on a black or dark brown wooden panel with the amber standing out three-dimensionally (see Fig. 10.10).

Another type of amber mosaic covers the entire panel with amber. After all pieces are fitted together and cemented into place, the whole slab is smoothed and polished at one time. In the past mosaics of this type were used for wall panels or cabinets and to cover the side of a box or jewelry casket (see Fig. 10.11). Other materials of about the same hardness as amber were used in combination with amber to provide greater versatility in design. For example, ivory, mother of pearl, and tortoiseshell were used, especially in the manufacture of attractive chessboards. Rare chessboards are still made entirely of amber encrusted over a solid board. Variations in the amber colors make the light and dark chessboard playing board squares. The darker pieces needed for the contrasting chessmen and squares are heated to produce an even dark coloring from yellow amber (see Figs. 10.12 and 10.13).

INDUSTRIAL PRODUCTION OF AMBER OBJECTS

In the early 1900s, the prevalent belief that amber possessed germicidal properties led to a vast amount of smokers' articles being manufactured of amber. Nearly one-half of the total production in the Baltic region was devoted to making cigar and cigarette holders and pipe mouthpieces. Vienna alone used up to 40 percent of the annual yield of East Prussian amber, and its factories were world famous for the high quality of such goods produced in that city. Smokers' articles were also made in Königsberg (Kaliningrad) and Memel (Klaipeda). Many mouthpieces and holders were made on lathes and pierced with machine drills, but some special shapes were made by sawing and hand-shaping. Pipe stems were made from slender amber strips that were drilled and then immersed in hot linseed oil to render them clearer and pliable enough to be bent into the curved

Figure 10.11. Contemporary amber box constructed of opaque rectangular amber tiles over Lucite. Made in Poland. Author's Collection.

Figure 10.12 *Chessboard made from light and dark Dominican amber. The chessmen are set with pearls, and the 18 k gold crowns are set with rubies and sapphires. The kings are 17.7 cm (2.75 in.) in height, the pawns are 24.6 mm (1.5 in.) in height. The chessboard is 67.8 cm x 67.8 cm (10.5 x 10.5 in.) Courtesy Jeffrey and Christian Lambujon Collection.*

Figure 10.13. *Chessboard and pieces made from Baltic amber exhibited in Tucson 2004. Made by the Russian amber artists who were assisting in the reconstruction of the Amber Room. The chess pieces are made of light yellow opaque Baltic amber and darker brownish-yellow translucent Baltic amber. The board is made with amber encrusted over a solid board.*

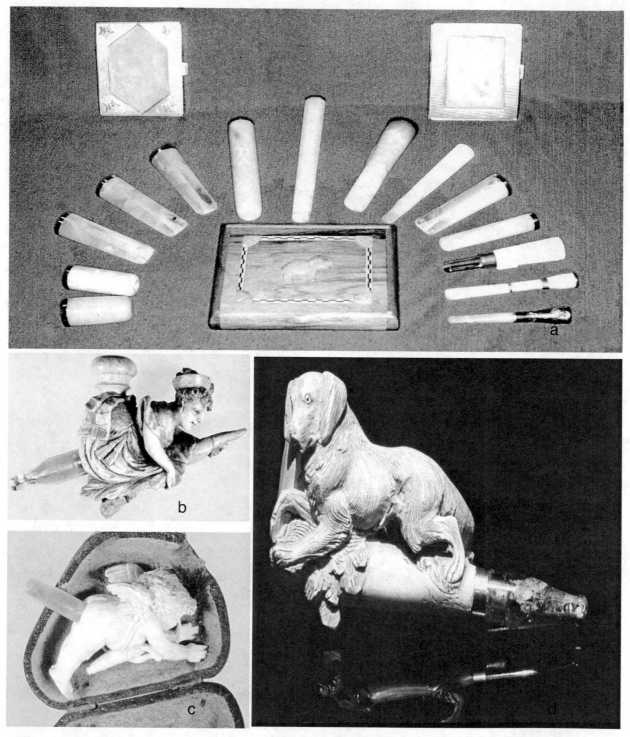

Figure 10.14. Smokers supplies, made of pressed amber: a, Cigarette holders. Courtesy Kuhn's Baltic Amber Collection; b, c, d, Meerschaum pipes with amber stems. Courtesy Burgess-Jastak Collection.

forms desired for mouthpieces (see Fig. 10.14).

Other articles made from amber, such as small carvings and beads, were specialties of factories in Gdańsk, Berlin, and Słupsk. During the early 1900s, transparent golden corals (or faceted beads) were in great demand in the United States. These beautiful hand-crafted beads can be found in antique shops today in a golden amber color. In most of Europe, however, opaque bastard yellow amber in the shape of olive or round beads was generally considered most valuable, while the French preferred water-clear beads. China and Korea produced round beads for mandarins and, in 1904; China imported approximately $25,000 worth of rough amber for this purpose alone. The round amber beads were believed to have a special power to heal, thus amber prayer beads, used to count the prayers, were considered to have more healing power. The word bead, itself, is derived from the English word *bede*, which means prayer (Rice 1977, 8–12).

During the 1950's and 60's, small firms in Poland attempted to use small natural amber pieces by heating and pressing them together with varying degrees of results. However, pressing outside an airtight environment and with insufficient pressure (only between 200-300 kg per cm2) often lead to dark red tint or insufficient clarity so the natural shapes of the nuggets had clearly visible borders at the joints. This lead to a process still used today in which natural irregular amber material is softened at 150-200°C into a cabochon or diamond shape using a single nugget to form the resulting molded stone. Because raw amber was becoming rather scarce in about 2002, attempts to press amber grains at 140° C or close to the softening point with along with other fossil resins such as young amber, gedanite, or copal to fill the spaces. Since the filling resins have a lower melting point, they ooze into the spaces between the amber fragments binding them together. The amber fragments remain multicolored producing a dapped appearance, thus the name "dapped" method. If the filler material used is a finely ground amber powder and pressed under high pressure, it remains white and opaque appearing like mosaic forms around the pieces (Gierlowski 2005, 8)

Table 10.2	Standard Bead Styles
1. Olives	Elongated elliptical
2. Zotten	Barrel-shaped, cylindrical; slightly rounded, almost plane at the two ends
3. Grecken	Like zotten, but shorter
4. Beads proper	Spherical
5. Coral	Faceted beads
6. Horse-corals	Flat, clear, faceted at the two ends

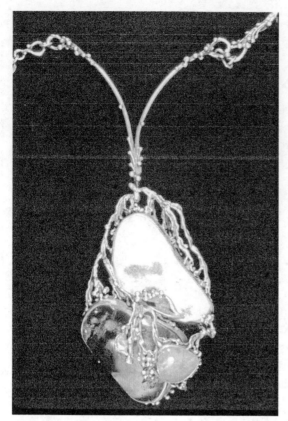

Figure 10.15. Fused silver work and amber, chrysoprase, and other semiprecious stones made by Janusz Góralski, Polish artist. Courtesy Kuhn's Baltic Amber Collection.

COMMENTS FROM AND ABOUT SELECTED ARTISTS WORKING IN AMBER

Orno Cooperative, Warsaw, Poland. An artist working at the Orno Cooperative in 1977 described his feelings about amber thus (Szejnert 1977):

To me, amber seems closer to human nature than any precious stone. It is warm in color and the touch, gay, luminous, and full of surprises. . . . It gives me pleasure to find remains of ancient life when polishing a piece of amber. . . . What can be more beautiful than a lump of sunshine given the minimum of polish not to spoil its natural charm and set in silver which contrasts so well with its color and substance?

Janusz Góralski. Janusz Góralski, another noted Polish artist of the 1980s, was especially known for his beautiful melted silver work. He framed his amber with silver, styled in massive baroque shapes, complimented with extravagantly sized, polished lumps of amber. Contrasting with the smooth amber were fused silver shell shapes, floral designs, seaweed, and other metaphorically designed shapes representing the sea from which amber comes. But, the amber was always the center, beautifully selected "symbol for the associations of the world under the sea" (Grabowska 1983, 30; see Fig. 10.15).

Malbork Castle Museum Artist. The artist at the Malbork Castle Museum amber shop eagerly demonstrated the equipment in his local shop across the river from the museum. He had equipment for hand drilling with a flexible drill, a motorized armature for using buffing wheels and grinding wheels, with a dust cover and vacuum vent. His workshop was a very sophisticated lapidary shop that included machinery such as an oven with shelves for slowly heating amber and an autoclave for raising pressure and releasing pressure. He also had a kiln used for lost-wax jewelry designing (see Fig. 10.16).

Today in Poland, amber is produced not only in large production facilities, but also by independent craftsmen in small shops, many of which are located along St. Mary's Street in Gdańsk. The artists exemplify great skill for working and techniques for finishing amber. Many contemporary artists in Poland create elaborate fused silver work in the jewelry that is also adorned by amber along with other

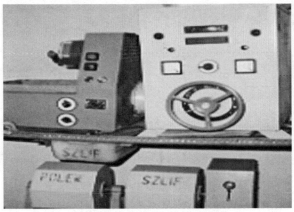

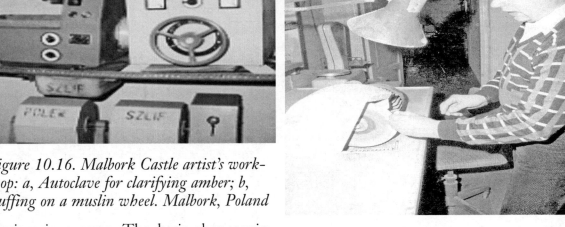

Figure 10.16. Malbork Castle artist's workshop: a, Autoclave for clarifying amber; b, Buffing on a muslin wheel. Malbork, Poland

semiprecious gems. The basic shop equipment includes flexible-shaft electric drills, motorized grinding wheels, muslin buffing wheels, with vacuum dust shields to remove the amber dust while grinding. Interestingly, they do not use constant water coolants as do many lapidaries in the United States who grind hard stones.

Kycler Workshop, of Gdańsk, Poland. One such shop owner, Zdzisław Kycler, explained the reason why running water is not used. "If the amber piece is not pressed too long or hard against the wheel, there is no need for constant running water. When the piece is smoothed and shaped on the sanding wheels, the final polish is applied." Kycler demonstrated the polishing phase, using a muslin buff that he used only for amber and a white silver polish purchased from Bennington, England. Intricate carving was done with the flexible shaft drill, using different sanding tips and buffing tips. During carving, the craftsman washed away excess amber dust from the crevices by submerging the amber piece into lukewarm sudsy water and rubbed gently with a soft toothbrush. When asked about how he produced the darker brown and sun spangles, he showed how it could be done simply by waving the butane soldering torch across the bottom of a rough piece. Then, the burned crust was ground off. This shop has produced many artistic

Figure 10.17. Kycler workshop in Gdańsk, Poland.

Figure 10.18. Bishop's cross, fused frame of stainless steel, adorned with 29 amber nuggets, made by Marion Owczarski, Polish artist, St. Mary's College, Orchard Lake Schools, Michigan.

twentieth-century carvings, some of which are included in the Burgess-Jastak Collection now housed at the University of Delaware Gallery, Wilmington, Delaware (Rice 1984, 16–21; see Fig. 10.17).

Marion Owczarski, from Warsaw, now of Orchard Lake, Michigan. Marion Owczarski, artist-in-residence at St. Mary's College, Orchard Lake Schools in Michigan, is internationally known for his stainless steel sculpture and stained glass church windows. He began as a child in war-torn Poland carving simple images in amber. He incorporated amber into his earliest works as a student at Poland's prestigious Academy of Fine Arts. According to Owczarski, "The stone's permanence, its age, and the fact it generates a spark in response to friction reminds me of my country, Poland. I combine amber with wood and stainless steel because the ages—old petrified resin evokes the past but demonstrates how well beauty resists time. Stainless steel, on the other hand, is the medium of the future, suggesting strength." Marion Owczarski continues to use "the tear of the sun," as he sometimes refers to amber, in his chalices and crosses where he combines amber and wood with the stainless steel. Owczarski created stainless steel crosses for all of the Polish bishops meeting at a convention at Orchard Lake Schools. One of the crosses is pictured in Figure 10.18.

Marion Owczarski says his favorite polish to produce a high luster on amber is tin oxide, talc, and paraffin used on a clean muslin buff. He sometimes works amber in a manner similar to techniques for carving wood sculptures in the folk art tradition. The natural markings are left on the piece and carving is done in more of an impressionistic or stylized manner. This technique gives the impression that the piece of material dictated what the theme of the sculpture would be and the artist simply helped it to materialize. Carving must be done carefully so as not to chip away unintended pieces because amber is brittle and bending or turning cutting implements will result in an

Figure 10.19. Stainless steel chalice with amber teardrop, signed M. OwczARSKI, WARSZAWA. *Marion Owczarski is a Polish artist working at St. Mary's College, Orchard Lake Schools, Michigan.*

unwanted chip.

One of Marion Owczarski's trademark pieces is his chalice constructed of hammered stainless steel adorned with an amber teardrop. He uses various colored stones to decorate them. The amber teardrop at the base of the pedestal of the chalice symbolizes the strength of the materials from which the art is produced. The forged steel bowl of the chalice is mounted on an inverted cone-shaped base of hand-hammered steel with a high polish. These unique pieces are signed M. Owczarski (see Fig. 10.19).

Norman Fortin, Toronto, Canada. One of Canada's outstanding artists, Norman Fortin, also is an artistic mentor of indigenous sculptors and jewelers in the Caribbean. He has mastered a technical fluency that spans a broad spectrum of media from ivory to precious stones and metals. He says he uses the medium that best speaks for the subject matter. He sculpts in amber and bronze and creates commissioned pieces of jewelry. A one kilogram piece of Baltic opaque yellow amber was purchased from Andrew Jaskow, of Sweden, at the Tucson Gem and Mineral Show, Arizona USA, and given to Fortin to create what he thought would be an appropriate sculpture for that lump of amber. The result was a figure of the head of a dreamy-eyed young girl with a flower deeply carved in relief at her forehead, long curly hair with garlands of flowers interwoven in the design. The artist tried not to take away too much amber during carving to keep the final sculpture weighing as close to 1 kilogram as possible to emphasize the large size of the amber piece. The back of the piece was polished to show natural markings of the crust to best advantage. The sculpture was mounted on a black and pink round marble base supported by a brass shaft with black coral at the base. At the bottom of the marble base is a Canadian copyright seal that reads Canadian Designer Copyright, Norman Fortin (see Fig. 10.20).

In his sculptures, Fortin uses ideas related to the early history of amber as his theme when working with amber. An other example of his talent was when a large lump of amber purchased in Poland was given to him with only the directions to sculpt what he determined

a

Figure 10.20. a, Carving of a young girl with flowing hair and garlands of flowers; b, Marble base has the copyright seal of Norman Fortin, Canada Designer.

b

Figure 10.21. Sculpture by Norman Fortin of symbolic amber themes: a, Front view; b, Side view; c, Back view. (Size: 5.7 cm x 8.25 x 13.9 cm [2.25 in. x 3.25 in. x 5.5 in.], Wt. 275 g)

would be right. The rough material had many inclusions of lignite, pyrite, and pitting from decayed wood material. The artist had to drill out the lignite and pyrite inclusions, thus he created a sculpture carved on all sides with deep piercing and intricate designs. The final sculpture begins with a beautiful sunburst, symbolic of amber's sun-worship connection. As the piece is turned, new themes are encountered, such as a Greek ship sailing on a sea with waves, shells, fish, and sea creatures to remind the viewer of amber's connection with the sea and the high value placed on amber by early Mediterranean cultures. The new side portrays tropical leaves surrounding a monk with a bird and monkey on his shoulder symbolizing amber's religious connections and its origin from living plants in a primeval forest. Next, at the feet of the monk is an octopus with tentacles reaching back to the sea, bringing us back to understanding amber's many uses. However, in the midst of this flora and fauna, amber reigns, a symbol for association with the world under the sea. The finished amber sculpture has a high polish and weighs 275 grams (see Fig. 10.21).

Debra Orvis, designer and owner of Time Traveling Trade Co. of Bay City, Michigan. Debra Orvis works amber from the Dominican Republic, Chiapas, Mexico, Poland, and Kaliningrad. She creates beautiful pendants and brooches using transparent amber that she etches in reverse, using the *verre églomisé* technique. Using her flexible shaft drill, she carves on the back of the transparent amber piece, producing detailed horses, dragons, seahorses, birds, and other elegant scenes. Finally, the engraved areas are filled with white acrylic to bring out the scene when viewed through the transparent amber.

Orvis has experimented with various methods of treating amber in her own jewelry workshop and offers special tips for the home workshop. Her first recommendation for amateur amber workers is the best polish that can be purchased locally, which is the white rouge stick, a wax-based polish for polishing plastics from a department store, such as Sears and Roebuck in USA. She has found this produces the best finished polish on amber. Many times the natural indentations in rough amber are left in the finished piece, but often crevices become filled with the polish and it is difficult to remove even with a toothbrush and detergent bath. To avoid white particles in the crevices, she prepares her polish by adding a black carbon ink. First, she melts the wax-based polish and adds a small amount of 100 percent carbon black ink. The polish is then allowed to harden again. It is now ready to use on a muslin buff, which is kept for use only with this particular polish on amber.

Orvis has tried various methods for clarifying amber in her workshop. She uses Cannola oil (a rapeseed oil with the same refractive index as amber) or virgin olive oil to induce a caramel color in opaque amber pendants. These natural amber pieces are unique. The oil is heated in a deep fat fryer to a temperature of about 135–148°C (275–300°F). The opaque amber is "deep fried," but

Figure 10.22. Channel inlay, using silver and different shades of amber to create the amber lady bug. Made by the author.

watched carefully until the amber reaches the desired color. If left in the hot oil too long, the amber piece begins to clarify, because the oil penetrates deeper into the porous material filling the air bubbles.

CUTTING AND POLISHING AMBER

Most important in working amber is selecting rough material suitable for producing the desired finished product. Each piece of raw amber should be carefully considered in the design so natural inclusions and even portions of the crust may be incorporated into the detail of the object produced. The crust may be ground off to display the true amber colors beneath the surface, or portions of the crust may be left untouched to provide a deeper coloration to such surfaces.

Shelly amber, formed in concentric layers, is weakly adherent along layer boundaries and, if drilling is required, the possibility of fracturing is more likely when the piece is drilled in the same plane as the layers. Rough pieces of this variety are better used in larger blocks requiring little drilling.

HAND METHODS OF WORKING AMBER

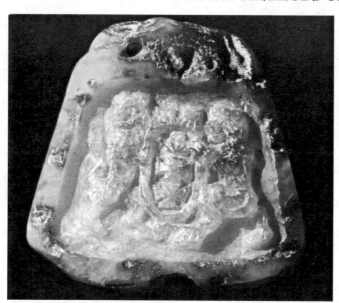

Figure 10.23. Relief carving of Seal of Gdańsk in Baltic amber. Courtesy Burgess-Jastak Collection.

Many artists work amber using simple hand tools for cutting, shaping, and polishing. The first step is to decide on the shape of the finished piece and to cut any large unneeded material away with a jeweler's saw or hacksaw, using slow, even sawing strokes. Great care must be taken to avoid any wobbling or bending of the saw, as pressure on the amber piece may cause it to chip or perhaps fracture completely.

After the desired shape has been roughed out, files are used to further shape the piece. A coarse rasp file may be used to remove thick crust, but its use is only recommended when thick material needs to be quickly removed. Generally, it is safer to use flat files in several degrees of coarseness for removing material and smoothing the surface. Begin shaping by making even strokes with the file and crossing these by strokes at right angles to eliminate scratch lines. Files readily clog as work proceeds, and a file card (file cleaner) should be kept handy and used frequently to keep the teeth clean. Small needle files of different shapes are employed to remove material from depressions or other irregularities in the surface, and a small triangular file is particularly useful. However, care must be taken to avoid the wedging of small files in crevices, as this easily causes chipping or even results in the whole piece splitting in two. After the shape is roughed out with files, the piece is ready for sanding.

All surfaces must now be further smoothed to remove all file scratches. For this purpose, sandpapers of decreasing coarseness are used and prepared as follows. Make the first of several boards with a coarse-grit paper, such as sandpaper 220, emery paper 220, or coarse-grit aluminum oxide paper. Tack the sandpaper around an 8 x 10 x 0.5 inch board. Holding the piece of amber in your fingers, rub the lump over the sanding board, using a scrubbing circular motion and varying movements so as to sand all surfaces of the piece. Repeat this process, using a medium-grit (400) sanding board. The final sanding step requires fine-grit aluminum oxide paper or sandpaper 600.

When working any of the ambers or fossil resins, hand sanding can be done dry, since the work is generally slow and heat does not build up enough to soften the amber. However, copal, or the so-called "African amber," is far more brittle and softer than the fossil ambers and, for this reason, should be sanded wet, using a waterproof sandpaper. Also, the lower melting point of copal results in the material becoming gummy during dry sanding; wet sanding prevents both temperature rise and excessive clogging of the sandpaper.

Upon completion of sanding, closely examine the piece to be sure all fine scratches are removed before polishing. It is very exasperating to polish a specimen and then discover that one or two deeper scratches were overlooked. If this happens, go back to sanding until such scratches are eliminated and the entire surface has a uniform frosted appearance.

The last step is polishing. Various suggestions by different professional lapidaries have been made regarding methods for buffing and polishing amber. Some use tripoli or rottenstone as the polishing agent, with oil as a lubricant; others recommend aluminum oxide, tin oxide, Linde A, or rouge. Tin oxide and Linde A may be applied as a paste made up with water, whereas rouge is applied dry. It is said that in Mexico, dry cigarette ashes were used as a polishing agent. Whichever agent is chosen, the following instructions serve as a general guide for polishing amber.

MECHANICAL POLISHING

Amber can be polished using power-driven lapidary equipment, but because of its fragile nature as compared to mineral gemstones, much more care is required during all stages. Amber will not stand any kind of high-speed friction grinding, and some specimens under a state of strain may shatter. Overheating from too much friction causes upper layers to melt and produces a rippled or "orange peel" surface. If the amber approaches its melting temperature for an excessive period, volatile substances escape and the piece is left weakened and more brittle. For these reasons, when power-driven wheels or laps are used, the speed should be slowed to around 100 revolutions per minute and large amounts of water should be supplied to abrasive or polishing wheels to avoid heat build up. Such reduced speed is especially important when shaping pendants and cabochons. Working on several pieces at one time may be necessary because the pieces build up static electric charges, which can be felt as tremors in the hand. If this continues, the pieces may shatter. By alternating from one piece to another, the electrical charge dissipates. Felt wheels are excellent for the finish polish, with any of the polish agents mentioned previously, but with water as the lubricant. Never use a felt wheel that has been used for polishing hard minerals, since scratches may develop on the amber (Practical Shop Notes 1915). Avoid using diamond laps, as they cause the surface to become dull and cloudy, giving an unpleasant appearance. Linde A, water, and a felt wheel rotating at a slow speed provide a combination that produces a finished piece with a fine luster. A Polish artist suggests using tin oxide, talc, and paraffin mixed to form a waxy paste polish (see Fig. 10.22; 10.23).

DRILLING AMBER

Amber may be drilled with a hand drill or an electric bead drill. When a hand drill is used, hold the amber object securely in the fingers to avoid splitting the piece. Using a vise on amber causes chipping and is not recommended. A hand drill may be used dry, as the speed can be varied to avoid excessive frictional heating. If an electric drill is used, however, a drill stand or commercially designed bead drill rig is necessary, because the amber must be submerged under water to avoid overheating while being drilled. Commercial bead drills not only have a stand, but also are designed so the drill bit can be moved in and out of the piece frequently, thus avoiding overheating. WARNING! Do not attempt to hold an electric drill while perforating a bead that is submerged in water!

When piercing, it is good practice to drill only halfway

Table 10.3 Faceting Guides			
Cutting Angles	Cutting Lap	Cutting Speed	Polishing
Culet 43º Crown 42º	Fine-extra fine Speed: normal	100 RPM	Lap: wax Agent: Linde A

through the material from one side, then turn the piece over to complete the hole. If the drill clogs or sticks in the amber, add oil to the bit to loosen it. Tugging or twisting usually results in chipping.

FACETING AMBER

Amber is easily faceted if the lap speed is reduced. Vargas and Vargas (1969, 259) suggest the use of a wax lap with copious amounts of lubricant and Linde A as the polish. Cutting may be slow, resulting from clogging of the lap with amber particles. The use of more water helps flush away such debris. Vargas and Vargas provide the information in Table 10.3 for faceting amber in the manner used for mineral gemstones, that is, in brilliant or step-cut styles.

AMBER ART ,ARTICLES

One of the major problems faced by museum curators and collectors of antique amber articles as well as rare fossil collections is the preservation of amber. Amber is affected by changes in temperature and by oxidation as it is exposed to the atmosphere over long periods of time. The outer surface of amber when exposed to air gradually begins to oxidize and decompose. At first a delicate crazing of the outer surface develops that is composed of a fine network of lines and the piece begins to lose its polish. As this continues, the edges may begin to crumble. Pieces may tend to redden or darken as they age. Museum curators who house antique amber articles find this a constant problem in attempting to preserve these pieces of art for the future. In regard to masterpieces of medieval guild craftsmen, amber pieces have been found to crumble, especially near glued surfaces. The question for museums is, Should the articles be restored and how can they be conserved so no further deterioration takes place?

Fraquet (1987, 70) describes the difference between restoration and conservation of amber objects. The process of restoration "includes the addition of melted, color-blended material with the hope to deceive and achieve a higher sale price," and conservation is described as "repairs undertaken with the premise that all work should be able to be dismounted at a future occasion when perhaps

Figure 10.24. Amber art by modern Polish artists, S. G. and L. Piro, as exhibited in Malbork Castle Amber Museum, Poland.

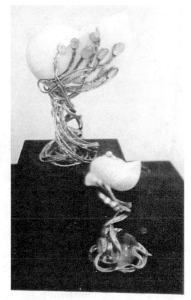

Figure 10.25. Amber art exhibited in Malbork Castle.

even better forms of conservation become available." Various methods have been used to prevent or at least retard this process. Some museums store specimens in water, glycerin, or mineral oil, which reduces the contact with air, similar to the preservation effect seawater and burial in clays had during amber's formation. Small rare fossil specimens are sometimes embedded in Canadian balsam, clarite, or synthetic resins (polystyrenes). In the past, specimens were preserved by placing an air-resistant wax or resin coating on the outer surface of the amber (Poinar 1992, 12). Objects have been coated with xylene solutions of Canada balsam, amber varnish, turpentine solutions of Dammar, shellac, and a variety of synthetic resins.

Beck (1982*a*), in discussing conservation, cautioned against the technique, described by Klebs, of preserving amber inclusions by submersion in water in which various preservatives such as ethanol were added to prevent fouling. Later it was discovered that this technique rendered amber porous and opaque on long exposure—thus, this is *not* recommended. The same problem occurs with ether and formaldehyde, both of which are deleterious to amber. A better technique of the past was to immerse the specimen in pure paraffin oil.

Prevention of deterioration of amber art requires that the room temperature be kept rather constant without overheating or extreme cold, and enough moisture must be maintained in the atmosphere. As to displaying the amber artifacts, placing them in airtight containers away from direct sunlight or a heat source is important. The exclusion of air is important, thus, airtight boxes or exhibit cases filled with nitrogen would constitute a perfect environment for valuable amber carvings according to Beck. Malbork Castle Amber Museum, Poland exhibits modernistic sculptures with silver and amber (see Fig. 10.24a &b).

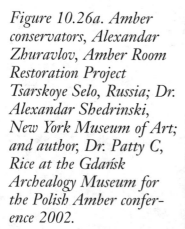

Figure 10.26a. Amber conservators, Alexandar Zhuravlov, Amber Room Restoration Project Tsarskoye Selo, Russia; Dr. Alexandar Shedrinski, New York Museum of Art; and author, Dr. Patty C, Rice at the Gdańsk Archealogy Museum for the Polish Amber conference 2002.

CARE OF AMBER JEWELRY

Amber beads, rings and other forms of jewelry will retain their original luster and splendor indefinitely if a few simple precautions are observed. Because of the softness of amber, its brittleness, and its susceptibility to attack by various chemicals and oxidation, amber jewelry pieces do require some special care in handling and storing.

Care of Beads

The first precaution is usually taken by the manufacturer by stringing the beads on silk or linen thread with knots between each bead to prevent mutual rubbing and chipping. Knots also provide a safeguard in the unfortunate circumstance of the string breaking, as only one bead will fall, leaving the rest safely knotted on the string.

Amber beads should not be stored where they will rub against metal or other pieces of jewelry. Soft flannel or velvet pouches made with drawstring tops are best for storing and protecting each individual item.

To remove dust and perspiration from amber beads, simply wipe them with a soft flannel cloth dampened with clean lukewarm water. They should then be dried carefully and rubbed lightly with clear olive oil, then rubbed with a soft cloth to remove excess oil and to restore the polish.

Avoid placing amber pieces in any strong solutions, especially ammonia, strong soaps, or detergents, and do not allow amber to come in contact with perfume or hair spray. Commercial jewelry cleaning solutions should not be used on amber.

All the foregoing can result in permanently dulling the polish. Furthermore, because some hydrocarbon compounds form a dull white coating on amber, it is advisable to avoid contact with common kitchen substances such as lard, salad oil, or butter. It is a good idea to remove amber jewelry while cooking to avoid excessive heat, as well as contact with these substances. Never use hot water for cleaning amber.

Care of Rings

Special care should be taken when wearing amber rings, which should always be removed from the fingers before the hands are immersed in anything but plain water. Avoid bumping rings against any hard surface.

Care of Carvings and Art Objects

All such objects should be handled gently to avoid knocks and chipping. Dust may be removed with a soft cloth or feather duster. Avoid

placing the object too near a heating duct, in direct sunlight, or in extremely warm areas. Carvings are best exhibited and protected in glass showcases; however, if showcases are lighted, be sure to have adequate ventilation to avoid excessive heat from the electric light.

Repair of Amber Objects

Broken articles may be repaired by coating the surfaces to be rejoined with linseed oil, then gently heating the surfaces over a low flame or charcoal fire. When the pieces are warm, press the matching surfaces together and continue to hold the article over the heat until the amber softens.

Epoxy glues typically used by lapidaries may be used for repair of fractured pieces. Very small amounts should be used with great care to prevent epoxy from coating areas other than the fractured surface. Hardened glue should not be removed from amber with glue solvents, as they will attack the amber.

Beck (1982a) recommends repairs to amber objects be made by moistening the fracture surfaces with aqueous potassium hydroxide, then pressing the fragments together. This same procedure is recommended to test for authentic amber because amber imitations will not stick together. He suggests gluing amber with alcoholic solutions of amber or of rosin.

Removal of Chemically Formed Coatings

If amber surfaces have come into contact with alcohol, the resulting white coating can be polished off using water and tripoli applied with the ball of the hand by rubbing over the piece. If a polishing wheel is used, it is important to use a new felt disk or one used only for polishing amber. After use, the polish compound is removed, using cold or lukewarm water.

Golden amber has retained its beauty for at least 40 million years; with the observance of these simple handling procedures, amber articles will retain their beauty.

Beck (1982a) states that:

> Any conservation treatment with natural or synthetic materials will introduce into an amber object contaminants that must interfere with its future authentication or provenance analysis. When some sort of treatment is absolutely necessary to prevent the decomposition of an amber find, paraffin is the only substance that can be used with safety. It is unique in that it does not absorb radiation in the principal diagnostic region of provenance analysis by infrared spectroscopy.

MAKE YOUR OWN POLYBERN

The embedding of small chips, nodules, and grains of raw amber into plastic or synthetic resin is a process easily accomplished by the amateur to create attractive beads or other ornaments. This method makes possible the utilization of waste material or pieces too small for other purposes, and the material can be readily molded into attractive ornaments. Synthetic resin compounds can be purchased at craft stores and hobby shops and are cast in molds of polyethylene, glass, or ceramic. For any casting plastic, be sure to follow the manufacturer's directions, as each type of resin requires a slightly different procedure. Practice with some inferior amber before attempting work with more expensive pieces.

The following six procedures are similar to those used in commercial production of polybern in the Baltic states of Poland, Lithuania, Kaliningrad, Russia, and other areas where this type of processed amber product is made.

1. Select amber chips you wish to embed according to the size of your mold. Small chips are best. Make sure they are free of dirt and sand.
2. The mold should be clean and have a highly polished wax finish.
3. Mix the resin in disposable paper cups, adding a yellow pigment, if desired, to match the color of the amber. Do not inhale fumes or let the plastic touch the skin.
4. Pour a thin layer of plastic resin into the mold and allow it to set.
5. Add a layer of fresh plastic resin to the top of the first layer, then embed the amber chips. Allow this layer to harden before proceeding.
6. Fill the mold with a third layer of resin and leave undisturbed until the plastic solidifies. Follow the manufacturer's instructions for removal of the casting from the mold.

Figure 10.26. Director of the Amber Room Restoration Project Tsarskoye Selo, Russia, Boris Igdalov, carving amber at the Tucson Amber and Organic Gem Show. Boris Igdalvo demonstrated techniques used by Russian amber artists who worked on the designs of the amber room. Old photographs were used to reproduce the amber art pieces exactly like the ones created in the 1600's. Because of the size of the amber pieces needed, much pressed amber was used for the restoration.

ANCIENT DNA RESEARCH USING FOSSILS
IN AMBER

Today the primary recognition of amber is no longer as a medium for art and jewelry; it is now known especially for its ancient fossil inclusions, brought to the public eye by the movie Jurassic Park in 1993 and more recently in 1997 by its sequel, The Lost World. Steven Spielberg based his film on Michael Crichton's 1990 novel, but this was not the origin of the idea. As early as 1977 a paleobiologist, Charles Pellegrino, from Rockville Center, New York, examined a piece of 85-million-year-old New Jersey amber with a perfectly preserved fly. Also being an author, he wrote in Omni (1979, 38–40), "Three more decades of technological advance and we may be able to extract and read DNA from flies' stomachs, where, if we are lucky, we will find the blood and skin of dinosaurs." Pellegrino suggested that if pieces of the genetic code were missing, which up to then had been the case in DNA recovered from ancient tissues, the missing genes could be reconstructed by borrowing from genes of modern relatives of the dinosaurs—such as today's reptiles (he did not know some dinosaurs were now considered to be more closely related to birds). He stated, "We could insert these (chromosomes) into a cell nucleus, provide a yolk and an eggshell, and hatch our own dinosaur"(Lewin 1996, 202). This was the basis for the fantasies of the movies, but more importantly, also the basis for the new science of ancient DNA research.

In 1984 the science began with Allan Wilson, from the University of California, Berkeley, who was working with chromosomes and set out to solve two problems: (1) The poor physical state of DNA in tissue that had been dead for hundreds, let alone millions of years; and (2) The development of a technique for recovering the ancient DNA. The Berkeley laboratory had already experimented with extracting DNA and macromolecules of protein from dead tissue and was looking at the possibility of obtaining amino acid sequences to shed light on evolutionary questions. However, researchers had been discouraged by the degradation or fragmentation of the protein molecules when found in aged tissues. For example, they attempted to extract DNA from a dried muscle of a quagga, an animal similar to a zebra, from an Amsterdam zoo that had died in 1883, but had been stored in the basement of the Museum of Natural History in Mainz, Germany (Lewin 1997, 203; Note: an article, "Faded Genes," in *Scientific American* 1985, stated the quagga had died 140 years ago which would indicate a date of 1844). Even though there had been much degradation of the mitochondria, the DNA was intact enough to be compared with the DNA extracted from a modern zebra and a modern horse. They found that the quagga, which in scientific literature had been compared anatomically with these two animals, presented a molecular analysis that showed no differences in DNA sequence between it and the plains zebra (Higuchi, Bowman, Freiberger, Ryder, and Wilson. 1984a, 282–4). This was the first demonstration that clonable DNA sequence information could be recovered from remains of extinct species (Lewin 1996, 124–5). They began speculating on the possibility of retrieving DNA from insects trapped in amber (Poinar 1993, 34–8).

In addition to Higuchi and Wilson, early pioneers in DNA research from amber inclusions were Poinar and Hess, of the University of California, Berkeley, who in 1982, examined the ultra-structure of tissue in a gnat estimated to be about 40 million years old from Baltic amber using an electron microscope. Poinar and Hess reported not only recognizable striated muscle fibers, but also cell nuclei, ribosomes, lipid droplets, endoplastic reticulum, and mitochondria (Poinar and Poinar 1992 and 1999, 194). They also found similar inclusions in tissues of an adult braconid wasp in Canadian Cretaceous amber (70–80 ma; Poinar and Poinar 1994, 144–7).

While working at the Cetus Corporation, Kary Mullis developed the polymerase chain reaction (PCR) technique that permits tiny amounts of DNA to multiply and reproduce to larger quantities for subsequent sequencing that can be studied The work of Mullis, who won the 1993 Nobel Prize in chemistry for his invention of the procedure of PCR, was to be pivotal in the extraction process of DNA from ancient insects in amber because only very small amounts of DNA are typically obtained from embalmed insects. Earlier in 1970 methods had been developed to use "restriction" enzymes to cut the DNA strand, which improved the effectiveness of isolating DNA. However, with such small quantities, it was necessary to boost the amount of material by adding a bacterium on the natural piece of DNA. Next, the bacteria were cultured and as they divided the inserted DNA were also copied. The PRC simplifies the technique of isolating and amplification of the DNA samples and being a cyclic process, it doubles the target DNA with each sequence. One molecule can be processed through 30 cycles of PCR yielding about a billion (Lewin 1994, 124). After heating, primers are added to produce a "DNA polymerase." Researchers now had modern scientific tools and techniques to attempt to isolate DNA from fossil insects in amber. The crucial point was avoiding contamination (Lewin 1996, 124–5).

Higuchi and Wilson (1984, 1557) removed tissue from body cavities of eight insects embalmed in Dominican amber. They found that the remains seemed to contain DNA templates capable of directing the copying of radioactive complementary DNA. To do this an exogenous primer, made from a purified single strand of the DNA multiplied with the assistance of an enzyme when bacteria such as *Escherichia coli* was added. Though this sounds very technical at this point, a description of the use of primers and enzyme polymerase in DNA extraction will be explained further when describing DeSalle's work.

They added phosphates, such as P32 labeled nucleotide triphosphates, being radioactive, which would attach to DNA and in ultraviolet (UV) light indicate DNAs presence by producing a red glow. Higuchi and Wilson's (1984, 1557) results indicated that DNA was present, but tests to determine if the DNA was from the insect tissue in amber or from contamination were not performed (Poinar 1992,

Fig.11.1. Raul Cano, George Poinar, and George Poinar, Jr. in the lab with DNA model in Berkeley, California. This team was one of the ones who successfully extracted DNA in 1992. Photo courtesy G. Poinar.

272). Higuchi et al. (1984, 282–4) stated: "If the long-term survival of DNA proves to be a general phenomenon, several fields including paleontology, evolutionary biology, archeology, and forensic science, may benefit."

In 1992 successful extraction and evaluation of the genomic DNA from stingless bees, *Proplebeia dominiana*, from Dominican amber was announced by Cano, Poinar, and Poinar (1992, 249–51; see Fig. 11.2).

Though this type of research is done in specialized scientific laboratories, a generalized description of the process follows that was adapted from the description by Rob DeSalle (DeSalle and Lindley 1997, 19–38) of the American Museum of Natural History, who, working with David Grimaldi, who also isolated DNA from fossils in amber in 1992.

According to DeSalle, to begin an experiment, the researcher, attempting to isolate ancient DNA from an insect embalmed in amber, must first find a transparent piece of amber containing a fossil inclusion that is clearly identifiable. Early researchers usually selected pieces about the size of walnuts with fossils about the size of mosquitoes. (DeSalle and Lindley 1997, 19-38)

The next step is to release the fossil from the amber without destroying the insect itself. Crushing or pounding would obviously destroy the fossil. In addition, it would be impossible to free the insect without picking up other microscopic pieces of pollen, midges, and small organisms from the past that were also embalmed in the amber, which, of course, would contaminate the results because each has its own DNA. Trying to dissolve the amber could cause the insect to float apart during the process and solvents would destroy parts of the inclusions. Chipping away pieces of amber risks damaging the insect with the chips and exposing the organism, damaging the specimen.

The researchers had to devise a way to cut the amber in half without disturbing or contaminating the insect, because once the amber was opened and the fossil exposed, airborne DNA from modern flies, viruses, bacteria, skin flakes, and other debris would land on the specimen. The first technique was to use a small circular saw on the end of a flexible shaft drill, similar to a Dermal used by some amber carvers. The small saw was carefully manipulated around the circumference of the amber piece without cutting too close to the insect. The amber was cut to the point where it could be broken apart, but not into the insect. Thus, the fossil was still intact in amber. When finished, the piece was carefully washed to remove dust from sawing and other DNA contaminants with alcohol according to DeSalle and Lindley (1997, 19–38).

These researchers suggest to actually obtain DNA, the next step of extracting the insect had to be done where contamination could be eliminated, e.g., no outside DNA would fall upon the specimen, and the DNA contained within the insect would be preserved. This required special equipment such as UV light and a box that would eliminate outside air. UV light affects the structure of DNA by altering the molecules' chemical components. Therefore, shining UV light on a potentially contaminated object would destroy any DNA on it. Based on this quality, a Plexiglas box, about the size of a boot box, was equipped with a UV light. It also was equipped with a positive airflow, with filtered air being forced into the box at a pressure slightly greater than atmospheric pressure. This caused air to flow out of the box when it was opened, rather than outside air flowing in. This prevented contaminants from entering. Rubber gloves were sealed to the inside wall of the box so the researcher could place his hands in the gloves without opening it to manipulate the objects inside (DeSalle and Lindley 1997, 19–38).

DeSalle and Lindley explain that all equipment to extract the insect, such as instruments, chemicals, test tubes, and caps, were placed into this box. Then UV photons were used to destroy the DNA on the equipment and the insides of the box. Since Plexiglas is opaque to UV, objects outside the box were not affected. After the UV light was switched off, the partially sawed amber piece was placed into the box. It was important that the box itself was opened as little as possible even though there was a positive airflow. Most important of all was the entire lab was kept as sterile as possible and away from other experiments to ensure that no contaminants were introduced into the project. Tools for amber DNA research were kept completely separate from any other equipment, away from the preparation area, and away from storage rooms for any other amber research.

To begin actually extracting the fossil insect from its amber casing, Rob DeSalle and Lindley describe immersing the scored amber nugget into super cold liquid nitrogen. This process makes amber more brittle and easier to break. The researcher continued to work with his hands in rubber gloves while using surgical forceps, carefully avoiding tearing the gloves as he gently broke away the amber from the insect inclusion. This procedure required great care so the insect would not be destroyed during the process. Remember, the insect had been entombed in this casing for 40 to 95 million years (depending on the type of amber with which the scientist was working). The amber's preserving quality was the very substance that kept the DNA intact and preserved this long. Exposure to air would render the insect vulnerable to bacteria and contamination—the cause of decay and decomposition (DeSalle and Lindley 1997, 19–38)

Prior to opening the amber, plastic test tubes were placed in the contamination-free box so the specimen could be placed in a sterile tube and sealed before being removed from the sterile environment. Researchers usually work in pairs so the assistant can record while the other one has his hands in the rubber gloves in the box. Test tubes are numbered in order to record what was placed in each tube.

The insects are very fragile so they actually had to be pulled apart as they were extracted from the brittle amber. Thus, the researcher would call out exactly what piece of tissue he was placing in each tube, that is, a wing in test tube one, a leg in test tube two, and so forth. The assistant recorded the information to prevent confusion developing later. Rob DeSalle suggests practicing pulling apart a modern insect [a dead one, of course] to get the feel of the procedure so the fossil will not be completely destroyed during the process. Once the test tubes were tightly sealed, the plastic box could be opened. All instruments used were then cleaned and sterilized to prevent unknown bacteria from the insect and amber from contaminating the next project.

Even the sealed test tubes were kept in an uncontaminated place to prevent unknown bacteria from entering the project. At this point in the procedure, the amazing ability of amber to preserve the insect was obvious as the researchers examined pieces in the test tubes. The parts were as intact as modern-day insects, including details of the digestive system, thorax, muscles, eyes, and so forth. The amber-producing resin caused the insect to dry out quickly, but it appears some chemical fluid in the original resin must have quickly soaked into the body, embalming the tissues. These well-preserved tissues could now be examined under an electron microscope to identify striated tissue of the muscle or venation of the wings. However, the piece of tissue used to extract DNA during the next steps of the procedure would be destroyed by continuation of the required manipulations.

Previously standardized conditions had been developed for working with DNA and were followed when the ancient DNA science began. These were first developed so the public would not worry about contaminating today's environment with bacteria and viruses from the past. But these same conditions are also important to keep today's organisms from contaminating the work in progress, thus, resulting in wrong conclusions based on material not present in the original tissues that were intended to be examined. Therefore, researchers continue the next step of the experiment in a different laboratory, with a separate entrance away from other laboratories and with a positive airflow. This lab was equipped with plenty of UV lights to destroy DNA. The

UV lights were kept on when not working with the amber fossil sample. Again the scientists wore gloves, a lab coat, as well as a mask with a respirator.

Proceeding with the experiment according to DeSalle, the researcher takes only a small portion of the specimen so the rest is preserved for further study if something goes wrong with the initial experiment. The DNA is inside the cells, which must be broken open. Since the cell membranes are made up of lipoproteins, a detergent solution can be used to break up the molecules. The soap dissolves the fatty proteins. Also, an enzyme is added because proteins such as deoxynuclease that result from the lipid breakdown may destroy the DNA. It is the cell structure that keeps DNA, contained within, away from the degrading molecules. Now the test tube contains a slushy fluid – a mixture of tissue and chemicals in addition to the DNA.

Next DeSalle describes adding phenol, an organic solvent, to the mixture. He advised that care needed to be taken at this point because phenol is a dangerous material and can burn the skin and even cause liver cancer, but it is used to separate DNA from the rest of the material in the test tube. The phenol will dissolve fatty molecules from the cell structure and will result in DNA molecules floating around in the water away from the soapy solution. Some proteins will not dissolve in either phenol or water and will form a cloudy layer between the two. The scientist shakes the test tube to thoroughly mix the contents, and places the tube in a centrifuge to spin for a short time. This forces the water and phenol to separate. Water containing DNA comes to the top, next is the milky layer of proteins, then the phenol layer on the bottom. A pipette was used to suck the water layer from the top of the test tube. Now as mentioned, the water contains the DNA, but since DNA falls apart in water, it can't be left in this condition for long. If it were to be stored, it would have to be frozen. Another purification step must be followed to eliminate any contamination that may have occurred in the water layer containing the DNA.

According to DeSalle (1997) the method used since 1990 was to separate the layer of water containing the DNA from the phenol layer. After discarding the phenol material, the water was placed in a new test tube with ethanol or alcohol and a small amount of a salt. This caused the DNA to precipitate out of the water and ethanol, forming a white precipitate. Then all that is needed is a regular lab centrifuge to separate the cloudy white precipitate from the solution. The DNA collects in the bottom of the test tube and the water and ethanol mixture is poured off. A drop of distilled water may be added to the test tube if only a small amount of DNA was collected.

DeSalle and Lindley (1997, 34) used Mullis' PCR procedure to increase the amount of DNA for later examination. The precipitate of DNA, with the distilled water added, now must be added to a specially prepared mixture of chemicals that work in the PCR machine to reproduce the DNA present. The complex DNA molecule is composed of discrete chemical units of four different types that are bound to another complementary chain. These form a spiral with each unit linked in a special sequence to a complementary unit in a double helix arranged in opposite directions forming a double-stranded DNA. Since DNA in the test tube has broken down over the pursuing ages, the material in the test tube will be broken into smaller fragments or pieces of these double helices specific to the insect from which it came. An enzyme is needed to help reproduce these DNA molecules. The machine is used to heat the test tube almost to the boiling point to cause the double helical fragments to dissociate from each other, but keep the individual strands intact. When the machine lowers the temperature to 75°C, the mixture with the enzyme will reform the double helix, fashioning the correct adjacent strand from the free-floating monomers in the mixture. What the enzyme polymerase does is manufacture the right DNA sequence to pair up with a single DNA strand already present. With each cycle the DNA is doubled. The machine can complete as many as 30 cycles in a couple of hours producing a billion times the DNA of the original sample.

According to DeSalle, the PCR machine needed can be as simple as metal hotplates with holes that accommodate 50 or more test tubes equipped with a small computer that can be programmed to set the number of cycles to go through and the rate at which to perform the cycles. However, more elaborate machines pick up the test tubes and dip them into hot and cool water or blow hot and cool air over them. The PCR is more efficient when temperature changes are as fast as possible. The enzyme, called polymerase, comes in many versions, but the most effective is one that can withstand high temperatures such as the version from a one-celled organism, called *Thermos aquaticus*, which lives in hot water. It was first found in hot springs in Yellowstone National Park surviving in water that was almost boiling. The "Taq polymerase," as it is called, can withstand the high temperatures needed for the PCR in order to break the DNA strands apart. The Taq polymerase is inactive at the high temperatures, but becomes active again when the test tubes are cooled. This is why the cycle from hot to cool is important.

Another ingredient necessary to assist the Taq polymerase to start working is called a "primer" and consists of a single strand of DNA, usually about 15 to 20 bases long. But the single strand DNA primer must match the DNA that is being reproduced. The primer will be able to attach itself to another single strand of DNA only if it has the complementary sequence. Therefore, the researchers use primers that match known insect sequences. The modern sequence primers are used to attempt to amplify DNA from the fossil insect that is 30 to 125 million years old. If the researcher has the correct primer, he or she can amplify tiny amounts of DNA from ancient insects or dried blood or bacteria.

After the correct primer is added to the test tubes in the PCR machine, the machine is heated to break apart the DNA double helices. It then cools the solution down to about 50°C—cool enough to allow the primers to hook up to the longer strands of separate DNA, making the beginnings of new double helices. Then, the PCR machine brings the temperature back up to about 75°C, which is the temperature Taq polymerase functions at most efficiently. The enzyme can then work at creating more double-helical DNA strands, making a copy of the original. Now, if no contaminants have entered the experiment, the DNA in the test tube retrieved from the PCR machine will be the DNA from the fossil insect, not from the researcher himself or the surrounding environment.

For a more detailed explanation of the DNA molecule and how to "reconstruct a dinosaur," see DeSalle's book *The Science of Jurassic Park and the Lost World or How to Build a Dinosaur* (DeSalle and Lindley 1997).

Now that two different laboratories at opposite ends of the United States had demonstrated that DNA could be isolated, other scientists were encouraged to pursue the search of ancient DNA from insects in different varieties and ages of amber. Andrew Ross (Ross 1998, 32-3), curator of fossil arthropods for the Natural History Museum of London, reports that scientists at the London Museum have tried to repeat the experiments to obtain DNA from fossil insects in Dominican amber and have found no DNA, therefore, casting doubt on earlier reports. Ross asserts that insects found in Baltic amber, particularly, are often found completely hollow without any internal tissue preserved because bacteria and enzymes continue to degrade the gut of the insect from the inside. Even though this is often the case, conversations with George Poinar suggest that contamination during failed experiments may be the cause of finding no DNA.

While other researchers discuss the controversy of DNA preservation in amber, such as Gutierrez and Marin (1998) who point out that the most ancient DNA recovered from an amber-preserved specimen may not be as ancient as it seems. Dr. Poinar provides a set of criteria to be followed if success is expected (Poinar 1999, 195). Poinar states that the amber fossils have a specialized situation because the tissue is protected by continuous contact with resin for millions of years. He also points out that only insect inclusions that were completely immersed in the resin after contact, that is not being exposed on the surface of the resin for hours during which time decay takes place, are likely to retain extractable DNA (Poinar 1999, 196). In addition, he stresses the procedure for eliminating contamination must be followed with extreme care. Poinar points out that DNA extracted from the Dominican amber fossil of a stingless bee (Poinar et al, 1992, the extinct termite (DeSalle et al 1992), an algarroba leaf (Poinar et al 1994) and a weevil in Lebanese amber (Cano et al 1994) provided answers to questions related to bio-geography, evolution and phylogenetic relationships.

For further reading on this controversy, also see *Ancient DNA from amber fossil bees?* by Walden and Robertson (1997) or *Problems of reproducibility - Does geological ancient DNA survive in amber-preserved insects?* by Walden and Robertson (1997).

Figure 11.2 Dominican amber fossil with two enhydra or bubbles with air and water. Does this air and water originate from he era when the Hymenaea protera was growing in the Dominican region or has osmosis changed the gas content so it no longer represents a sample of the original air. Amber is a porous material. This question is still debatable. Ross describes these large bubbles projecting from the bodies of termites in Dominican amber as being common and consisting of methane produced by the bacteria, in the gut of the termite, that helped it to digest its food.

CONCLUSIONS

Oh ant, you were passing by a mighty tree
When a drop of amber fell down unto thee.
So typical when you live you're just a poor soul
When you're dead you're worth your price in gold.
MARTIAL (AD 88). Epigrams, book VI

Amber is a major player in the development of human cultures and a valuable paleoenvironmental marker. This versatile gemstone and scientific marvel is appreciated for its simplicity and complexity, its beauty and utility (see Fig. 12.1). This book compiles the history of amber and amber research to provide a wide spectrum of facts related to amber. As a medium for the connoisseur, jewelry collector, bead collector, chess and sculpture collector, as well as the scientific researcher, amber requires an interdisciplinary approach to appreciate this beautiful mysterious gemstone. As the result of scientific research, fossils within amber have received wide attention even in the popular world. The study of amber fossils has provided new information such as the first appearance of particular groups of insects in the fossil records, new dates for recognizing symbiotic associations, and a view of microevolution in insect groups. Paleoentomologists continue to study fossil inclusions to identify new species, the number of different species, the abundance of each species, and their relationship to the habitat.

Questions such as; What did the forests look like during the time of the amber-producing trees? What insects became extinct when mass extinction of life on Earth occurred as a result of past catastrophic events? still need to be answered. For example, when a massive aster-

Figure 12.1 Amber casket and round amber beads made by 21st century Tsarskoe Selo Palace craftsmen after the style of the 18th century Amber Guild masters. These artistic techniques were lost for many years, but now are being revived in Russia, Lithuania, Latvia, Poland, and the Ukraine.

oid smashed into the Earth about 65 million years ago, perhaps causing the destruction of the two orders of dinosaurs, why didn't the 17 orders of insects found in amber occurring in the Cretaceous become extinct? This question encouraged more study of the Cretaceous amber insects and during the past ten years, scientists researched and documented Cretaceous fossiliferous amber from New Jersey (Grimaldi 1998), Lebanon (Poinar 2001) and reexamined fossils from Myanmar (Cruickshank 2003). The reassigning of the amber bearing strata of the mines in Hukawng Valley, Myanmar, by Cruickshank revealed that the paleoenvironment in which the Burmese amber was produced may have been the warmest tropical climate of all of the Cretaceous amber forests (Grimaldi and Engle 2002).

For example, Engel and Grimaldi (2004) documented new sub family, genus, and species of primitive earwigs in Middle Cretaceous amber from mines in Myanmar. They describe the oldest definitive fossil, naming it *Burmapygia* n.sp., on record. (superfamily, Pygidicranidae,) In 1992 Poinar described parasitic wasps (family Serphitidae) found in Canadian and Siberian amber as no longer in existing at the end of the Cretaceous age (p. 274). We now have documentation of prehistoric ants living together in colonies in New Jersey 92-million-year-old amber fossils (Agosi, Grimaldi, and Carpenter 1998). Even though this social structure was developed in ancient times, about 50 million years earlier than previously thought, ants were not very plentiful until the Eocene age. Dr. E. O. Wilson, a biologist from Harvard, told the Washington Post (1998) "We know that ants were stinging dinosaurs."

Another area for scientific study is the question of providing the exact age of amber. The geologist's method of examining the strata where amber is found is restricted because amber has been redeposited and is located in secondary layers. Methods attempted for dating amber include infrared, mass, and nuclear magnetic resonance spectroscopy. However, according to Poinar (1992. 275), using nuclear magnetic resonance spectroscopy appears to be the most promising because it measures the exomethane peaks of fossilized resins that are considered to peak in inverse proportion to the age. Then again, this method was challenged by Grimaldi (1995, 203–17), who disputed the age indicated for some Dominican amber by comparing fossil insect evidence in some amber specimens.

Actually, the oldest fossil resins date back to the Carboniferous period about 320 million years ago and differ physically and chemically from other fossil resins known. Scientific analysis showed they were produced by seed ferns (pteridosperms). Permian age (260 Ma) limestone in the western piedmont area of Chekarda River in the Ural Mountains yield microscopic lenses of resin. Triassic period fossil resins have been found in Bavaria encasing primitive bacteria, protozoa, fungal spores, algae, protozoa and amoeba-like organisms (Poinar 1992; Platt 1995). Poinar (2001) explored the amber of Lebanon describing it as the oldest amber in the world containing insect remains. The study of Cretaceous fossiliferous resins provided evidence of the first appearance of eusocial termites, bees, and vespid wasps. This study has implications for interpreting distributions of modern day ants (Grimaldi and Agosti 2000).

    Amber and copal were produced from many different kinds of trees - from tropical broad-leafed trees found in the Dominican Republic to a variety of different conifers. Since the main characteristic used to identify Baltic amber was the presence of amber acid or succinic acid content, which ranges from three to eight percent, the pine source as the original amber tree has been brought into question. Generally, pine tree resins do not contain succinic acid. Therefore, designating of amber producing trees as *Pinus succinifera* by early scientists in the Baltic region was imprecise, even though it was a collective term for many extinct trees thought to have produced an unnaturally large secretion of resin. The term was coined based on the botanical inclusions studied by earlier botanists. Based on these inclusions, it is known that the climate must have been sub-tropical because the amber forest included palm trees, cypresses, magnolias, rhododendrons, tea shrubs, oaks, maples, horse-chestnuts, mistletoe and cinnamon trees with a ground cover of heathers, mosses, lichens and fungi.

    Paleobotanists continue to search for affinities with recent tree resins. There has been a lengthy debate about the origin of the Baltic amber tree. Botanists have put forth three most likely species of living trees that could be related to the ancient source tree that produced the succinite amber in Baltic regions. These candidates are the *Cedrus atlantica*, or the cedar now growing in the Atlas Mountains in North Africa; the *Pseudolarix wheri* or larch now growing in Canada; or the *Agathis australia*. Recent chemical analysis indicated it was more similar to the resin produced by monkey-puzzle trees (family Araucariaceae). Today Kauri pine, also a member of this family, continues to grow in New Zealand and Australia. The main problem is no araucarian inclusions have yet been found in Baltic amber. However, recent evidence indicates a pine origin because studies have recently shown some trees within the pine family may also produce succinic acid. Both plant and insect inclusions indicate that Baltic amber was produced in an environment that was subtropical and supported a mixture of different trees (Ross 1998, 18).

    Dominican amber, on the other hand, has been shown to be related to the *Hymenaea*. Poinar described the amber producing tree as *Hymenaea protera* (1991b) a leguminous tree now extinct. The flora and fauna structures found in the amber resemble the present-day trees, commonly called "algarroba." These inclusions support the data found by infrared spectrometry studies by Schlee and Glochner (1978), nuclear magnetic resonance studies by Lambert, Fry and

Poinar (1985) and the pyrolysis mass spectrometry studies by Poinar and Haverkamp in 1985. Francis Hueber and Jean Langenheim (1986) compared the Dominican amber with resin from the present day leguminous tree species of *Hymenaea verrucosa*. This did not present the same dilemma as Baltic amber because structures found in the resin corresponded with the technical analysis. For a fascinating description of the Dominican amber-producing forest one needs to read George Poinar and Roberta Poinar's book, *The Amber Forest-A Reconstruction of a Vanished World* (1999).

Another question in the minds of amber "aficionados" is; How does resin become amber? Many researchers describe the original resin being produced in abundance because it was either a defensive mechanism against fungal attack, or perhaps insect infestations, or caused by radical climactic change. It is important to realize that it was the resin from vascular tissue that changed into amber, not the "sap." It is not the same as sap, which is a watery juice which transports mineral salts and sugars through the plant, generally a hardwood tree. In contrast, resin is a sticky vascular tissue that protects the tree from insect invasion and disease. It is exuded in tear drop shapes, stalactites , or blobs that flowed or dripped down the trunk or filled cracks in the bark of the tree.

Thus, flowing from the vascular tissue from some kinds of trees, the resin began to go through a number of stages to become amber. First, as the amberization began eons ago the resin becames insoluble in water, then the sticky clumps were carried by moving water and then buried in the sediments of an oxygen poor environment. Once resin, is released from the tree, it will decompose if left exposed to the atmosphere. During the first stage, the molecular structure within the resin began to "cross chain link" with one another, called a polymerization process. There is no exact time ascribed to this process, but it generally takes several million years in a continuum starting from the fresh resin until it becomes amber. At first, this makes the resin hard and brittle as opposed to its soft sticky state. This hardened state, called copal, is an intermediate stage and is not as hard or durable as true amber.

Copal deposits today are found in large quantities mainly in the southern hemisphere in regions such as Africa, Australia, New Zealand, and South America. Colombia in South America has extensive copal deposits. Large gem shows often have dealers selling this resin as fossiliferous amber, but the insects are recent. Poinar's testing of this resin has shown much of it is less than 250 years old. Madagascar and Kenya also have mines producing highly fossiliferous copal.

After polymerization begins, the next stage is the evaporation. Volatile oils, called terpenes, slowly evaporate out of the amber. It may take millions of years before this process changes the copal into a solid structure with longer chains of molecules approaching the structure of amber. Some research indicates these stages need to take place in an environment that is anaerobic, or even sustained for a period time immersed in seawater. As trees fell and decayed, the viscous resin hardened and was carried into the rivers. This river-borne resin reached seas where it lay in a salt-water environment under-going a series of changes that transformed it into amber. According to Dr. Kos-mowska-Ceranowicz (1996, 3) amber from the Sambian Peninsula and Poland is actually in a secondary deposit. She describes the original amber forest as being located further South, stating that identification of the Baltic Sea as the origination is mistaken, because the Baltic is a mere ten thousand years old, while the amber itself is at least forty million years old. Thus, amber today is considered to be redeposited by the Tertiary seas making it much older than the formation, especially so when found in alluvial sands along Baltic shores.

Even today if amber is exposed to air for several years, it undergoes oxidation. The surface, even if fashioned into jewelry, darkens and changes color. If natural pieces were buried in a sand dune where they are aerated, they form a crust over the surface and continue to darken. This oxidation process is attributed to one of the reasons why the oldest amber is usually found in small pieces weighing only a few grams. For example, Lebanese deposits, dating back 125 million years, usually are found in minuscule sizes (Poinar 2001). Hence, natural amber is a continually changing gemstone, with all the processes still going on that helped the original sticky resin set hard enough so it can be worked and polished.

Baltic amber, only about 40 to 50 million years old, is found in sizes and shapes ranging from tear drop to icicles or stalactites to large blocks. The largest known piece of Baltic amber ever found in Poland weighed 21.5 pounds (9.75 kg.) and was found in 1860 near Szczecin (German Stettin) Poland, on the Odra (Oder) River. The piece is now in the Natural History Museum of Berlin, Germany. Many large pieces of Paleogene amber have been documented with the largest confirmed piece coming from Sarawak, Malaysia (Borneo)-, and weighed 150 pounds (68 kg.).

During the viscous stage a wide range of life became encased in amber—not only the small insects too weak to pull out of the sticky resin, but also small lizards, frogs, spiders and other arthropods. Though small lizards had been found in Dominican amber, it was not until 1997 that an authenticated lizard was reported in Baltic amber (Kosmowska-Ceranowicz, Kulicka, and Gierłowska 1997). Evidence of birds and mammalian life are evident from feathers, hairs from rodents, and paw imprints. Baltic amber has also encased mammalian molars, perhaps from a pig that had died and laid face down in a bed of resin. It appeared as though the resin seeped around the decaying jaw preserving the teeth from decay.

In 1996 a Dominican amber mine produced a specimen of amber

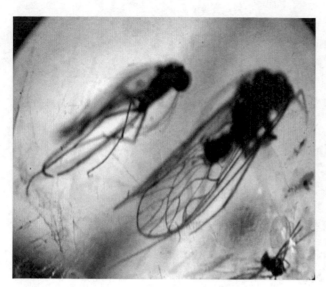

Figure 12.2. Winged insects in Domincan amber. Insects make up about 85% of all the fauna inclusions in amber. Length 3.7 mm

including the spine and ribs of a mouse (MacPhee and Grimaldi 1996). This discovery completely changed the theory of the population of the West Indian Islands by land animals. Poinar (1999) described the source of the bones as coming from a small insectivore belonging to a now extinct genus *Nesophontes*. In 2002 the first English edition of Weitschat and Wichard's *Atlas of Plants and Animals in Baltic Amber*, originally published in German (1998), became available identifying the variety of fossils previously found in Baltic amber. They describe fauna inclusions, size, habitat and behavior as selective agents and attribute the predominance of insects as a result of selective fossilization. It also may be an indicator that there was a high frequency of insects in the paleobiosphere of the amber forest. Based on Baltic fossil inclusions insects comprise 85 percent of the total fossil fauna in amber. The next most frequently found non-insect arthropod were mainly Arachnida, or spiders, taking up 12 percent of the animal section. This leaves only 3 percent for gastropods, worms and vertebrates (Weitschat and Wichard 2002). If the reader is looking for systematic taxonomic reference for fossils in amber, this is the source book to refer to and is way beyond the scope of this volume. It even includes an index of scientific names and taxonomic terms. For a simplified method for the amateur to identify the fossils within his/her own specimen, Ross (1998) provides a "Key to Identification" for Arthropods, bugs, and winged insects. Another book by Poinar (1999), *The Amber Forest*, describes the Dominican amber producing forest.

Currently scientists, knowing amber's origin as fossil resin from prehistoric trees, now classify it as an organic gemstone or as a mineraloid. Dr. Kosmowska-Ceranowicz (1996) points out the processes involved in transforming resin into amber have chemical equations that explain how the volatile particles in fresh resin were lost as it solidified - isometric processes, polymerization, and oxidization   assisted by bacteria all contributed to the whole process. Even with all this knowledge, exact times or ages when the resin changes to copal and then to amber has not been identified in a clear-cut chemical model.

In 1995 Anderson and Crelling at the American Chemical Society Symposium proposed a new classification and nomenclature of fossil resins based on their chemical components, but many European scientists do not accept this suggestion. We still have the geological term *resinites* being used by geologists and petrologists. To distinguish Baltic amber, it is the succinic acid content separating amber as *succinite* and those fossilized resins with less than three percent succinic acid have the name *retinite* assigned to it. However, researchers tend to use Anderson's classification when describing new fossil resins.

Petrologists continue to find fossil resins in their search for fossil fuels. Thus, amber presence something for all. Amber deposits of different ages occur in many places all over the earth. Many of these deposits only have trace amounts of amber. Only a few locations have deposits large enough to be mined and used commercially.

Archaeologists continue to study amber artifacts from archaeological sites to determine old trade routes leading from the Baltic regions to England as early at 10,000 BC and to the Mediterranean as early as 1,600 BC. Based on archaeological finds that revealed amber had a significant roll in European culture, an Italian-Polish Commission for Amber Route Studies was organized in Rome in 1972. This commission saw two trends in current research into amber: 1, the significance and role of amber in culture, both human and social life, and 2, studies of amber's geological and natural aspects. All of the latest methods and techniques of analysis aid in the scientific investigations and interdisciplinary character of amber research (Tabaczyńska 1997).

The amber medium itself also has provided an area for scientific research such as the process that gives amber its unique preservation qualities. Different types of amber preserve the ancient organisms in different ways. Dominican amber is reported to be the best preservative because the fossil usually still has the internal tissues preserved in perfect condition. see Fig. 11.3 and 12.3 Often Baltic amber insects are not as well preserved and hard-shelled bugs may be hollow with decayed insides or there may be a white coating around the insect. Borneo, Burmese and Mexican amber insects are less preserved and may be incomplete and distorted (Ross 1998, 24). Scientists continue to sort out fakes such as the so called "Piltdown Fly," a carved fake discovered in the collections of the Natural History Museum, London, that in 1966 cast doubts on Baltic amber reflecting the fauna of prehistoric time because living species were preserved in it. Then, in 1993 it was reexamined using modern scientific equipment. It was discovered the fly had been inserted into an amber piece, which had been hollowed out and glued back together (Ross 1998, 5). Currently, on the market one can buy a falsified fossilized lizard in amber for $100. The base is usually real amber, but the lizard inserted similarly to the "Piltdown Fly" fake.

To the general population, the role of amber as an

Figure 12.3. Wood-boring beetle (Coleoptera: Platypodidae) in Dominican amber. Wood particals and beetle droppings are also in the piece. More than 70 families of beetles have been recorded in dominican amber. Even though these beetles probably bored into the Hymenaea protera trees that procduced the Dominican amber, they are rare in Baltic amber. Length 3.2mm.

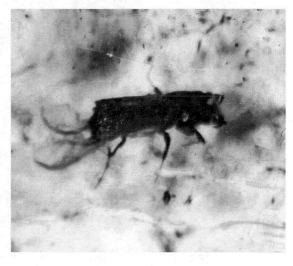

Figure 12.4. Amber altar called "Our Lady of the Labor People," in St. Brygida Church in Gdańsk. illustration from the M. Drapikowski's publication 2001, Polski Jubiler, No. 11-25.

artistic medium is how amber actually touches and influences lives. Amber has been a precious stone since the Phoenician traders. It has found its place in folklore and for medicinal values. Amber cosmetics are being manufactured in Poland based on the belief that the amber acid will create a youthful complexion. As far-fetched as it may seem, during World War II Russian scientists experimented using amber as a conductor in some rockets. Large amber creations were thought to be a thing of the past, only preserved in sculptures created by 18th century amber guild craftsmen, however, in May of 2003 the opening of the reconstructed Amber Room by Russian amber artisans brought many admiring tourists to St. Petersburg to view this magnificent work and wonder – an exact copy of the room commissions in 1701 by Frederick 1 King in Prussia. The only difference from the original is that some of the large carvings today are now made from fine quality pressed amber and dyed to enhance the color.

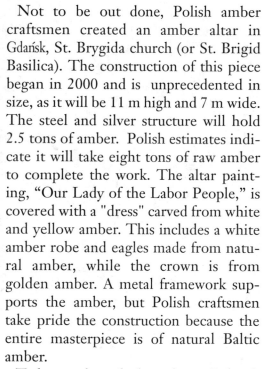

Not to be out done, Polish amber craftsmen created an amber altar in Gdańsk, St. Brygida church (or St. Brigid Basilica). The construction of this piece began in 2000 and is unprecedented in size, as it will be 11 m high and 7 m wide. The steel and silver structure will hold 2.5 tons of amber. Polish estimates indicate it will take eight tons of raw amber to complete the work. The altar painting, "Our Lady of the Labor People," is covered with a "dress" carved from white and yellow amber. This includes a white amber robe and eagles made from natural amber, while the crown is from golden amber. A metal framework supports the amber, but Polish craftsmen take pride the construction because the entire masterpiece is of natural Baltic amber.

Today amber dealers from Poland, Lithuania, Ukraine, and Russia are found selling amber articles at gem and mineral shows throughout Europe and the United states. Elite high fashion models in Poland adorn themselves with amber created in exquisitely modern styles in the Amber Association of Poland's Amberif Fashion Shows. Amber not only enhances the beauty of the wearers, but mystical beliefs say it helps one be joyful, happy and lighten the burdens of life. Just as in ancient times, amber is used to tap into the power of the sun, bringing success, abundance, healing, vitality and joy. Also when amber is carved with special markings, particularly one's initials, it is thought to be a protective talisman.

Figure 12.5. Amber altar called "Our Lady of the Labor People," in St, Brygida Church in Gdańsk.. Folds of the robes flow outside the painting. When finished they will be part of a three-dimensional composition filling the presbytery. Courtesy Gabriela Gierlowska.

**Peasants in the Kurpie Region
of Northern Poland wearing
amber at the Archealogy
Museum in Poland.**

APPENDICES

APPENDIX A
CLASSIFICATION SYSTEM
FOR FOSSIL RESINS

In 1995, Ken B. Anderson, Chemistry Division of the Argonne National Laboratory, and John C. Crelling, Department of Geology of the Southern Illinois University organized an American Chemical Society symposium of *Amber, Resinite, and Fossil Resins.* The symposium and resulting book brought together researchers, studying amber from many different fields such as biology, geology, petrology and chemistry. Their reports focused on fossil resin research, such as chemical studies of amber, including structural characterization, isotopic composition, maturation studies, resinite-derived oils, and amino acid distributions. Rather than mentioning the older European classification, the authors described amber among sedimentary organic products, yet being unique for its exceptional preservation (including exceptional preservation of materials included with it); for its value as an "organic gemstone"; and for the extraordinary role it has played in human history.

The authors report the current state of the art in scientific (especially chemical) studies of amber. Their chemical properties are a consequence of both their biological origins and the geological environment into which they were deposited and in which they have subsequently matured. Using modern analytical techniques, researchers are now able to address questions that could not be addressed in the past.

The first problem encountered was a difficulty with the nomenclature of the materials, amber, resinite and fossil resin as well as the term "resin" being used alone only for modern samples.

The authors define amber, resinite and fossil resins as: *"solid, discrete organic materials found in coals and other sediments as macroscopic and microscopic particles, which are derived from the resins of higher plants."* They explain two principal aspects of their definition being (i) fossil resins must be solid, discrete bodies, and (ii) that they must be derived from the resins of higher plants. However, past researchers used a variety of names of fossil resins or amber, making it difficult to compare current research results. In particular, European researchers, who discovered much of the earlier recorded amber, based the names on locality, discoverer, etc. Therefore, amber finds are documented using such names as Glessite, Beckerite, Stantienite, Rumanite, Chemawinite, Burmite, Simetite, Bitterfeldite, Settlingite, Walchowite, Schraufite, Schlierseeite, etc. Many of these names are entrenched in popular usage and in scientific literature making it impossible to restrict their usage.

Summarized and reprinted with permission from Ken B. Anderson and John C. Crelling, eds. *Amber, Resinite and Fossil Resins* (Washington, D.C.: American Chemical Society, 1995) ix–xiii. Copyright 1995 American Chemical Society.

Classification of amber

As an attempt to rationalize the nomenclature of fossil and recent resins and to provide a logical basis for comparison of amber for the purposes of scientific discussion, a classification system, based on the structural characteristics of amber itself has been proposed. This classification system has not yet been universally adopted, however, a number of workers who are presently using this system appear to find it useful. The bases for classification using this system are described in the Table I below.

Table I Classification System for Fossil Resins

<u>*Class I*</u> The macromolecular structures of all Class I resinites are derived from polymers of labdanoid diterpenes, including labdatriene carboxylic acids, alcohols and hydrocarbons

Class Ia Derived from/based on polymers and copolymers of labdanoid diterpenes having the regular [1S,4aR, 5S, 8aR] configuration, normally including, but not limited to, communic acid and communol and incorporating significant amounts of succinic acid.

Class Ib Derived from/based on polymers and copolymers of labdanoid diterpenes having the regular [1S, 4aR, 5S, 8aR] configuration, often including but not limited to communic acid, communol, and biformene (and related isomers.) Succinic acid is absent.

Class Ic Derived from/based on polymers and copolymers, of labdanoid diterpenes having the enantio [1S, 4aS, 5R, 8aS] configuration, including but not limited to ozic acid, ozol, and enantio biformenes (and related isomers.)

<u>*Class II*</u> Derived from/based on polymers of bicyclic sesquiterpenoid hydrocarbons, especially cadinene, and related isomers.

<u>*Class III*</u> Natural (fossil) polystyrene

<u>*Class IV*</u> Fundamental structural character is apparently non-polymeric, especially incorporating sesquiterpenoids based on the cedrane carbon skeleton.

<u>*Class V*</u>   Non-polymeric diterpenoid carboxylic acid, especially based on the abietane, pirmarane and iso-pimarane carbon skeletons."

APPENDIX B
COLLECTIONS OF FOSSILIFEROUS AMBER
IN INSTITUTIONS OPEN TO THE PUBLIC

Location	Institution	Source of Amber	Number of Pieces
Amsterdam, Holland	National Museum, amber vase with a cover decorated with enamel, colored stones and gold rings	Baltic	Due to close til 2008
Beijing, China	Geological Museum	Chinese	2,000
Berkeley, California	University of California	Mexican Alaskan	3,000 200
Berlin, Germany	Humboldt University, Natural History/Paleontology Museum, largest lump of amber ever found in Poland (9750 grams, found 1890 on Oder River.)	Baltic	20,000
Cambridge, Massachusetts, USA	Harvard University, Museum of Comparative Zoology	Baltic Canadian	16,000 5,000
Chicago, Illinois, USA	Field Museum of Natural History	Baltic	2,600
Copenhagen, Denmark	Geologist Museum, amber found in Denmark and Greenland. Houses largest piece found in Denmark	Retinite Greenland	
Copenhagen, Denmark	Rosenborg, crown jewels of the royal family, fifteenth to nineteenth centuries, including amber chandeliers.	Baltic	
Copenhagen, Denmark	Zoological Museum	Baltic	7,600
Gainesville, Florida	Florida State Collection of Arthropods, Florida Department of Agriculture	Dominican	3,500
Göttingen, Germany	Institute for Geology and Paleontology	Baltic Total	11,000 17,000
Halle, Germany	State Museum	Baltic	1,000
Hamburg, Germany	Geological Museum	Baltic	4,000

Data from: G. O. Poinar, Jr., *Life in Amber* (Palo Alto: Stanford University Press, 1992) 66; S. Ritzkowski, "Geschichte der Bernsteinsammlung der Albertus-Universität zu Königsberg i. Pr. In *Bernstein Tränen der Götter*, R. Slotta and M. Ganzelewski, eds. (Bochum: Deutsches Bergbau Museum Bochum,1997) 296; D.Grimaldi, *Amber, Window to the Past* (N.Y.: Harry N Abrams, Inc., 1996) 40, 50. 2004 various sites such as 3dotstudio.com/amber-home.html www.emporia.edu/earth-sci/amber/museum.html and direct correspondence with Museums.

Hollviken, Sweden	Swedish Amber Museum	Baltic	
Kaliningrad, Russia	Amber Museum, amber jewelry, boxes, copies of parts of Catherine the Great's amber room Location of world's biggest amber mine.	Baltic	6,000
Lawrence, Kansas	University of Kansas Nat. His. Museum - entomology also smaller representation New Jersey, Arkansas, Wyoming Lebanese, Alaskan (Engel)	Baltic Burmese Dominican Mexican	
London, England	British Museum of Natural History, includes the large 15 kg lump of amber from Burma, specimen of retinite in sandstone from New Zealand	Baltic Burmese Sicilian	25,000 300 50
Ludwigsburg, Germany	State Museum	Baltic	2,000
Malbork, Poland founded 1965	Malbork Castle Museum Old Amber Guilds historical artifacts: caskets, altars, table utensils, candlesticks, jewelry, sculpted figurines	Baltic	
Matakoe, New Zealand	The Kauri Museum includes exhibit on Kauri industry, many large pieces and carvings	Kauri	many
Moscow, Russia	Geological Institute	Baltic Sibirian	5,000 4,000
New York, New York	American Museum of Natural History, includes a traveling amber exhibition www,amnh.org/exhibitions/amber/	Baltic Dominican New Jersey many more	500 2,000 ++n
Ottawa, Ontario, Canada	Biosystematics Research Institute	Canadian	300
Paris, France	National Museum of Natural History	Baltic	2,000
Palanga, Lithuania	Museum of Amber, amber from archeological sites, numerous fossiliferous amber, natural forms and modern items Claims to have largest collection in world.	Baltic neolithic Jurassic insects	25,000 15,000
Puerto Plata, Dominican Republic	Dominican Amber Museum	Dominican amber	
Ribnitz-Damgarten, Germany	Deutsches Bernsteinmuseum both a museum and workshop to learn how to polish amber.	Baltic	

Location	Museum / Description	Type	Collection
Skagen, Denmark	Skagen Amber Museum Denmark's largest collection of insect inclusions, old Art Nouveau amber jewelry	Baltic	
Skive, Denmark of prehistoric amber beads in	Skive Museum, largest find Denmark dating from 2500–2200 B.C. largest prehistoric find of Denmark amber	Baltic	13,000 beads 50 neck- laces
St. Petersburg, Russia	Zoological Institute	Baltic	25,000
Stuttgart, Germany	State Museum of Natural History, includes Cretaceous age amber from Lebanon, Oligocene Dominican amber, Miocene amber from Chiapas, Mexico	Baltic Dominican	2,500 4,600 Total collection 10,000
Vilnius, Lithuania branch also in Nida, Lithuania established in 1998 -2002	Amber Museum Gallery - Gintaro galerija Muziejus amber from artists' perspective sculpted by Lithuanian jewelers and folk artists.	Baltic	many
Vilnius, Lithuania founded 1992	Davainis' Amber Sculpture Museum contains information on formation of amber and amber inclusions	Baltic	many
Wilmington, Delaware	University of Delaware Gallery Contains the Burgess-Jastak Amber Collection, carvings & jewelry	Baltic	250
Warsaw, Poland	Muzeum Ziemi (Museum of Earth), scientific amber display regarding origins, environments, life in amber and synthesis of present-day studies	Baltic Ukrainian Bitterfeld and other fos- sil resins of the world	25,000 Total collection 29,500
Washington, D.C.	Smithsonian Institution also has Kansas amber	Dominican	5,500

(Other museums may have amber, but these are the ones I was able to
locate information about in literature or on the internet.)

APPENDIX C
CHARTS FOR COMPARING AMBER
WITH COMMON SUBSTITUTES

Reprinted with permission from John Sinkankas, *Gemstone and Mineral Data Book*, (Tucson: Geoscience Press, 1972).

Table of Specific Gravity

1.03	Amber, copalx
1.05	Amber, copal, polystyrene
1.06	Amber, copal
1.10	Amber, copal, meerschaum, jet
1.18	Acrylic plastics (Plexiglas, Lucite, perspex) jet, meerschaum
1.19	Acrylics, jet, meerschaum
1.20	Jet, meerschaum
1.25	Bakelite (amber imitation thermosetting plastic), jet
1.26	Tortoiseshell, jet, Bakelite
1.29	Tortoiseshell, jet, Bakelite, cellulose acetate plastics
1.30	Tortoiseshell, jet, Bakelite, cellulose acetate
1.32	Tortoiseshell, jet, Bakelite, cellulose acetate, casein plastics
1.35	Casein plastics, tortoiseshell, jet, cellulose acetate
1.36	Cellulose nitrate plastics (celluloids), cellulose acetate, jet
1.38	Cellulose nitrate, cellulose acetate, vegetable ivory, jet
1.40	Cellulose nitrate, cellulose acetate, vegetable ivory, mineral coal
1.42	Cellulose nitrate, ivory

Refractive Indices

1.49	Polypropylene, acrylic, cellulose acetate
1.50	Celluloid, polyethylene, cellulose acetate
1.53	Nylon, polyethylene
1.539	Amber, glass
1.54	Amber, copal, glass, polyethylene
1.545	Amber, glass
1.55	Casein, Bakelite
1.56	Polystyrene-acrylonitrile, casein, Bakelite
1.57	Polystyrene-acrylonitrile, Bakelite
1.60	Ekanite, glass
1.64	Bakelite, glass
1.66	Bakelite, glass

APPENDIX D
AMBER FLOW CHART: AN OVERVIEW OF THE EXTRACTION AND USES OF AMBER

Amber flow chart: An overview of the extraction and uses of amber.

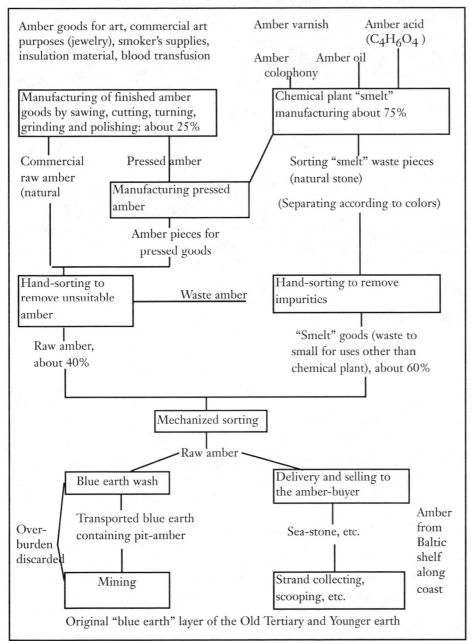

Reprinted from K. Andrée, *Der Bernstein und seine Bedeutung in Naturund Geisteswissenschaften, Kunst und Kunstgewerve, Technik, Industrie und Handel* (Königsberg: Albertus-Universität).

APPENDIX E
INSTITUTE & SOME MUSEUMS URLS:
A NON-PROFIT ORGANIZATION
DEDICATED TO THE STUDY OF AMBER
GOALS
- To develop a data bank on fossils in amber.
- To prepare exhibits and educational material for schools and museums.
- To acquire rare and unusual amber specimens for study and safekeeping.
- To characterize amber from different geographical regions.
- To provide funds to specialists for scientific studies.
SERVICES
- Identification of amber fossils. Authentification of amber material
- Evaluation & Certification of rare and unusual amber fossils
- Procure specific amber fossils for others
- Lectures on amber and amber fossils

For further information contact: G. O. Poinar, President
4405 NW Queens
Corvallis, Oregon 79330 USA

Poinar often found on:
http://oregonstate.edu/dept/ncs/newsearch/2003amber.htm

MUSEUM RESOURCES URLs:
American Museum of Natural History at New York D. Grimaldi
  http://www.amnh.org/exhibitions/amber/

Natural History Museum - entomology,Univ. of Kansas, M. Engle
  http://nhm.ku.edu

London Museum of Natural History, Ross, A.
  http://www.nhm.ac.uk/science/intro/palaeo/project1/

Paleontological Museum at Oslo, Norway
  http://www.toyen.uio.no/palmus/galleri/montre/english/
  m_rav_e.htm

Swedish Amber Museum, L. Brost
  http://www.brost.se

Palanga Amber Museum
  http://www.pgm.lt/Gintaras/ amber _fossil .en.htm

The Kauri Museum, Matakohe, New Zealand
  http://www.kauri-museum.com/

INTERNET RESOURCES:

(Internet resources are often changing and sometimes are not reliable. However, since the internet is so convenient, I am including some of the sites I have found very helpful educationally. One can find many resources in addition to the ones listed by using the following key words in a search: amber, amber jewelry, fossil resins, amber mining.)

Amber:Arboreal Gold (Jewels of the Past)
    http://www.uky.edu/ArtsSciences/Geology/webdogs/amber/ welcome.html

Amber Gallery
    http://www.ambergallery.com/

Amber Reading Room, Andzia's Amber Jewelry
    http://www.amberjewelry.com/

Ambere from Lithuania
    http://www.litamber.lt/

Arnold, V.
    http://home.t-online.de/home/Arnold-Heide/ebinstrt.htm

GemRocks: Amber
    www.cst.cmich.edu/users/dietr1rv/ amber .htm

Americawest, Lundberg, D. The Amber Listserver
    http://www.americawest.com/listserv/amber.html
    http://www.americawest.com/

Earth Science
    http://www.emporia.edu/earthsci/amber/amber/.htm

Meyer. M.
    http://www.3dotstudio.com/amberhome.html

Mizdgiris, K. & M. Nida, Lietuva.
    http://www.ambergallery.It/english/muziejus-kas_yra_gintaras.htm

Polish Mall
    http://www.polishmall.com/cgi-bin/merchant2/merchant.mv

Platt, G. Amber Home
    http://www.gplatt.demon.co.uk/

Encyclopedic Theosophical University Press
    http://www.org/pasadena/etgloss/am-ani.htm

Ward-Aber, S. World of Amber
    http://www.emporia.edu/earthsci/amber/amber.htm

Rice, Patty C.
    http://homepage.mac.com/pattyrice1/ambertreelady/Menu16.html

APPENDIX F
PHOTOGRAPHIC REFERENCES

Page front piece double spread. Malbork Castle, Poland. Courtesy Jane Wojtan, Friends Polish Arts .

Page 286 Figure 6.17. A mermithid nematode emerging from an ant in Baltic amber. Description of nematode published in "Poinar, Jr., G. 2002. First fossil record of nematode parasitism of ants; a 40 million year tale, *Parasitology* 125:457-459.

Page 286 Figure 6.17b. A mermithid nematode emerging from an planthopper in Baltic amber. Published in "Poinar, Jr., G. 2002. *Heydenius brownii* sp.n. (Nematoda: Mermithidae) parasitising a planthopper (Homoptera: Achilidae) in Baltic amber. *Nematology* 3:753-757."

Page 286 Figure 6.17c. A braconid wasp larva emerging from its host ant in Baltic amber. Published in "Poinar, Jr., G. O. & Miller, J. 2002. First fossil record of endoparasitism of adult ants (Formicidae: Hymenoptera) by Braconidae (Hymenoptera). *Ann. Entomol. Soc. Amer*. 95:41-43.

Page 286 Figure 6.17d. A rare wind scorpion in Baltic amber. Not yet published Poinar, Jr., G. 2004.

Page 286 Figure 6.17e. A rare tanyderid fly, *Macrochile spectrum*, in Baltic amber, Known from earlier works. Poinar, Jr., G. 2004.

Page 293 Figure 7.3. Nodules of Chemawinite from Cedar Lake, Manitoba. Courtesy of the Canadian Museum of Natural History, Toronto, Canada.

Page 297 Figure 7.6. "Frey's wasp" the famous ant, *Sphecomyrma freyi*, fossilized in New Jersey amber, providing a link between wasps and ants. Courtesy Harvard University Museum of Comparative Zoology.

Page 302 Figure 7.7. Lebanese amber weevil *Lebanorhinus succinus*. Poinar Amber Collection. Photo by G. Poinar.

Page 320 Figure 7.26. A new weevil in Dominican amber. Description published in "Poinar, Jr., G. & Voisin, J. F. 2003. A Dominican amber weevil, *Velatis dominicana* gen. n., sp. n. and key to the genera of the Anchonini (Molytinae, Curculionidae). *Nouv. Revu Ent*. (N.S.) 19:373-381."

Page 322 Figure 7.27. Dominican amber inclusions: Gecko Courtesy Amber Fox, Richard Fox.

Page 324 Figure 7.29. A flower of the extinct palm, *Trithinix dominicana* in Dominican amber. Published in "Poinar, Jr., G. O. 2002. Fossil palm flowers in Dominican and Baltic amber. *Botanical Journal of the Linnaen Society* 139:361-367."

Page 325 Figure 7.30. An extinct termite, Mastotermese electrodominicus, the basis of David Grimaldi's research. Courtesy Natural History Museum, New York.

Page 329 Figure 7.33. The first puffball in amber. Published in "Poinar, Jr., G. O. 2001. Fossil puffballs (Gasteromycetes: Lycoperdales) in Mexican amber. *Historical Biology* 15: 199-122."

Page 346 Figure 7.44. The earliest known hymenomycete fungus in Burmese amber. Description published in "Poinar, Jr., G. O. & Brown, A. E. 2003. A non-gilled hymnenomycete in Cretaceous amber. *Mycological Research* 107:763-768.

Page 435 Figure 11.1. David Grimaldi of the American Museum of Natural History extinct Termite *(Mastotermes electrodominicus)* in Dominican amber 23 to 30 million years old. Courtesy D. Grimaldi copyright AMNH 1996.

REFERENCES AND BIBLIOGRAPHY

Aber, S. W., and Kosmowska-Ceranowicz, B. 2001a. Bursztyn i inne zywice kopalne swiata. Kredowe zywice kopalne Ameryki Polnocnej: cedaryt (czemawinit), jelinit. *Polski Jublier* , 2(13), p. 22-24

Aber, S. W., and Kosmowska-Ceranowicz, B. 2001b. Kansas amber: Historic review and new description. In A. Butrimas (Ed.) Baltic Amber *Proceedings of the International Interdisciplinary Conference: Baltic Amber in Natural Sciences, Archaeology, and Applied Arts.* Vilnius Academy of Fine Arts Press: Vilnius, Lithuania.

Acra, A. 1981. Lebanese amber encapsulates pre-history. *AUB Today.* Beirut: American University of Beirut. 12-3.

Acra, A., R. Milki, and F. Acra. 1972. The occurrence of amber in Lebanon. *Abstract of the Reunion of the Scientific Association for the Advancement of Sciences,* Beirut 4:76-7 (UNESCO).

Agosti, D., D. Grimaldi, and J. M. Carpenter. 1998. Oldest Known Ant Fossils Discovered. *Nature* January: 447.

Agricola, G. (Bauer, G.) 1546. *De Natura Fossillium (On the nature of fossils).* ix:479-480.

Alexander, A. E. 1976. Renewed interest in amber, organic compound, cited in The Gemologist's Corner. *The National Jeweler* (March):27.

Allen, J. 1976a. Amber and its substitutes, Part I: Historical aspects. *The Bead Journal* 2(Winter):4:15-9.

Allen, J. 1976b. Amber and its substitutes, Part II: Mineral analysis. *The Bead Journal* 2(Spring):4:11-2.

Allen, J. 1976c. Amber and its substitutes, Part III: Is it real? Testing amber. *The Bead Journal* 2(Spring):1:20-31.

Almazjuvelirexport. 1978. *Bernstein.* Moscow. 20 pp.

Almazjuvelirexport. 1989. *Bernstein.* Moscow. 44 pp.

Amber, substance of the sun. *Chemistry* 45:21-2.

Amber. 1933. *Manufacturing Jewelers* 92:3-21.

Anderson, B. W. 1971. *Gemstones for every man.* New York: Van Nostrand Reinhold.

Anderson, B. W. 1973. *Gem testing.* New York: Emerson Books.

Anderson, K. B. 1994. The nature and fate of natural resins in the geosphere. IV. Middle and Upper Cretaceous amber from the Taimyr Peninsula, Siberia—evidence for a new form of polylabdanoid of resinite and revision of the classification of Class I resinites. *Organic Geochemistry* 21(2):209-12.

Anderson, K. B., and R. E. Botto. 1993. The nature and fate of natural resins in the geosphere. III. Re-evaluation of the structure and composition of *Highgate Copalite* and *Glessite. Organic Geochemistry* 20:1027-38.

Anderson, K. B., and J. B. Crelling, eds. 1995. *Amber, resinite and fossil resins.* Washington, D.C.: American Chemical Society.

Andrée, K. 1924. Öst Preussens Bernstein und seine Bedeutung. *Ostdeutscher Naturwart* S3:183-9. Also in 1925, S3:120-134.

Andrée, K. 1929. Bernsteinforschung einst und jetzt. In *Bernstein-Forschungen.* 1:I-XXII.

Andrée, K. 1937. *Der Bernstein und seine Bedeutung in Natur und Geisterwissenschaften, Kunst und Kunstgewerbe, Technik, Industrie und Handel.* Königsberg: Grfe und Unzer Albertus-Universität.

Andrée, K. 1951. *Der Bernstein: Das Bernsteinland und sein Leben.* Stuttgart: Kosmos.

Anon. 1985. Faded Genes. *Scientific American* 252:67.

Ansulis, V. 1979. *Baltic Amber* in Lithuanian *Balatijas dzintars.* Riga: Izdernieciba "Liesma." 131pp.

Antropov, A. V. 1995. A new species of the genus Trypoxylon Latrielle (Hymenoptera, Sphecidae) from Dominican amber. *Paleontologicheskiy Zhurnal,* 1:125-8; *Paleontological Journal* 29(1):172-7.

Arem, J. 1977. *Color encyclopedia of gemstones.* New York: Van Nostrand Reinhold.

Arem, J. 1992. *Gems and jewelry,* 2nd ed. Tucson: Geoscience Press.

Aurifaber, Andreas. 1551. *Succini Historia, ein kurtzer gründlicher Bericht, woher der Agtstein oder Börnstein ursprünglich komme, dass er kein Baumhartz sey, sonder ein geschlecht des Bergwachs, und wie man ihn mannig-faltiglich in Artzneien möge gebrauchen. Durch Andream Aurifabrum Vratislautensem Medicum Königsberg in Preussen.* Königsberg: Hans Lufft; 2. Also in 1572. Königsberg: Johan Daubmann.

Axon, G. V. 1967. *The wonderful world of gems.* New York: Criterion Books.

Aycke, J. C. von. 1835. *Fragmente zur Naturgeschichte des Bernsteins.* Danzig: Nicda.

Bada, J. L., X. S. Wang, H. N. Poinar, S. Pääbo, G. O. Poinar, Jr. 1994. Amino acid racemization in amber entombed insects—Implication for DNA preservation. *Geochimica cosmochimica Acta* 58:3131-5.

Baer, J. W. 1937. Floating "gold" from the Baltic. *The Jewelers' Circular-Keystone,* pp. 82-6.

Bajeri, J. J. 1708. Oryktographia Noric . . . Noriberegae.

Bandel, K., and N. Vávra. 1981. Ein fossiles Harz aus der Unterkreide Jordaniens. *Neues Jahrbuch Geologie und*

Palaontologie Mb. 1:19–33.

Barfod, J., F. Jacobs, and S. Ritzkowski. 1989. *Bernstein Schätze in Niedersachsen.* Germany: Knorr & Hirth Verlag,
Seelze.

Baroni-Urbani, C., and J. B. Saunders. 1980. The fauna of the Dominican Republic amber: The present status of knowledge. *Proceedings of the Ninth Caribbean Geological Conference* (Santo Domingo, August 1980) 213–23.

Barrera, A. De. 1860. *Gems and jewels, their history, geography, chemistry and analysis.* London: Richard Bentley.

Barry, T. H. 1932. *Natural varnish resins.* London: Ernest Benn.

Barthel, M., and H. Hetzer. 1982. Bernstein-Inklusen aus dem Miozän des Bitterfelder Raumes. *Zeitschrift für angewandte Geologie* 28:314–36.

Basharkevitch, A. P., L. I. Ilkevitch, and L. I. Matruntchik. 1983. *Fossil resins of Belorussian "Polessya."* In Russian. Dokl. AN BSSR Belorussian Soviet Republic 27(7):644–5.

Bather, F. O. 1922. See Cockerell, 1922.

Bauer, J. 1974. *Field guide in color to minerals, rocks and precious stones.* London: Octopus Books.

Bauer, M. 1904. *Precious stones, II.* London: Charles Griffin. Reprint 1968. Mineola, N.Y.: Dover Publications (page references are to reprint edition).

Beard, A., and F. Rogers. 1940. *5000 years of gems and jewelry.* New York: Frederick A. Stokes.

Beck, C. W. 1965. The origin of amber found in Gough's Cave, Cheddar, Somerset. *Proceedings, University of Bristol Spelaecological Society* 10(3):272–6.

Beck, C. W. 1971. Archaeological chemistry. In *Science and archaeology,* R. Brill, ed. Cambridge, Mass.: The MIT Press.

Beck, C. W. 1982*a. Authentication and conservation of amber: Conflict of interests.* Washington, D.C.: Washington Congress for Science and Technology in the Service of Conservation.

Beck, C. W. 1982*b.* Physical methods used to determine the geological origin of amber and other fossil resins; some critical remarks: comment. *Physics & Chemistry of Minerals* 8(3):146–7.

Beck, C. W. 1983. The amber trade. *Savaria-Bulletin der Museen des Kohitats,* Vas 16. Internationales Kolloquim 1982. Bozsok-Szombathelly.

Beck, C. W. 1986. Spectroscopic investigations of amber. *Applied Sepctroscopy Review* 22:57–110.

Beck, C. W. 1995. The provenience analysis of amber. *American Journal of Archaeology* 99 (January):125–7.

Beck, C. W., and L. Y. Beck. 1995. Analysis and provenience of Minoan and Mycenaean amber, V. Pylos and Messenia. *Greek, Roman and Byzantine Studies* 36(Summer):119–35.

Beck, C. W., and Shennan, S. 1991. *Amber in prehistoric Britain.* Oxbow Books Limited 8:231.

Beck, C. W., M. Gerving, and E. Wilbur. 1967. *The provenience of archaeological amber artifacts: An annotated bibliography.* London: International Institute for Conservation of Historic and Artistic Works.

Beck, C. W., J. B. Lambert, and J. S. Frye. 1986. Beckerite. *Physics and Chemistry of Minerals* 13:441–3.

Beck, C. W., E. C. Stout, and B. Kosmowska-Ceranowicz. 1993. A large find of supposed amber from the Baltic Sea. *Geologiska Foreningens i Stockholm Forhandlingar* 115(2):145–50.

Beck, C. W., E. Wilbur, and S. Meret. 1964. Infra-red Spectra and the origin of amber. *Nature* 201:256–7.

Beck, C. W., E. Wilbur, S. Meret, D. Kossove, and K. Kermani. 1965. Infrared spectra of amber and identification of Baltic amber. *Archaeometry* 8:96–109.

Beckmann, F. 1859. *Ursprung und Bedeutung des Bernsteinnames Elektron.* Braunsberg: Zeltschrift Fürd Geschichte und Alterthumskunde Ermelands. Also in 1860, pp. 201–43.

Beckmann, F. 1860.*Über den altpreussisch-litauischen Bernsteinnamen Gentaras oder Gintaras.* Braunsberg: Zeitschrift für Geschichte und Altertumskunde Ermelands, 1:633–40.

Bellmann, E. 1913*a.* Amber. *The Mining Journal* 101:129.

Bellmann, E. 1913*b.* Recovery and treatment of amber at Palmnicken (East Prussia). *The Mining Journal* 102:122.

Berendt, G. C. 1830. *Die Insekten im Bernstein. Ein Beitrag Zur Thiergeschichte der Vorwelt.* Berlin/Danzig.

Berendt, G. C. 1845. *Die im Bernstein befindlichen organischen Reste der Vorwelt. I.* In H. R. Göppert and G. C.

Berendt. 1845. Der Bernstein und die in ihm befindlichen Pflanzenreste. Berlin-Danzig. Also in 1854.

Berendt, G. C. 1866. Erläuterungen zur geologischen Karte des West-Samlandes. Verbreitung und Lagerung der Tertiär-Formationen. (Geology of Kurisches Haff.) In German. *Schriften der Königlichen Physikalisch Ökonomischen Gesellschaft zu Königsberg.* 7:131–44.

Berendt, G. C. 1869. *Gologic des Kurischen Haffes und seiner Umgebung* (Geology of Kurisches Haff). Königsberg: Commission bei W. Koch.

Berner, R. A., and G. P. Landis. 1987. Chemical analysis of gaseous bubble inclusions in amber: The composition of ancient air? *American Journal of Science* 287(8):757–62.

Berner, R. A., and G. P. Landis, 1988. Gas bubbles in fossil amber as possible indicators of the major gas composition of ancient air. *Science* 239(4846):1406–9.

Berry, E. W. 1927. The Baltic amber deposits. *Science Monthly* 24:268–78.

Berry, E. W. 1930*a*. Ancestry of our trees. *Science Monthly.* 31:260–3.

Berry, E. W. 1930*b*. The past climates of the North Polar Region. *Smithsonian Miscellaneous Collection* 82(6).

Bishop, L. 2003. Conspiracy of Silence. *DEEPIMAGE.co.uk. http://www.deepimage.co.uk/wrecks/wilhelm-gustoff/wilhelm-gustloff_amberroom.htm*

Black, G. F. 1893. Scottish charms and amulets. In *Proceedings of the Scottish Antiquarian Society*, Vol. III. Edenburgh: Scottish Museum of Antiquities.

Błaszak, M. 1987. Bursztyn w osadach trzeciorzędowych w okolicy Moźdánowa koło Słupska. (The amber in the Tertiary sediments in the vicinity of Moźdánowa near Słupsk) In Polish. *Biuletyn Instytutu Geologicznego* 356:103–19.

Blavatsky, H. P. 1999. *Encyclopedic Theosophical Glossary:Am-Ani.* Theosophical University Press web site

Block, J. A. 1974. *The story of jewelry.* New York: William Morrow.

Blunck, G. 1929. Bakterieneinschlüsse im Bernstein. *Centralblatt für Mineralogie, Geologie und Palaontologie* B(11):554–5.

Bock, F. S. 1767. *Versuch einer kurzen Naturgeschichte des Preussischen Bernsteins und einer neuen wahrscheinlichen. Erklärung seines Ursprungs.* Königsberg: Zeise & Hartung.

Bolland, H. R., and W. L. Magowski. 1990. Neophyllobius succineus n.sp. from Baltic amber (Acari: Raphignathoidea: Camerobiidae). *Entomologische Berichten* 50(2):17–21.

Bölsche, W. 1927. *Im Bernsteinwald.* Berlin/Stuttgart: Kosmos Gesellschaft der Naturfreunde, Franckh'sche Verlalgsshandlung.

Bombicci, L. 1873, 1875. *Corso di Mineralogi Seconday edizione grandemente variata ed accrescuatam etc*, 2 volumes. Bologna.

Borglund, E., and J. Flauensgaard. 1968. *Working in plastic, bone, amber and horn.* New York: Reinhold.

Bortolotti, G. 1993. *L'oro del nord: commerci e diffusione dell'ambra nell'antichita* (Northern gold: Trade and spread of amber in antiquity). In Italian. *Universo* 73(4):442–58.

Bousfield, E. L., and G. O. Poinar Jr, 1994. A new terrestrial amphipod from Tertiary amber deposits of Chiapas Province, southern Mexico. *Historical Biology* 7(2):105–14.

Bradford, E. 1953. *Four centuries of European jewellerey.* Middlesex, England: Spring Books.

Brill, R., ed. 1971. *Science and archaeology.* Cambridge, Mass.: The MIT Press.

Broughton, P. L. 1974. Conceptual frameworks for geographic-botanical affinities of fossil resins. *Canadian Journal of Earth Science* 11(4):583–94.

Brouwer, P. A. 1959. *Economica Dominicano* 11:16.

Brouwer, S. B., and P. A. Brouwer. 1980. Geologia de la Region Ambarifera Oriental de la Republica Dominicana. *Proceedings of the Ninth Caribbean Geological Conference* (Santo Domingo, August 1980), 200–12.

Brown, B. V., and E. M. Pike. 1990. Three new fossil phorid flies (Diptera: Phoridae) from Canadian Late Cretaceous amber. *Canadian Journal of Earth Science* 27:845–8.

Brown, C. O. 1966. Geology of Central Dominican Republic. In *Caribbean Geological Investigations*, H. Hess, ed. Geological Society of America, 11–84.

Brues, C. T. 1951. Insects in amber. *Scientific American* 185(5):56–61.

Buddhue, J. D. 1935. Mexican amber. *Rocks and Minerals* 10:170–1.

Budge, E. A. W. 1961. *Amulets and talismans.* New York: University Books.

Budrys, R. 1974. *Gintaras. Bukletas.* [Yantar.] Lietuvos TSR Dailes muziejaus leidinys.

Buffum, A. 1896. *The tears of Heliades, or amber as a gem.* London: Sampson Low, Marston.

Burdukiewicz, J. M. 1981. Stanowisko kultury hamburskiej Siedlnica 17a, gm. Wroclow, Poland. *Slaskie Sprawozdania Archeologiczne.* Wroclow: 22:5–11.

Burleigh, R., and P. Whalley. 1983. On the relative age of amber and copal. *Journal of Natural History* 17:919–21.

Butenas, P. 1973*a*. *Gentaro Sneka.* Balandis: Karys 4:110–4.

Butenas, P. 1973*b*. *Gentaro Sneka.* Geguze: Karys 5:159–64.

Cammann, S. 1962. *Substance and symbols in Chinese toggles.* Philadelphia: University of Pennsylvania Press.

Cano, R. J. 1994. Bacillus DNA in amber: A window to ancient symbiotic relationships? *Features ASM News* 60(3):29–134.

Cano, R. J., and M. K. Borucki. 1995. Revival and identification of bacterial spores in 25- to 40-million-year-old Dominican amber. *Science* 268(5213):1060–4.

Cano, R. J., H. Poinar, and G. O. Poinar, Jr. 1992. Isolation and partial characterization of DNA from the bee *Proplebeia dominicana* (Apidae: Hymenoptera) in 25-40 million-year-old amber. *Medical Science Research* 20:249–51.

Cano, R. J., H. N. Poinar, D. Roubik, and G. O. Poinar, Jr. 1992. Enzymatic amplification of nucleotide sequencing of Protia 18f rRNA gene of the bee, *Proplebeia dominicana* (Apidae: Hymenoptera) isolated from 25-40 million-year-old Dominican Amber. *Medical Science Research* 20:619–22.

Cano, R. J., H. N. Poinar, N. J. Pieniazek, A. Acra, and G. O. Poinar, Jr. 1993. Amplification and sequencing of DNA from 120–135-million-year-old weevil. *Nature* 363:536–8.

Cano, R. J., H. N. Poinar, M. Boruchi, and G. Poinar. 1994. *Bacillus* DNA in fossil bees: an ancient symbiosis? *Applied Environmental Microbiology* 60:2164–7.

Carpenter, F. M. 1935. Fossil insects in Canadian amber. *University of Toronto Studies, Geological Series*, No. 38, 69.

Carpenter, F. M. 1938. Insects and arachnids from Canadian amber. *Introduction. University of Toronto Studies of Geological Survey* (Series 400), 7–13.

Carpenter, F. M,. and W. L. Brown, Jr. 1967. The first Mesozoic ants. *Science* 157:1038–40.

Carpenter, F. M., J. W. Folsom, E. O. Essig, A. C. Kinsey, C. T. Brues, M. W. Boesel, and H. E. Ewing. 1937. Insects and arachnids from Canadian amber. *University Toronto Studies.* Geological Series 40:7–62.

Caspary, R., and R. Klebs. 1907. Die Flora des Bernsteins und anderer fossiler Harze des ostpreussichen Tertiärs. In *Abhandlungen der Königlichen (Preussischen) Geologischen Landesanstalt*, Berlin.

Cherfas, J. 1991. Ancient DNA: Still busy after death. *Science* 253 (September 20):1354.

Chętnik, A. 1952. *Przemysli sztuka bursztyniarstwa nad Narwią* (Amber working art and industry on the River Narew). In Polish. Warsaw: Muzeum Ziemi.

Chętnik, A. 1964. Twórczość ludowa Kurpiów w dziedzinie sztuki. *Polska Sztuka Ludowa.* Wrocław 18(3):107–26.

Chętnik, Adam. 1981 Mały słownik odmian bursztynu polskiego. (A short terminological dictionary of varieties of Polish amber). Wstępem poprzedziła Barbara Kosmowska-Ceranowicz. Maszynopis przygotował do druku Zbigniew Wojcik.. *Prace Muzeum Ziemi* 34:31–8. English summary in Kosmowska-Ceronowicz 1990.

Chhibber, H. L. 1934. *The mineral resources of Burma.* London: Macmillan.

Childe, G. V. 1939. *The dawn of European civilization*, 3rd ed. New York: Knopf.

Cockerell, T. D. 1917. Some American fossil insects. *Proceedings U.S. National Museum* 51:89–106.

Cockerell, T. D. 1922. Fossils in Burmese Amber. *Nature* 109:713–4.

Cofta-Broniewska, A. 1997. *Amber in the material culture of the communities of the region of Kujawa during the period of the Roman influence.* . In: KOSMOWSKA-CERANOWICZ & H. PANER (ed) Investigations into amber. Proceedings of the International interdisciplinary Symposium "Baltic amber and other fossil resins 997 Urbs Gyddanyzc 1997 Gdańsk" 2-6 September 1997 Gdańsk : 157-175

Cokendolpher, J. C. 1986. A new species of fossil *Pellobunus* from Dominican Republic amber (Ar. achnida:Opiliones:Phalangodidae). *Caribbean Journal of Science.* 22(3–4):205–11.

Conti, B. A. 1996. Amber, the Dominican Republic produces world-class specimens. *Rocks and Gems* (November):38–42.

Conwentz, H. W. 1886a. Die Bernsteinfichte. *Berichte der Deutschen Botanischen.* Gesellshaft Berlin. 4:375–7.

Conwentz, H. W. 1886b. *Die flora des Bernsteins. Bd. 2, Die Angiospermen des Bernsteins.* Danzig: Commissions-Verlag von W. Engelman 1886, 1893. And in 1886, Naturforschende Gesellschaft.

Conwentz, H. W. 1890a. *Monographie der baltischen Bernsteinbäume.* Danzig: F. Bertling.

Conwentz, H. W. 1890b. *Monographie der baltischen Bernsteinbäume.* Vergleichende Untersuchungen über die Vegetationsorgane und Blüten, sowie über das Harz und die Krankheiten der Bernsteinbäume. Leipzig: Commissions-Verläg von Wilhelm Engelmann 151(18):141.

Cook, J. 1770. *The natural history of lac, amber, and myrrh. . . .* London: Sold by Mr. Woodfall.

Cruickshank, R. D. and Ko Ko. 2003. Geology of an amber locality in the Hukawng Valley, Northern Myanmar. *Journal of Asian Earth Sciences* 21:441-455

Cuba, J. D., ca. 1483–1491. *Hortus Sanitatus* (old herbal book). Cap. Ixx. Strassburg: Jean Pryss.

Czeczott, H. 1961. Składi wiek flory bursztynów bałtyckich (The flora of the Baltic amber and its age). In Polish. Czesc I. *Prace Muzeum Ziemi.* (Paleobotaniczne) Warsaw: Muzeum Ziemi 4:119–45.

D'Aulaire, E., and O. D'Aulaire. 1974. For-ever amber. *Reader's Digest* (reprint from Scandinavian edition). (December).

D'Aulaire, E., and O. D'Aulaire. 1978. Amber: gold of the north. *Scanorama Magazine* (January):61–4.

Dahlstrom, A., and L. Brost. 1996. *The amber book.* Tucson: Geoscience Press.

Dahms, P. 1901. Über das Vorkommen und die Verwendung des Bernsteins. *Zeitschrift für prakt. Geologie* 9:201–11.

Dall, W. H. 1870. *Alaska and its resources.* London: Sampson Lawson and Mareston.

Das, H. A. 1969. Examination of amber samples by nondestsructive activation analysis. *Radiochemical and Radioanalytical Letters* 1(4).

Davidov, C., and G. Dawes. 1988. *The Bakelite jewelry book.* New York: Abbeville Publishing.

DeNavarro, J. M. 1925. Prehistoric routes betweeen Northern Europe and Italy defined by the amber trade. *The Geographical Journal* 66(6 December):481–507.

Denef, K., G. Doll, D. Hausser, W. Karpe. 1990. Vorkommen und Nutzung von Begleitrohstoffen einschliesslich Bernstein im Raum Bitterfeld (Deposits and use of accompanying raw material, including amber, in the region of Bitterfeld) *Zeitschrift für Angewandte Geologie* 36(8):300–6.

DeSalle, R., and D. Lindey. 1997. *The science of Jurassic Park and the lost world or, how to build a dinosaur.* New York: Basic Books.

DeSalle, R., J. Gatesy, W. Wheeler, and D. Grimaldi. 1992. DNA sequences from a fossil termite in Oligo-Miocene amber and their phylogenetic implications. *Science* 257:1889-1882.

Desautels, P. E. 1977. *The gem kingdom.* New York: Random House.

Disney, R. H., and A. J. Ross. 1997. *Abaristophora and Puliciphora* (Diptera: Phoridae) from Dominican amber and revisionary notes on modern species. *European Journal of Entomology* 94(1):127–35.

Diving for amber. ca. 1890. *St. Paul's Magazine.*

Domke, W. 1952. Der erste sichere Fund eines Myxomyceten im baltischen Bernstein. *Mitteilungen Geologie Staatsinst.* Hamburg 21:152–62.

Durham, J. W. 1957. Amber through the ages. *Pacific Discovery* 10(2):3–5.

Dyl, S. J., and M. L. Wilson. 1995. *Nomenclature for mineral localities of the former Soviet Union,* C. Yagley, ed. Centerline, Mich.: The MMS Conglomerate.

Eberle, W., W. Hirdes, R. Muff, and M. Pelaez. 1980. The geology of the Cordillera Septentrional. *Proceedings of the Ninth Caribbean Geological Conference* (Santo Domingo, August 1980), 619–32.

Emelyanov, E. M., V. V. Sivkov, and S. M. Isachenko. 1997. New data from lithologic-geochemical research on Palaeogene deposits from the Primorsk Quarry of Kaliningrad Amber Factory, Sambian Peninsula, Russia. In *Baltic amber and other fossil resins,* B. Kosmowska-Ceranowicz, ed. Warsaw: Muzeum Ziemi, 11–2.

Encyclopedia Americana. International 1975 ed., s.v. "plastics."

Encyclopaedia Britannica. 1971 ed., s.v. "amber."

Encyclopaedia Britannica 1971 ed., s.v. "plastics."

Encyclopaedia Britannica Micropaedia (Ready Reference).1974 ed., s.v. "amber."

Encyclopedia Lituania. 1970 ed. s.v. "amber."

Engel, M.S. 1997. A new fossil bee from the Oligo-Miocene Dominican amber (Hymenoptera:Halictidae). *Apidologie* 28(2):97–102.

Engel, M.S. (2001) A monograph of Baltic amber bees and evolution of the Apoidea (Hymenoptera). *Bulletin American Museum of Natural History* 259: 1-192.

Erikson, J. M. 1969. *The Universal Bead.* New York: W. W. Norton.

Evans, J. 1976. *Magical jewels of the Middle Ages and Renaissance.* Mineola, N.Y.: Dover Publications.

Farrington, O. C. 1923. *Amber,* Department of Geology Leaflet No. 3. Chicago: Field Museum of Natural History.

Fecarotta, E. 1639–1805. *Dell'Ambra Siciliana, Testi di antichi autori siciliani.* Carmelo Erio Fiore, ed. Palarmo: Edizioni Boemi.

Fennah, R. G. 1987. A new genus and species of Cixiidae (Homoptera: Fulgoroidea) from Lower Cretaceous amber. *Journal of Natural History* 21(5)1237–40.

Fernie, W. T. 1907. *Precious stones: For curative wear and other remedial uses.* Bristol: John Wright.

Ferrara, F. 1805. *Memorie . . . Sopra l'Ambra Siciliana.* Palermo: Dalla Reale Stamperia.

Fielder, M. 1976. What is this gem called amber? *Lapidary Journal* 30(5):1244–9.

Fielding, W. J. 1945. *Strange superstitions and magical practices.* Philadelphia: Blakiston.

Fischman, J. 1993. Going for the old: ancient DNA draws a crowd. *Science* 262 (October 29):655–6.

Folinsbee, R. E., H. Baadsgaard, G. L. Cumming, and J. Nascimbene. 1964. Radiometric dating of the Bearpaw Sea. *Bulletin American Association of Petrology Geology* 48:525.

Fraas, O. 1878. Geologisches aus dem Libanon. *Jahreshefte des Vereins für vaterolandische Naturkunde.* Wurttemberg 34:257–391.

Frahm, J. P. 1993. Mosses in Dominican Amber. *Journal Hattori Botany Lab.* 74:249–59.

Frahm, J. P. 1996. New records of fossil mosses from Dominican amber. *Cryptogamie: Bryologie et Lichenologie* 17(3): 231–6.

Fraquet, H. 1987. *Amber.* London: Butterworths Gem Books.

Fraquet, H. R. 1982. Amber from the Dominican Republic. *Journal of Gemology* 18(4):321–33.

Frondel, J. W. 1967*a*. X-Ray diffraction study of some fossil and modern resins. *Science* 155:1411–3.

Frondel, J. W. 1967*b*. X-Ray diffraction study of some fossil elimis. *Nature* 215(5168):1360–1.

Frondel, J. W. 1968. Amber facts and fancies. *Economic Botany* 22(4):381–2.

Frondel, J. W. 1969. Fossil elemi species identified by thin-layer chromatography, *Die Naturewissenschaften* 56:280.

Fuhrmann, R., and R. Borsdorf. 1986. (The Lower Miocene amber of Bitterfeld). In German. *Zeitschrift der Angewandte Geologie* 32(12):309–16.

Garty, J., C. Giele, W. E. Krumbein. 1982. On the occurrence of pyrite in a lichen-like inclusion in Eocene amber (Baltic). *Palaeogeography, Palaeoclimatology, Palaeoecology* 39(1–2):139–47.

Gawęda, M. T. 1995. Bursztyn (Jantar)—lecznicze złoto Bałtyku. In *Medycyna ludowa. Woda żywa I martwa. Ziółka*. Ciechanów, Poland: Wydawnictwo Drukarz, 156–9.

Gemological Institute of America. 1971. *Manual for colored stones*. Los Angeles: Gemological Institute of America.

Gerhardt, P. 1995. *Technical comments—bacteria revived from amber-fossilized bees after 25+ million years.* Typescript. Michigan State University, Microbiology Professor Emeritus.

Ghiurca, V. 1990. New considerations on Romanian amber. In *Sixth meeting on amber and amber-bearing sediments proceedings* (Warsaw 20–21 October) : Report Summaries. Prace Muzeum Ziemi 41 p. 158.

Gierłowska,G. 2002. *Amber in therapdutics.* Gdańsk, Poland: "Bursztynowa Hossa" publishing House.

Gierłowska, G. 2003. *Guide to amber imitations.* Gdańsk, Poland: "Bursztynowa Hossa" publishing House.

Gierłowska, G. 2004. *The beauty of amber.* Gdańsk, Poland: "Bursztynowa Hossa" publishing House.

Gierławski, W. 1997. Gdańsk's craftsmen, workshops and the management of amber resources in the twentieth century. In *Baltic amber and other fossil resins*, B. Kosmowska-Ceranowicz, ed. Warsaw: Museum of the Earth.

Gimbutas, M. 1963. *The Balts*. New York: Frederick A. Praeger.

Goebel, S. 1558, 1582, etc. *De Succino Libri II, quorum prior theologicuss, posterior de Succini origine agit.* cum corolario C. Gesneri 1565–66 De omni reram fossilium genere. . . . Francofurti [Frankfort].

Goldsmith, E. 1879. Proceedings, Academy of Natural Science Philadelphia.

Göppert, H. R. 1853. Über Bernsteinflora und dem Bernstein. In *Bericht über die zur Bekanntmachung, geeigneten Verhandlungen der Köngisberg 1*. Berlin: Preussischen Akademie der Wissenschaften zu Berlin, 450–76.

Göppert, H. R. and G. C. Berendt. 1845. *Die Bernstein und die in ihm befindichen Pflanzenreste der Vorwelt.* Vol. 1. Berlin.

Göppert, H. R. and A. Menge. 1883 1886. *Die Flora des Bernsteins und ihre Beziehungen zur Flora der Tertiärformation und der Oegenwart.* Vol. 1. Danzig: Naturforschende Gesellschaft.

Gough, L. J., and J. S. Mills. 1972. The composition of succinite (Baltic amber). *Nature* 239:527–8.

Grabowska, J. 1983. *Polish amber.* Warsaw: Interpress.

Green, R. L. 1962. *Myths of the Norsemen.* London: Bodley Head.

Grimaldi, D. A. 1993. The care and study of fossiliferous amber. *Curator* 36:31–49.

Grimaldi, D. A. 1995. The age of Dominican amber. In *Amber, resinite and fossil resins*, K. B. Anderson, and J. C. Crelling, eds. Washington, D.C.:American Chemical Society, 203–17.

Grimaldi, D. A. 1996*a*. *Amber: Window to the past.* New York: Harry N. Abrams.

Grimaldi, D. A. 1996*b*. Captured in amber, the exquisitely preserved tissues in insects in amber reveal some genetic secrets of evolution. *Scientific American* (April):84–91.

Grimaldi, D. A. and G. R. Case. 1995. A feather in amber from the Upper Cretaceous of New Jersey. *American Museum Novitates* 3126: 1-6.

Grimaldi. D. A. and D. Agosti. 2000. A formicine in New Jersey Cretaceous amber (Hymenoptera: formicidae) and early evolution of the ants. *Proceedings of the National Academy of Sciences, USA* 97: 13678-13683.

Grimaldi, D. A., A. Shedrinsky, A. Ross, and N. Baer. 1994. Forgeries of fossils in "amber": History, identification, and case studies. *Curator, the Museum Journal* 37(4): 251–73.

Grimaldi. D. A., J. A. Lillegraven, T. W. Wampler, D. Bookwalter, and A. Shedrinsky. 2000a. Amber from Upper Cretaceous through Paleocene strata of the Hanna Basin, Wyoming, with evidence for source and taphonomy of fossil resins. *Rocky Mountain Geology* 35: 1263-204.

Grimaldi, D. A., M. S. Engel, and P. C. Nascimbene. 2002. Fossiliferous Cretaceous amber from Myanmar (Burma): its rediscovery, biotic diversity, and paleontological significance. *American Museum Novitate*, 3361: 1-72.

Gudynas, P. and S. Pinkus. 1967. *The Palanga Museum of Amber.* Vilnius: Mintis Books.

Gudynas, P., J. Kuzminskis, J. Balcikonis, S. Bernotiene, K Kairluketyte, and I. Vaineikyte. 1969. *Lietuviu Liaudies Menas Drabuzia.* Vilnius: Vaga.

Gutierrez, G. and A. Marin. 1998. The most ancient DNA recovered from an amber-preserved specimen may not be as ancient as it seems. *Molecular Biology & Evolution* 15: 926-929.

Hackman, W., and I. Valovirta. 1996. Barnstenssnackans egendomliga parasiter av slaktet *Leucochloridium* (Trematoda: Digenea), livscykel och forekomsti i Finland (Unusual parasite of an Amber snail, *Leucochloridium* [Trematoda: Digenea]: Lifecycle and occurrence in Finland). In Finnish. *Memoranda Societatis pro Fauna et Flora Fennica* 72(1):13-6.

Haczewski, S. 1838. *O Bursztynie.* Lwów: Sylwan 14:191-251.

Haddow, J. G. 1891. *Amber, all about it.* Liverpool: Cope's Smoke Room Booklets, No. 7.

Hamamoto, T., and K. Horikoshi, 1994. Characterization of a bacterium isolated from amber. *Biodiversity & Conservation* 3(7):567-72.

Hartknoch, M. C. 1684. *Alt und Neues Preussen oder Preussischer Historien.* Frankfurt-Leipzig: Zwey Theile.

Hartmann, P. J. 1677. *Succini Prussici physica et civilis historia cum demonstratione.* Frankfort: Hallervord.

Heer, O. 1859. *Die Tertiäre Flora der Schweiz (1855–1859).* Winterhur 3:327–50.

Heer, O. 1869. *Miocän baltische Flora.* Series Beiträge zur Naturkunde Preussens. Königsberg: Koch 2:1–4.

Heie, O. E., and E. M. Pike. 1992. New aphids in Cretaceous amber from Alberta (Insecta, Homoptera) *Canadian Entomologist* 124(6):1027–53.

Heie, O. E., and P. Wegierek. 1998 A list of fossil aphids (Homoptera: Aphidinea). *Annals of the Upper Silesian Museum in Bytom. Entomology.* 8/9:159-192.

Helm, O. 1877. Notizen über die chemische und physikalische Beshaffenheit des Bernsteins. *Reichardt's Archiv der Pharmacy.* 98(3): 2211, 229-246.

Helm, O. 1881. Mittheilungen über Bernstein. 2, Glessit, 3 Über sicilianischen und rumanischen Bernstein. *Schriften der Naturforschenden Gesellschaft zu Danzig* 5:291–6.

Helm, O. 1885. Über die Herkunft des in der alten Königsgräbern von Mykena gefundenen Bernstein und über den Bernsteinsäuregehalt verschiedener fossiler Harze. *Schriften der Naturforschenden Gesellschaft zu Danzig* 2:234–9.

Helm, O. 1892a. Mittheilungen über Bernstein. 14, Über Rumanite ein in Rumanien vorkommendes fossiles Harz, 15, Über den Succinit und die ihm verwandten fossilen Harze (1890, 1892). *Schriften der Naturforschenden Gesellschaft zu Danzig* 7:186–295.

Helm, O. 1892b. On a new fossil, amber-like resin occurring in Burma, from Upper Burma. *Records of the Geological Survey of India* 25:180–1.

Helm, O. 1896. Mitteilungen über Bernstein 127, Über den Gedanit, Succinit und eine Abaart des letztern, den sogenannten murber Bernstein. *Schriften der Naturforschenden Gesellschaft zu Danzig* 9:51–7.

Hennig, W. 1965. Der Acalyptratae des Baltischen Bernsteins und ihre Bedentung für die Erforschung der phylogenetischen Entwicklung die ser Dipteren—Gruppe. *Stuttgarter Beiträge zur Naturkunde* (Series A) 145:1–215.

Hennig, W. 1966. *Fannia scalaris* Fabaricius, eine rezente Art im Baltischen Bernstein? (Diptera: Muscidae). *Stuttgarter Beiträge zur Naturkunde* (Series A) 150:1–12.

Henwood, A. A. 1992a. Exceptional preservation of dipteran flight muscle and the taphonomy of insects in amber. *Palaios* 7(2):203–13.

Henwood, A. A. 1992b. Soft-part preservation of beetles in Tertiary amber from the Dominican Republic. *Paleontology* 35(4):901–12.

Henwood, A. A. 1993a. Still life in amber. *New Scientist* 1859:31–4.

Henwood, A. A. 1993b. Ecology and taphonomy of Dominican Republic amber and its inclusions. *Lethaia* 26(3):237–45.

Henwood, A. 1993c. Recent plant resins and the taphonomy of organisms in amber: A review. *Modern Geology* 19(1):35–59.

Hermanni, D. (Herman, Daniels). 1583. *De Rana et lacerta: succino Prussiaco insitis. (Poem)* Cracow.

Hibbett, D. S., D. Grimaldi, and M. J. Donoghue. 1997. Cretaceous mushrooms in amber. *Nature* 377: 487.

Higuchi, R. G., and A. C. Wilson. 1984. Recovery of DNA from extinct species. *Federation Proceedings* 43:1557.

Higuchi, R. G., B. Bowman, M. Freeberger, O. A. Ryder, and A. C. Wilson. 1984a. DNA sequences from the Quagga, an extinct member of the horse family. *Nature* 312: 282–4.

Hilts, P. 1996. Expedition to far New Jersey finds trove of amber fossils. *New York Times,* 30 January.

Hodges, D. M. 1962. *Healing stones.* Hiwatha, Iowa: Pyramid Publisher of Iowa.

Hong, Y.-Ch. 1981. *Eocene fossil Diptera (Insecta) in amber of Fushun coalfield.* Beijing: Geological Publishing House.

Hong, Y.-Ch, T.-C. Yang, S. T. Wang, S. E. Wang, Y. K. Li, M. R. Sun, H.-C. Sun, and N.-C. Tu. 1974.

Stratigraphy and paleontology of Fushun coalfield, Liaoning Province. *Acta Geologica Sinica* 2:113–49.

Hopfenberg, H. B., L. C. Witchey, and G. O. Poinar, Jr. 1988. Is the air in amber ancient? *Science* 241:717–24.

Horibe, Y., and H. Craig. 1987. Trapped gas in amber: A paleobotanical and geochemical inquiry. *Eos* 68:1513.

Hornaday, W. D. 1915. Kauri gum deposits of New Zealand. *Mining Press* 110:181–2.

Hotz, R. L. 1993. Scientists find dinosaur-era DNA. *Detroit News,* 10 June.

Hueber, F. M., and J. Langenheim. 1986. Dominican amber tree had African ancestors. *Geotimes* 31: 8–10.

Hunger, R. 1977. *Magic of amber.* London: N.A.G. Press.

Infopedia. 1995. The Funk & Wagnalls New Encyclopedia, (CD-ROM) Future Vision Multimedia, Inc.

Iturralde-Vinent, M.A., and R. D. E MacPhee. 1996. Age and paleogeographical origin of Dominican amber. *Science* 273(5283):1850–2.

Jagodziński, M. 1982. Neolityczne pracownie bursztyniarskie na Żuławach. *Z Otchłani Wieków.* Poznań 1–3:9–18.

Jakubowski, M. 1914. Amber fishers and mines. *The Jewelers' Circular-Weekly* (23 December) 102:1–22.

Jewell, W. B. 1931. Geology and mineral resources of Hardin County, Tennessee. *Tennessee Divison of Geology Bulletin* 37:94–5.

Jobes, G. 1961. *Dictionary of mythology, folklore and symbols, Part 1.* New York: Scarecrow Press.

John, J. F. 1816. *Naturgeschichte des Succins, oder des sogenannten Bernsteins.* Cologne: Theodor Franz Thiriart.

Kasparyan, D. P. 1995. A review of the ichneumon flies of Townesitinae subfam. nov. (Hymenoptera, Ichneumonidae) from the Baltic amber. *Paleontologicheskiy Zhurnal.* Vol. 4, 1994, pp. 86–96; *Paleontological Journal* 28(4):114–26.

Katinas, V.I. 1971. *Amber and amber-bearing deposits of the southern Baltic area. (Jantar i jantarenosnyje otlozhenija Juzhnoj Pribaltiki.)* Vilnius: "Mintis." In Russian.

Katinas, V. 1983. *Baltijos gintaras.* Vilnius: Mokslas.

Katinas, V. 1987. *The amberbearing deposits of the southern Baltic area.* Vilnius: "Mintis":156 (In Russian)

Kaunhoven, F. 1913. Der Bernstein in Ostpreussen. *Jahrbuch der Preussischen Geologischen.* Berlin: Landesanstalt 34(2):1–80.

Keferstein, C. 1849. *Mineralogia Polyglotta.* Halle, Germany.

Keilbach, R. 1937. *Neue Forschungen über samländische Bernsteineinschlüsse. Der Naturforscher.* Berlin 13:398–400.

Kemp, E. C. 1925. *American consular commerce reports.* Danzig: American Consulate (27 April).

Kennedy, G. 1967. The fundamentals of gemstone carving. *Lapidary Journal* 68–73.

Kharin, G. S. 1995. Geological conditions of the amber-bearing deposits originating in the Baltic region. *Amber and fossils* 1(1):47–53.

King, A. A. 1961. Texas gemstones. *Report of Investigations No. 42.* Texas State Geology Department.

Kircher, G. 1950. Amber inclusions. (Submarine Bernsteineinschlusse) *Endeavour* London 9:70–5.

Klebs, R. 1880. *Der Bernstein. Seine Gewinnung, Geschichte und geologische Bedeutung.* Erläuterung und Katalog der Bernstein-Sammlung der Firma Stantien & Becker. Königsberg: Deuch von G Laudien.

Klebs, R. 1882. *Der Bernsteinschmuck der Steinzeit von der Baggerei bei Schwarzort und anderen Lokalitäten Preussens.* Königsberg: Schriften der Physikalisch-Ökonomischen Gesellschaft zu Königsberg in Preussa

Klebs, R. 1889. *Aufstellung und Katalog des Bernstein-Museums von Stantien und Becker*—Nebst einer kurzen geschichte des Bernsteins. Königsberg: Hartung.

Klebs, R. 1897. Cedarit, ein neues bernsteinähnliches fossiles Harz Canadas und sein Vergleich mit anderen fossilen Harzen. *Jahrbuch der Königlich preussischen geologischen Landesanstalt und Bergakademie zu Berlin für das Jahr 1896,* 199–230.

Klebs, R. 1910. *Der Bernstein und seine Bedeutung für Ostpreussen.* Königsberg: Naturforschung und Medizin 38–52.

Knowlton, F. H. 1896. American amber-producing tree. *Science* 3:582–4.

Koivula, J. I. 1993. Golden windows/fossils. *Lapidary Journal* 47(4):22–7.

Kolberg, A. 1878. *Pytheas. Geography-history. Erörtrungen über der Bernsteinland der altesten Zeit.* Braunsberg: Zeitschrift für Geschichte und Altertumskunde Ermelands 6:442–520.

Kolendo, J. 1990. Napływ bursztynu z Północy na tereny imperium rzymskiego w I–VI w.n.e. *Prace Muzeum Ziemi.* Warsaw 41:91–100.

Komarow, W. L. 1943. Proiskhozhdenie rastenii. In Russian. Referred to in 1974 *Amber in Poland, a guide to the exhibiton,* A. Zalewska, ed. Warsaw: Wsydawnictwa Geologiczne.

Kornilovitch, N. 1903. Has the structure of striated muscle of insects in amber been preserved? (In Russian) Sicgribukas ku stryjtdyra oioereczbioikisak-tych myszc u nasekomych, wstrechaushchisa w iskopaimom yantarew? *Protokoly Obszczestva Estestvoispytateley pri Imperatorskom Urewskom Universitete.* Yurev: Yurev University 13(2):198–206. See in: Amber in nature. 1984.

Koshil, I. M., I. S. Vasilishin, and V. I. Panchenko. 1993. Bernstein aus der Ukraine, *Lapis*. Munich 18:34–7.

Kośko, A. 1991. Specyfika rozwoju kulturowego społeczeństwa Nizu Polski w dobie schylkowego neolitu i wczesnej epoki brazu. Zarys problematyki (The distinctive features of the cultural development of the society of the plains of Poland toward the close of the Neolithic Period and in the early Bronze Age. A summary of the subject). In Polish.

In *Lubelskie Materiały Archeologiczne*. Lublin 6:23–37.

Kosmowska-Ceranowicz, B. 1978. *La Genesi e le Varieta Delle Resine Fossili. Ambra Oro Del Nord*. Venice: Alfieri.

Kosmowska-Ceranowicz, B. 1986. Bernsteinfunde und Bernsteinlagerstätten in Polen. *Zeitschrift d. Deutsche Gemmologische Gesselschaft*. Stuttgart 35 (1/2) s. 21-26.

Kosmowska-Ceranowicz, B. 1991. Zarys wiadomości o leczeniu bursztynem. In: A. Szyma_ski (ed.) Biomineralizacja i biomateria_y. Warszawa PWN 1991, s. 152-162.

Kosmowska-Ceranowicz, B. 1994. Bursztyn z Borneo—największe na świecie złoże żywicy kopalnej. *Przegląd Geologiczny* 7:576–8.

Kosmowska-Ceranowicz, B. 1995. Das Bernsteineinfuhrende Tertiär des Chłapowo-Samland Delta. *Klassische Fundstellen der Paläontologie Bd. III*, Goldschneck-Verlag Korb,180-190.

Kosmowska-Ceranowicz, B. 1996. Bernstein die Lagerstatte und ihre Entstehung. In: M.Ganzelewski & R. Slotta (ed.) – *Bernstein – Tränen der Götter. Bochum.161-168*.

Kosmowska-Ceranowicz, B. 1996 *(and 1997 second ed., 1998 third ed., 2001 fourth ed.).Amber, treasure of the ancient seas.* Warsaw: Oficyna Wydawnicza "Sadyba" Warsaw.

Kosmowska-Ceranowicz, B. 1999. Succinite and some other fossil resins in Poland and Europe (Deposits, finds, features and differences in IRS). *Proceedings of the World Congress on Amber Inclusions*. In Estudios Del Museo de Ciencias Naturales de Alava 14(2):74-117 Spain.

Kosmowska-Ceranowicz, B., ed. 2001. *The amber treasure trove*. Warsaw: Oficyna Wydawnicza "Sadyba."

Kosmowska-Ceranowicz, B., and T. Konart. 1989. (1995 second ed.)*Tajemnice Burszynu*. (The Mysteries of Amber). Warsaw: Wydawnictwo Sport I Turystyka.

Kosmowska-Ceranowicz, B., and G. R. Krumbiegel. 1988. Die unter miozäne Bernstein und andere fossile Harze von Bitterfeld, DDR. *Proceedings sixth annual symposium on amber and amber-bearing sediments*. B. Kosmowska-Ceranowicz, ed. Warsaw: Muzeum Ziemi, 17–20.

Kosmowska-Ceranowicz, B., and G. R. Krumbiegel. 1989*a*. Geologie und Geschichte des Bitterfelder Bernsteins und andere fossiler Harze. *Hallesches Jahrbuch für Geowissenschaften*. Gotha.14:1–25.

Kosmowska-Ceranowicz, B., and G. R. Krumbiegel. 1989*b*. Der Bitterfelder Bernstein: Geschichte, Geologie und Genese. *Fundgrube*. Popularwissenschaftliche zeitshrift für Geologie Mineralogie Palaontologie Bergbaugeschichte 25:34–9.

Kosmowska-Ceranowicz, B., and G. R. Krumbiegel. 1990. Bursztyn bitterfeldzki-saksonski I inne zywice kopalne zokolic Halle/NRD. In *Przegląd Geologiczny* 38(9):394–400.

Kosmowska-Ceranowicz B., Migaszewski Z.1988. Petrography and IRS of fossil black resins, jets and black amber jewellery. In: The sixth meeting on amber and amber-bearing sediments. Warsaw, Museum of the Earth 1988, s. 9-10.

Kosmowska-Ceranowicz B., Migaszewski Z.1988. *O czarnym bursztynie i gagacie*. Przegl_d Geologiczny 7 s. 413-421.

Kosmowska-Ceranowicz B., Migaszewski Z.1990. *Petrography and IRS of fossil black resins, jets and black amber jewellery*. Prace Muzeum Ziemi, 41 s. 156-157.

Kosmowska-Ceranowicz, ed. English version, pp.9–10. Also in Polish. O czarnym bursztynie i gagacie. *Przegląd Geologiczny*. Warsaw: Muzeum Ziemi. 36(7):413–21.

Kosmowska-Ceranowicz, B., and T. Pietrzak. 1982. *Znaleziska i dawne kopalnie bursztynu w Polsce* (Findings of amber and ancient amber mining in Poland). In Polish. Katalog Opracowania dokumentacyjne Muzeum Ziemi. Warsaw: Wydawnictwa Geologiczne 6:132.

Kosmowska-Ceranowicz, B., E. Sontag, G. Jabłoński, and J. Bielak. 2000. *Succinum. electron, glaesum, agtstein, bernstein, bursxtyn, bärnsten, gintaras, jantar, rav, rafr, ambra, ámbar, ambre jaune, amber.* Gdańsk, Poland:Amber Association of Poland (Stowarzyszenie Bursztynnikow we Polsxce) 16.

Kosmowska-Ceranowicz, B., R. Kulicka, K. Leciejewicz, P. Mierzejewski, and T. Pietrzak. 1984. *Amber in nature. Guide and catalogue of the exhibition*. Warsaw: Wydawnictwa Geologiczne.

Kosmowska-Ceranowicz, B., G. Kociszewska-Musiał, T. Musiał, and Müller. 1990. The Amber-bearing Tertiary sediment near Parczew. [Bursztynoneśne osadytyzeciorze dowe okouc Parczewa. In Polish. *Prace Muzeum Ziemi* 41:21–35.

Kostiashova, Z. 1997*a*. The production of jewelry, artistic and fancy goods at the Kaliningrad Amber Factory (1945–1996). In *Baltic amber and other fossil resins*, B. Kosmowska-Ceranowicz, ed. Warsaw: Muzeum Ziemi, 44–5.

Kostiashova, Z. 1997b. *Russian Amber Company at Jantarnyj, Kaliningrad, Russia.* Oral lecture given at the symposium on Baltic amber and other fossil resins (2–6 September), held in Archeological Museum, Gdańsk, Poland.

Kostyniuk, M. 1961. Nowozelandzka kauri a pochodzenie bursztynu. *Wszechświat* 10:263–6.

Koziorowska, L. 1988. Comparative studies on microelements in amber. Warsaw: Central Laboratory of the Institute of the History of Material Culture, Polish Academy of Sciences. In *Proceedings of the sixth meeting on amber and amber-bearing sediments* (20–21 October) Warsaw: Muzeum Ziemi, 11–2.

Kozlowski, R. 1951. O naukowym znaczeniu badań bursztynu. *Wiadomosci.* Warsaw: Muzeum Ziemi 5(2)446–451.

Kraus, E. H. 1939. *Gems and gem materials.* New York: McGraw-Hill.

Krumbiegel, G. 1997. Beckerite from the Goitsche Mine near Bitterfeld (Saxon, Germany). In *Baltic amber and other fossil resins,* B. Kosmowska-Ceranowicz, ed. Warsaw: Muzeum Ziemi, 38–9.

Krumbiegel, G. and B. Krumbiegel. 1996. Bernstein-Fossile Harze aus aller Welt. Zweite Auflage, Goldschneck-Verlag, Korb. Also in *Fossilien, Sonderband 7,* Weinstadt 1994.

Krumbiegel, G. & B. Krumbiegel. 2001. *Faszination Bernstein.* Germany: Goldschneck-Verlag.

Krylov, A. 1992. *Amber and fossils.* Kaliningrad: Museum of the World Ocean.

Krylov, A. D. 1997. Cover letter by President/Director of Amber Program, Sea Venture Bureau, Ltd. Kaliningrad, Russia.

Krzemiński, W. 1992a. Limoniidae (Diptera, Nematocera) from Dominican amber. I. Genus Molophilus Curtis, 1833. *Acta Zoologica Cracoviensia* 35(1):107–11.

Krzemiński, W. 1992b. Fossil Tipulomorpha (Diptera, Nematocera) from Baltic amber (Upper Eocene). Revision of the genus Helius Lepeletier et Serville (Limoniidae) *Acta Zoologica Cracoviensia* 35(3):597–601.

Kucharska, M., and A. Kwaitkowski. 1979. Thin layer chromatography of amber samples. *Journal of Chromatography* 189:482–4.

Kucharska, M., and A. Kwiatkowski. 1978. Research methods for the chemical composition of amber; and problems concerning the origin of amber. In Polish. *Prace Muzeum Ziemi* 29:149–56.

Kula, L. J. 1966. *Kultura wschodniopomorska na Pomorzu Gdańskim.* Tom I. Worclaw: Materiały. 556 pp.

Kulicka, R. 1996a. Traces of animal life captured in amber. In *Amber, treasure of the ancient seas,* B. Kosmowska-Ceranowicz, ed. Warsaw: Muzeum Ziemi.

Kulicka, R. 1996b. The use of amber in art. In *Amber, treasure of the ancient seas,* B. Kosmowska-Ceranowicz, ed. Warsaw: Muzeum Ziemi, 22–3.

Kulicka, R. 1997. *Meng tertiaria w bursztynie baltyckim ze zbiorow.* Warsaw: Muzeum Ziemi.

Kulicka, R. 2001. New genera and species of Strepsiptera from the Baltic Amber. Warsaw:*Prace Muzeum Ziemi* 46: 3-16.

Kulicka, R. and Z. Sikorska-Piwowska. 1999. Mammalian Ichnites in Amber. In Kosmowska-Ceranowicz ed. *Investigations into Amber, Proceedings of the International Interdisciplinary Symposium: Baltic Amber and Other Fossil Resins.* Gdańsk, Poland: Arachaeological Museum in Gdański, Museum of the Earth, Polilsh Academy of Sciences

Kumuszki, Cz. A. 1993. An amber recipe. *Lapidary Journal* 47(4):27.

Kunz, G. F. 1971. *The curious lore of precious stones.* Mineola, N.Y.: Dover Publications.

Kupryjanowicz, J. 2001. Spiders (Araneae) in Baltic amber in the collections of the Museum of the Earth, Warsaw. *Prace Muzeum Ziemi* 46: 59-64.

Koziorowska, L. 1988. Comparative studies on microelements in amber. *Central Laboratory of the Institute of the History of Material Culture.* Warsaw: Polish Academy of Sciences 11.

Kwiatkowska, K. 1996. Amber in the archaeological record. In *Amber, treasure of the ancient seas,* B. Kosmowska-Ceranowicz (ed.), p. 16-20. (also: 1997 second ed., 1998 third ed., 2001 fourth ed.). Warsaw: Oficyna Wydawnicza "Sadyba."

Lallemand, F. 1981–1983. Pytheas et la route de l'etain et de l'ambre. (Pytheas and the tin and amber route). *Bulletin Societe de Geographie de Marseille* 85(14):35–46.

Lambert, J. B., and J. S. Frye. 1982. Carbon functonalities in amber. *Science* 217:55–7.

Lambert, J. B., C. W. Beck, and J. S. Frye. 1988. Analysis of European amber by carbon-13 nuclear magnetic resonance spectroscopy. *Archaeometry* 30:248–63.

Lambert, J. B., J. S. Frye, and G. O. Poinar, Jr. 1985. Amber from the Dominican Republic: Analysis of the nuclear magnetic resonance spectroscopy. *Archaeometry* 27:43–51.

Lambert, J. B., J. S. Frye, and G. O. Poinar. 1990. Analysis of North American amber by carbon-13 NMR spectroscopy. *Geoarchaeometry* 25:43–52.

Lambert, J. B., S. C. Johnson, and G. O. Poinar. 1996. Nuclear magnetic resonance characterization of

Cretaceous amber. *Archaeometry* 38:325-35.

Langenheim, J. H. 1964. Present status of botanical studies of ambers. *Botanical Museum Leaflets.* Cambridge, Mass: Harvard University Press. 20(8)225–87.

Langenheim, J. H. 1966. Botanical source of amber from Chiapas, Mexico. *Ciencias* 24:201–10.

Langenheim, J. H., 1968. Catalogue of infrared spectra of fossil resins (ambers) 1. North and South America. *Botanical Museum Leaflets, Harvard University* 22:65–120.

Langenheim, J. H. 1969. Amber, a botanical inquiry. *Science* 163(3872):1157–64.

Langenheim, J. H. 1973. Leguminous resin-producing trees in Africa and South America. In *Tropical forest ecosystems in Africa and South America: A comparative review.* Washington, D.C.: Smithsonian Institution Press, 89–104.

Langenheim, J. H. 1995. Biology of amber-producing trees: Focus on case studies of *Hymenaea* and *Agathis.* In *Amber, resinite and fossil resins.* K. B. Anderson and J. C. Crelling, eds. Washington, D.C.:American Chemical Society, 1–31.

Langenheim, J. H., and C. W. Beck. 1965. Infrared spectra as a means of determining botanical sources of amber. *Science* 149:52–5.

Langenheim, J. H., and C. W. Beck. 1968. Catalogue of infrared spectra of fossil resins (ambers) in North and South America. *Botanical Museum Leaflets, Harvard University* 22(3):65–120.

Langenheim, J. H., and Y. T. Lee. 1974. Reinstatement of the genus *Hymenaea* (Leguminosae: Caesalpinioidcac) in Africa. *Brittonia* 26:3–21.

Langenheim, J. H., Y. T. Lee, and S. S. Martin. 1973. An evolutionary and ecological perspective of Amozonia *Hylaea* species of *Hymenaea* (Leguminosae: Caesalpinioidaea). *Acta Amazonica* 3:5–38.

Langenheim, R. L., Jr., C. J. Smiley, and J. Gray. 1960. Cretaceous amber from the arctic coastal plain of Alaska. *Bulletin of the Geological Society of America.* 71:1345–56.

Langenheim, Jr., R. L., Buddhue, J. D., & Jelinek, G. 1965. Age and occurrence of the fossil resins Bacalite, Kansasite, and Jelinite. *Journal of Paleontology* 39(2):283-287.

LaPlace, M. M. J. 1996. L'ecphrasis de la parole d'apparat dans l'Electrum et le De domo de Lucien, et la représentation des deux styles d'une esthetique inspiree de Pindare et de Platon. *The Journal of Hellenic Studies* 116:158–65.

Larsson, S. 1978. *Baltic amber: A palaeobiological study. Entomonograph No. 1.* Klampenborg, Denmark: Scandinavian Science Press.

Laufer, B. 1907. Historical jottings on amber in Asia. *Memoirs of the American Anthropological Association.* Lancaster, Pa. 1:3, reprinted in 1964. Methobad, N.Y.: Kaus Reprint Co.

Lawrence, P. N. 1985. Ten species of Collembda from Baltic amber. *Prace Muzeum Ziemi.* Warsaw: Muzeum Ziemi. 37:101–4.

Leach, M. 1949. *Standard dictionary of folklore, mythology and legend.* New York: Funk & Wagnalls.

Lebez, D. 1968. The analysis of archeological amber from the Baltic Sea by thin layer chromatography. *Journal of Chromatography* 33:544–7.

Leciejewicz, K. 1996. Varieties of Baltic amber. In *Amber, treasure of the ancient seas.* B. Kosmowska-Ceranowicz, ed. p. 13-16 .(also: 1997 second ed., 1998 third ed., 2001 fourth ed.). Warsaw: Oficyna Wydawnicza "Sadyba."

Lethbridge, T. C. 1979. In *The magic of amber,* R. Hunger, ed. Radnor, Pa.: Chilton, 28.

Lewin, R. 1996. *Patterns in evolution, the new molecular view.* New York: Scientific American Library.

Lcy, W. 1938. The story of amber. *Natural History* (May):351–77.

Ley, W. 1951. *Dragons in amber.* New York: Viking Press.

Liddicoat, R. T. 1969. *Handbook of gem identification.* Los Angeles: Gemological Institute of America.

Litwin, R. J., and S. R. Ash. 1991. First early Mesozoic amber in the Western Hemisphere. *Geology* 19(3):273–6.

Lomonosov, M. V. 1954. (*Complete Collection of Works VI,* Title is in Russian) Moscow-Leningrad.

Lože, I. 1969*a.* Novji centr obrabotki jantarja epochi neolita v Vostocnoj Pribaltike. In *Sovetskaja Geologaija Moskva,* 3:124–134. (Work in progress referred to in Gudynas and Pinkus 1967, 42–4.)

Lože, I. 1969*b.* Senakas dzintara rot as Baltijas juras piekrasste. *"Padomju Latvijas Sieviete"* 11(19):1–15.

Lože, I. 1970. *Sesno Ticejumu un tradiciju atspogulojums akmens laikmete maksla Austrumbaltija Arheologija un etnografia* IX. Riga, Latvia.

Lucas, A. 1934. *Ancient Egyptian materials in industries.* London: Edward Arnold.

Iuka, L. J. 1966. *Kultura wschodniopomorska na Pomorzu Gdańskim.* Tom I. Worclaw: Materiały.

Luschen, H. 1968. *Die Namen der Steine.* Munich: Ott Verlag.

MacFall, R. P. 1975. *Gem hunter's guide,* 5th ed. New York: Thomas C. Crowell.

Magnus, Olaus. 1539. *Carta Marina*. Venice.

Magnus, Olaus. 1555. *Historia de gentibus Septentrionalibus earumque diversis statibus conditionibus* (History of the Nordic People). Ramae.

Maldonado Capriles, J., and G. O. Poinar, Jr. 1995. *Stomatomiris*, a new fossil mirid genus in Dominican amber (Heteroptera: Miridae). *Caribbean Journal of Science* 31(3–4):281–3.

Manufacturing Jewellers. 1933. Amber 92, 3:21.

Mazurowski, R. F. 1983. Bursztyn w epoce kamienia na ziemiach polskich. (Amber in the Polish territories the Stone Age.) Materiały Starożytne i Wczesnośredniowieczne. Warsaw. 5:7–134.

Mazurowski, R. F. 1984. Amber treatment workshops of the Rzucewo-Culture in Żuławy. *Przegląd Archeologiczne.*

Wroclaw. 32:5–60.xxxxxII4 `er

Mazurowski, R. F. 1987*a*. Nowe badania nad osadnictwem ludności kultury rzucewskiej w Suchaczu, woj. Elbąskie, w latach 1980–1983. In *Badania archeologiczne w województwie elbąskim w latach 1980–1983*. Malbork 141–176.

Mazurowski, R. F. 1987*b*. Badania żuławskiego rejonu bursztyniarskiego ludności kultury rzucewskiej, Niedźwiedziówka, stanowisko 1–3. In *Badania archeologiczne w województwie elbąskim w latach 1980–1983*. Malbork 70–119.

McAlpine, J. F., and J. E. H. Martin. 1969. Canadian amber: A paleontological treasure chest. *The Canadian Entomologist* (August) 101:819–38.

McDonald. L. S. 1940. *Jewels and gems*. New York: Thomas C. Crowell.

Menge, A. 1855. Über die Scheerenspinnen, Cherneridae. *Schriften der Naturforschenden Gesellshaft*. Danzig 5:1–41.

Michener, C. D., and G. Poinar, Jr. 1996. The known bee fauna of the Dominican amber. *Journal of the Kansas Entomological Society* 69(4):353–61.

Mierzejewski, P. 1976. On application of scanning electron microscope to the study of organic inclusions from the Baltic amber. *Rocznik Polskiego Towarzystwa Geologicznego* Kraków. 46:291–5.

Mierzejewski, P. 1978. Electron microscopy study of the milky impurities covering arthropod inclusions in Baltic amber. *Prace Muzeum Ziemi* 28:79–84.

Miller, A. M. 1988. *Gems and jewelry appraising*. New York: Van Nostrand Reinhold.

Miller, A. M., and J. Sinkankas. 1994. *Standard catalog of gem values*, 2nd ed. Tucson: Geoscience Press.

Mills, J. S., R. White, and L. J. Gough. 1984, 1985. The chemical composition of Baltic amber. *Chemical Geology* 47(1–2):15–39.

Mizgiris, M. 2000. *Amber Museum-gallery Morphology*. Lithuania: Gintaro Galerija Muziejus. www.amber-gallery.Lt/english/ muziejus-morfologija.htm.

Mosini, V., and R. Samperi. 1985. Correlations between Baltic amber and *Pinus* resins. *Phytochemistry* 24:859–61.

Mumford, J. K. 1924. *The Bakelite story*. New York: Robert L. Stillson.

Münster, S. 1559. *Cosmos Universalis* (Description of the world). Basle: I.

Murgoci, G. 1903. *Gisements du Succin de Roumanie avec an apercu Sur Les Resines-Fosiles*. Bucharest: l'Imprimerie de l'Etat.

Murgoci, G. 1924. *Les Ambres Roumanins*. Correspondance Economique de Roumanie; Bulletin officiel du Ministere de l'Industrie et du Commerce, VI-e annee, No. 6.

Mustoe, G. E. 1985. Eocene amber from the Pacific coast of North America. *Geological Society of America Bulletin* 96(12):1530–6.

Narune, J. 1961. *Ambarita (Amberella) Leyenda*. New York: Council of Lithuanian Women.

Nash, P. 1996. Flowers in amber: Letter to the Editor. *New York Times*, 8 February.

Nel, A., and E. Jarzembowski. 1997. New fossil *Sisyridae* and *Nevrorthidae* (Insecta: Neuroptera) from Eocene Baltic amber and Upper Miocene of France. *European Journal of Entomology* 94(2):287–94.

Nissenbaum, A. 1991. Amber and the direct observation of paleomicrobiota. In *Diversity of environmental bio-geochemistry*, J. Berthelin, ed. Elsevier: Developments in Geochemistry. 6:15–21.

Nissenbaum, A., and A. Horowitz. 1992. The Levantine amber belt. *Journal of African Earth Sciences* 14(2):295–300.

Niwiński, K. 1997. Amber in ancient Egypt. : B. Kosmowska-Ceranowicz & H. Paner (ed.) Investigations into amber. Proceedings of the International Interdisciplinary Symposium *Baltic amber and other fossil resins 997 Urbs Gyddanyzc – 1997* Gdańsk" 2-6 September 1997. Gdańsk, p. 115-119.

Noetling, F. 1892. Preliminary report on the economic resources of amber and jade mines in Upper Burma. *Records of the Geological Survey of India* 25:130–5.

O'Keefe, S., T. Pike, and G. Poinar. 1997. *Palaeoleptochromus schaufussi* (gen.nov., sp.nov.), a new antlike stone beetle (Coleoptera: Scydmaenidae) from Canadian Cretaceous amber. *Canadian Entomologist.* 129(3):379–85.

Oppert, J. 1876. *L'Amber Jaune Chez les Assyriens.* Paris: F. Vieweg.

Pääbo, S. 1993. Ancient DNA. *Scientific American* 269(November):86–92.

Paclt, J. A. 1953. A system of caustolites. *Tschermaks Mineralogische und Petrographische Mitteilungen* 3(4):332–47.

Palmer, D. 1992. The DNA that stays forever in amber. *New Scientist* 1843:15.

Patterson, J. E. 2005. Wonderful Article. Rocks & Minerals 80 (1):9.

Parsons, C. J. 1969. *Practical gem knowledge for the amateur.* San Diego: Lapidary Journal.

Pęczalska. A. 1982. *Złoto północy. Opowieści o bursztynie.* Katowice, Poland: Wydawnictwo "Śląsk."

Pelka, O. 1920. *Bernstein.* Berlin: R. C. Schmidt.

Pellagrino, C. 1979. Dinosauer capsule. *Omni* 7:38–40.

Penrose, R. A. F., Jr. 1912. Kauri gum mining in New Zealand. *Journal of Geology* 20:1.

Petar, A. V. 1934. Amber. *U.S. Bureau Mines Information Circular 6789.* Washington, D.C.: United States Bureau of Mines.

Petrified tree tears. 1973. *Science Digest,* (August): 86–7.

Pielińska, A. 1996. The origins of amber. In *Amber, treasure of the ancient seas,* B. Kosmowska-Ceranowicz, ed. p. 9-10 .(also: 1997 second ed., 1998 third ed., 2001 fourth ed.) Warsaw: Oficyna Wydawnicza "Sadyba." .

Pilaitis, G. 2002. Second wind for the sun-stone. *Lietuvos Rytas* on *www.baltkurs.com/english/spring/17amber.htm*

Piwocki, M. and I. Olkowicz-Paprocka, 1987. Litostratygrafia paleogenu, perspektywy i metodyka poszukiwań bursztynu w północnej Polsce. (Lithostratigraphy of the Palaeogene: methods and outlooks of amber prospecting in northern Poland). In Polish. *Biuletyn Instytutu Geologicznego* 356:7–28.

Pliny, the Elder. *The Natural History of Pliny,* Book 37, Chapter 10, Translated by D. E. Eichholz. 1962. Cambridge, Mass.: Harvard University Press.

Pliny, the Elder. *The Natural History of Pliny.* Book 37, Chapter 12; Translated by John Bostock, and H. T. Riley. 1857. London: Henry G. Bohn. VI:402–3.

Poinar, G. O., Jr. 1977. Fossil nematodes from Mexican amber. *Nematologica* 23(2):232–8.

Poinar, G. O., Jr. 1981. Fossil dauer rhabditoid nematodes.*Nematologica* 27:466–7.

Poinar, G. O., Jr. 1982*a.* Amber—True or False? *Gems and Minerals* (April) 534:80–4.

Poinar, G. O., Jr. 1982*b.* Sealed in amber. *Natural History* 91:26–32.

Poinar, G. O., Jr. 1983. *The Natural History of Nemmatodes.* Englewood Cliffs, N.J.: Prentice-Hall.

Poinar, G. O., Jr. 1984*a.* First fossil record of parasitism by insect parasitic Tylenchida (Allantonematidae: Nematoda). *Journal of Parasitology* 70:306–8.

Poinar, G. O., Jr. 1984*b. Heydenius dominicus* n.sp. (Nematoda: Mermithidae), a fossil parasite from the Dominican Republic. *Journal of Nematology* 16:371–5.

Poinar, G. O., Jr. 1984*c.* Fossil evidence of nematode parasitism. *Revue Nematology* 7:210–3.

Poinar, G. O., Jr. 1985*a.* Fossil evidence of insect parasitism by mites. *International Journal of Acarology* 11:37–8.

Poinar, G. O., Jr. 1985*b.* Tertiary forest in amber. *Pacific Horticulture* 46:39–41.

Poinar, G. O., Jr. 1987. Fossil evidence of spider parasitism by Ichneomonidae. *Journal of Arachnology* 14:399–400.

Poinar, G. O., Jr. 1988*a.* Hair in Dominican amber: Evidence for Tertiary land mammals in the Antilles. *Experientia* 44:88–9.

Poinar, G. O., Jr. 1988*b. Zorotypus palaeus* n.sp., a fossil Zoraptera (Insecta) in Dominican amber. *Proceedings of the New York Entomology Society* 96:253–9.

Poinar, G. O., Jr. 1988*c.* The amber ark. *Natural History* 97:42–7.

Poinar, G. O., Jr. 1991*a. Praecoris dominicana* gen. N., sp.n. (Holoptilinae: Reduviidae: Hemiptera) from Dominican amber with an interpretation of past behavior based on functional morphology. *Entomologica Scandinavica* 22: 193–9.

Poinar, G. O., Jr. 1991*b. Hymenaea protera* sp.n. (Leguminoseae, Caesalpinioideae) from dominican amber has African affinities. *Experientia* 47:1075–82.

Poinar, G. O., Jr. 1991*c.* Resinites, with examples from New Zealand and Australia. *Fuel Processing Technology* 28:135–48.

Poinar, G. O., Jr. 1991*d.* The mycetophagous and entomophagous stages of *Iotonchium californicum* n. sp. (Iotonchiidae: Tylenchida). *Review de Nematologie* 14:565–80.

Poinar, G. O., Jr. 1992. *Life in amber.* Palo Alto: Stanford University Press.

Poinar, G. O., Jr. 1993*a.* Insects in amber. In *Annual review of entomology.* T. E. Mittler et al., eds., 38:145–59. Annual Reviews Inc.

Poinar, G. O., Jr. 1993*b.* Still life in amber. *The Sciences* (March/April):34–8.

Poinar, G. O., Jr. 1996*a*. Fossil velvet worms in Baltic and Dominican amber: Onychophoran evolution and biogeography. *Science* 273(5280):1370–1.

Poinar, G. O., Jr. 1996*b*. Older and wiser. *Lapidary Journal*. 49:523–56.

Poinar, G. O., Jr. 2001. Fossil puffballs (Gasteromycetes: Lycoperdales) in Mexican amber. *Historical Biology* 15:199-122.

Poinar, G. O., Jr. 2002. First fossil record of nematode parasitism of ants; a 40 million year tale. *Parasitology* 125:457-459.

Poinar, G. O., Jr. 2002. Fossil palm flowers in Dominican and Baltic amber. *Botanical Journal of the Linnean Society* 139:361-357.

Poinar, G. O., Jr. 2002. *Heydenius brownii* sp. n. (Nemaroda: Mermithidae) parasitising a planthopper (Homoptera: Achilidae) in Baltic amber. *Nematology* 3:753-757.

Poinar, G. O., Jr. 2004. *Journal of Petroleum Geology* 27(2) 207-209.

Poinar, G. O., Jr., and J. Brodzinsky. 1986. Fossil evidence of nematode (Tylenchida) parasitism in Staphylinidae
Coleoptera). *Nematologica* 32:353–5.

Poinar, G. O., Jr., and A. E. Brown. 2003. A non-gilled hymenomycete in Cretaceous amber. *Mycological Research* 107:763-768.

Poinar, G. O., Jr., and D. C. Cannatella. 1987. An Upper Eocene frog from the Dominican Republic and its implications for Caribbean biogeography. *Science* 237:1215–6.

Poinar, G. O., Jr., and J. T. Doyen. 1992. A termite bug, *Termitaradus protera*, n.sp. (Termitaphidae: Hemiptera), from Mexican amber. *Entomologica Scandinavica*.

Poinar, G. O., Jr., and D. Grimaldi. 1990. Fossil and extant macrochelid mites (Acari: Macrochelidae) phoretic on drosophilid flies (Diptedra: Drosophilidae). *Journal of the New York Entomological Society* 98:88–92.

Poinar, G. O., Jr., and J. Haverkamp. 1985. Use of pryolysis mass spectrometry in the identification of amber samples. *Journal of Baltic Studies* 16:210–21.

Poinar, G. O., Jr., and R. Hess. 1982. Ultrastructure of 40-million-year-old insect tissue. *Science* 215:1241–2.

Poinar, G. O., Jr., and R. Hess. 1985. Preservative qualities of recent and fossil resins: electron micrograph studies on tissue preserved in Baltic amber. *Journal of Baltic Studies* 16:222–30.

Poinar, G. O. Jr., and R. Milki. 2001. *Lebanese amber; the oldest insect ecosystem in fossilization resin*. Corvfallis, OR:Oregon State University Press.

Poinar, G. O. Jr., and J. Miller. 2002. First fossil record of endoparasitism of adult ants (Formicidae: Humenoptera) by Braconidae (Hymenoiptera). *Annals of the Entomological Society of America* 95(1):41-43.

Poinar, G. O., Jr., and R. Poinar. 1994. *The quest for life in amber*. Reading, Mass.: Addison Wesley.

Poinar, G. O. Jr., and R. Poinar. 1999. *The amber forest, a reconstruction of a vanished world*. Princeton, MA:Princeton University Press.

Poinar, G. O., Jr., and C. Ricci. 1992. Bdelloid rotifers in Dominican amber: Evidence for parthenogenetic continuity. *Experientia*.

Poinar, G. O., Jr., and B. Roth. 1991. Terrestrial snails (Gastropoda) in Dominican amber. *The Veliger* 34:253–8.

Poinar, G. O., Jr., and J. A. Santiago-Blay. 1989. A fossil solpugid, *Happlodontus proterus* n.gen., sp.n. (Arachnida: Solpugida) from Dominican amber. *Journal of the New York Entomology Society* 97:125–32.

Poinar, G. O., Jr., and R. Singer. 1990. Upper Eocene gilled mushroom from the Dominican Republic. *Science* 248:1099–1101.

Poinar, G. O., Jr., and G. M. Thomas. 1982. An entomophthoralean fungus from Dominican amber. *Mycologia* 74:332–34.

Poinar, G. O., Jr., and G. M. Thomas. 1984. A fossil entomogenouss fungus from Dominican amber. *Experientia* 40:578–9.

Poinar, G. O., Jr., and J. F. Voisin. 2003. A Dominican amber weevil, *Velatis dominicana* gen. n., sp. n. and key to the genera of the Anchonini (Molytinae, Curculionidae). *Nouv. Revue Entomology* (N.S.) 19:373-381.

Poinar, G. O., Jr., U. C. Bauer, and B. M. Waggoner. 1993. Terrestrial soft-bodied protists and other microorganisms in Triassic amber. *Science* 259(5092):222–4.

Poinar, G. O., Jr., H. N. Poinar, and R. J. Cano. 1994. DNA from amber inclusions. In *Ancient DNA*, B. Herrmann and S. Hummel, eds. New York: Springer-Verlag.

Poinar, G. O., Jr., A. E. Treat, and R. V. Southcott. 1991. Mite parasitism of moths: Examples of paleosymbiosis in Dominican amber. *Experientia* 47:210–2.

Poinar, G. O., Jr., K. Warheit, and J. Brodzinsky. 1985. A fossil feather in Dominican amber. *IRCS Medical Science* 13:927.

Poinar, H. N., Cano, R. J., and G. O. Poinar, Jr. 1994. DNA from extinct plant. *Nature* 363: 677.

Poinar, H. N., M. Hoss, J. L. Bada, and S. Pääbo. 1996. Amino acid racemization and the preservation of ancient DNA. *Science* 27:84–6.

Popov, Yu. A. 1992. A new genus of Emesinae from Dominican amber (Heteroptera: Reduviidae) *Acta Zoologica Cracoviensia.* 35(3):435–43.

Practical shop notes. 1915. *The Jewelers' Circular.*

Priest, K. G. 1995. Age of bacteria from amber: Technical comments. *Science* 270:2015–7.

Prockat, F. 1931. Amber mining in Germany. *Engineering and Mining Journal* 129:305–7.

Protescu, O. 1937. Etude geologique et paleobiologique de l'ambre Roumain. *Buletinul Societatii Romdane de Geologie* 3:1–46.

Raicevich, I. 1788. *Osservazione storiche, naturali e Politeche intorno la Valachia e Moldavia.* Naples: G. Raimondo.

Raimundas, S. 1978. Third symposium of the sciences and arts. *Drauagas Lithuanian Newspaper* (8 April): 1.

Rasnitsyn, A. P., and A. J. Ross. 2000. A preliminary list of arthropod families present in the Burmese amber collection at the Natural History Museum, London. *Bulletin of the Natural History Museum, London (Geology)*, 56:21-24.

Rath, M. 1971. Golden amber, the magnificent historian. *Lapidary Journal* 25:36–46.

Raukas, A. 1944. *Latvju Raksti Ornament Letter.* Riga, Latvia: Valstsspapkru Spiestuve.

Rebaux, M. 1880. Natural Amber. In French. *Annales de Chimie et de Physicque*, Paris.

Reed, A. H., and T. W. Collins. 1967. *New Zealand's forest kind—the kauri.* Wellington, Auckland: Consolidated Press Holding.

Reineking von Bock, G. 1981. *Bernstein.* München: Callwey.

Rice, P. C. 1977. Beads galore and more. *Jewelry Making Gems and Minerals* 475:8–12.

Rice, P. C. 1978. Amber is back. *Gems and Minerals* 488:14–7.

Rice, P. C. 1979. Amber of Santo Domingo—mining in the Dominican Republic. *Lapidary Journal* 210:1804–9.

Rice, P. C. 1980*a*. Personal visits to Dominican Mines with Salvador Brouwer, Dominican Geologist, to Sierra de Agua, Mina Los Payes, Arroyos de Mameyes, and Bayaguana; Martin Nelson, United States Peace Corp Geologist, to Palo Alto, Palo Quemado, La Toca in Palo Alto de La Cumbre.

Rice, P. C. 1980*b*. Personal communication with Pompilio Brouwer, past director general de minas y petroleo; Francis Heuber, Paleobotanist, Smithsonian Institution; Robert Woodruff, Entomologist, Florida Dept. Agriculture; Nelson Gil, Dean Mining and Engineering, University of Santiago; Jake Bodzinsky, America, Ins., Virginia/Dominican Republic.

Rice, P. C. 1980*c*. Amber and its mystical past. *Lapidary Journal* 310:1312–6.

Rice, P. C. 1980*d*. Nuggets of island sun. *American Way* 13(1):70–2, 75, 76.

Rice, P. C. 1981*a*. Alluring amber. *Lapidary Journal* 410:64–70.

Rice, P. C. 1981*b*. Amber mining in the Dominican Republic. *Rocks and Minerals* 56:145–52.

Rice, P. C. 1984. Amber—the past and present. *Jewelry Making Gems and Minerals* 566:16–21.

Rice, P. C., and H. Rice. 1980. Nuggets of the sun. *Americas* 32(10):39–42.

Rice, P. C. 1997. Personal visits to Baltic Mines with Barbara Kosmowska-Ceranowicz; Siegfried Ritzkowski; Norbert Vávra; Aleksander Krilov.

Rice, P. C. 2004. The 2003 Reopening of the Amber Room. *Rocks & Minerals* 79(5):30-312.

Rieppel, O. 1980. Green anole in Dominican Amber. *Nature* 286:486–7.

Rimantienė, R. 1992. Neolithic hunter-gatherers at Sventoji in Lithuania. *Antiquity* 66(June):367–76.

Ritzkowski, S. 1996. Geschichte der Bernsteinsammlung der Albertus-Universität zu Königisberg i. Pr. In Bernstein Tränen der Götter, R. Slotta and M. Ganzelewski, eds. Bochum: Deutsches Bergbau Museum Bochum, 293–8.

Rohde, A. 1942. Das Bernsteinzimmer Friedrich I von Zarskoje Selo zum Königsberg er Schloss. *Kunstrundschau.* 50:88–91.

Ross, A. J. 1997. Insects in amber. *Geology Today* 13(1):24–8.

Ross, A. J. 1998. *Amber in the Natural Time Capsule.* London:The Natural History Museum.

Ross, A. 1998b. *Amber.* Cambridge, MA:Harvard University Press.

Röttlander, R. C. A. 1974. Die Chemie des Bernsteins. *Chemie Unserer Zeit* 8:78–83.

Röttlander, R. C. A. 1970. On the formation of amber from *Pinus* resin. *Archaeometry* 12:35–51.

Rudler, F. M. 1890 (ca.) A piece of amber. In *Science for all.* London: Cassell, 213–8.

Runge, W. 1868. Der Bernstein in Ostpreussen. *2 Vorträge.* Berlin: Lüderitz.

Ruzic, R. H. 1973*a*. Amber of Chiapas, Mexico, Part 1. *Lapidary Journal* 27:1304–6.

Ruzic, R. H. 1973*b*. Amber of Chiapas, Mexico, Part 2. *Lapidary Journal* 27:1400–6.

Salmon, W. 1696. *Family dictionary*. London: British Museum Library.

Samples, C. C. 1905. Amber. *England Mining Journal* 80:250.

Sanderson, M. W., and T. H. Farr. 1960. Amber with insects and plant inclusions from the Dominincan Republic. *Science* 131:1313.

Saunders, W. B., R. H. Mapes, F. M. Carpenter, and W. C. Elsik. 1974. Fossiliferous amber from the Eocene (Claiborne) of the gulf coastal plain. *Geological Society of America Bulletin* 85:979–84.

Saurov, L., V. Akseonov, I. Morozow, I. Bagayev, and A. Ivanov. 2003. *The case of the Amber Room*: A Vitaly Akseonov film. DVD. Russia: Culture and Communications of the Tsarkoye Selo Museum Presevation.

Savkevich, S. S. 1970. *Jantar (Yantar)*. Leningrad: "Nedra." In Russian.

Savkevich. S. S. 1975. State of investigation and prospects for amber in the U.S.S.R. *International Geology Review* 17(August): 919–24

(Savkevich, S. S. 1980. (New developments in amber and other fossil resins: mineralogical studies) Russian with English abstract. In *Gem minerals (high pressure minerals 1978) Proceedings of the XI general meeting of IMA, Novosibirsk* V. V. Bukano, ed. Academy of Sciences of the USSR, 17–28.

Savkevich, S. S. 1981. Physical methods used to determine the geological origin of amber and other fossil resins: Some critical remarks. *Physics and Chemistry of Minerals* 7:1–4.

Savkevich, S. S. 1983. Processes of transformation of amber and some amber-like fossil resins in connection with their formation conditions and environments of occurrence In Russian. *Izvestiya Akademii Nauk SSSR, Seriya Geologicheskaya*, (December):96–106.

Savkevich, S. S. 1988. Yantar. In *Proceedings of the sixth meeting on amber and amber-bearing sediments* (20–21 October) Polish Academy of Sciences. Warsaw: Muzeum Ziemi.

Savkevich, S. S., and T. N. Popkova. 1973. New data on "amber" from the right bank area of the Kheta and the Khatanga Rivers. *Mineralogy, Doklady Adademii Nauk S.S.S.R.* 208(1–6):131–2.

Savkevich, S. S., and I. A. Shakhs. 1964. Infra-red absorption spectra of Baltic amber. *Journal of Applied Chemistry, USSR* 37:2227–9 and 2717–9.

Schawaller, W. 1991. The first Mesozoic pseudoscorpion, from Cretaceous Canadian amber. *Paleontology* 34(4):971–6.

Schimper, T. 1870, 1872. Traité de paléontologie végétale. Paris: J. B. Bailliere et fils. 2:377.

Schlee, D. 1980. *Bernstein-Raritaten. Farben, Strukturen, Fossilien, Handwerk*. Stuttgart: Staatliches Museum für Naturkunde.

Schlee, D. 1984*a*. *Bernstein-Neuigkeiten*. Stuttgarter: Beiträge zur Naturkunde, (Series C) 18:1–100.

Schlee, D. 1984*b*. Notizen über einigen Bernsteine, und Kopale aus aller Welt. *Stuttgarter Beiträge zur Naturkunde* (Series C) 18:299–337.

Schlee, D. 1986. *Der Bernsteinwald*. Stuttgart: Staatliches Museum fur Naturkunde.

Schlee, D. 1990. Das Bernstein-Kabinett. *Stuttgarter Beiträge zur Naturkunde* (Series C) 28:100 pp.

Schlee, D. and H. G. Dietrich. 1970. Insektenführender Bernstein aus der Unterkreide des Libanon. In *Neues Jahrbuch für Geologie und Palaeontologie*. Stuttgart 1:40–50.

Schlee, D., and W. Glöckner. 1978. Bernstein. *Stuttgarter Beiträge zur Naturkunde* (Series C) 8:72 pp.

Schliemann, H. 1878. *Mycenae.(Mykenae)* Leipzig: Wissenschaftliche Buchgesellschaft. Also New York: Darmstadt, 1964.

Schliemann, H. 1886. *Tiryns*. Leipzig: F. A. Brockhaus.

Schliephake, G. 1993. Beiträge zue Kenntnis fossiler Fransenflügler (Thysanoptera, Insecta) aus dem Bernstein des Tertiär. II. Beiträge: Aeolothripidae (Melanthripinae) und Thripidae (Dendrothripinae und Thripinae) (Contributions to the knowledge of fossil Thysanoptera [Insecta] from Tertiary Amber. Part II: Aeolothripidae [Melanthripinae] and Thripidae [Dendrothripinae and Thripinae]) *Zoologische Jahrbucher: Abteilung fur Systematik, Ökologie und Geographie der Tiere* 120(2):215–51.

Schlüter, T., and F. von Gnielinski. 1987. The East African copal: Its geologic, stratigraphic, palaeontologic, significance and comparison with fossil resins of similar age. *National Museums of Tanzania Occasional Paper* 8.

Schubert, K. 1961. Neue Untersuchungen über Bau und Leben der Bernsteinkiefern (Pinus succinifera [Conw.] emend.) *Beiheft zum Geologischen Jahrbuch*. Hanover 78:3–149.

Schwartz, J. 1998. Ants Already a Nuisance at Dinosaur Picnics. *Washington Post* Jan.29: A3

Schweisheimer, W. 1968. The medical power of jewels. *Jewelers' Circular-Keystone* 139(7):70–1, 93, 96.

Sendel, N. 1742. *Historia succinorum corpora aliena volventium et nature opera pictarum et caclator ex regis*. Lipsiae: apud I. F. Gleditschium and in Elbing/Dresden.

Serebritsky, A. J., G. A. Ilinski, S. S. Savkevich, and I. L. Zaikina. 1979. North Ukranian amber (USSR). In Russian. *Vestnik Leningradskogo Universiteta Geologiya-Geografiya* 12:34–43.

Service, R. F. 1996. Just how old is that DNA, anyway?—Molecular evolution. *Science* 272:810.

Shedrinsky, A. M. 1996. Amber: a lecture to the Society on May 14, 1996, An Interview with Dr. Alexander Shedrinsky. *The American Society of Jewelry Historians Newsletter* 19(3):5–26.

Shedrinsky, A. M., and N. S. Baer. 1995. The application of analytical pyrolysis to the study of cultural materials in *Applied Pyrolysis Handbook*, Thomas P. Wampler, ed. New York: Marcel Dekker.

Shedrinsky, A. M., D. Grimaldi, T. P. Wampler, and N. S. Baer. 1991. Amber and copal pyrolysis gas chromatographic (PyGC) studies of provenance. *Wiener Berichte über Naturwissenschaft in der Kunst 6/7/8* (1981–1991):37–63. Vienna Institute fur Silikatchemie und Archaometric an der Hochshule für Angewandte Kunst.

Shedrinsky, A. M., D. A. Grimaldi, J. J. Boon, and N. S. Baer. 1993. Application of pyrolysis-gas chromatography and pyrolysis-gas chromatography/mass spectrometry to the unmasking of amber forgeries. *Journal of Analytical and Applied Pyrolysis* 25:77–95.

Sherratt, Andrew. 1995. Electric gold: Re-opening the amber route. Review article Amber in Archaeology: Proceedings of the Second International Conference on Amber in Archaeology: Liblice 1990, by Curt W. Beck and Jan Bouzek, in collaboration with Dagmar Dreslerova, *Antiquity* 69 (March):200–3.

Shinaq, R,, and K. Bandel. 1998. The flora of an estuarine channel margin in the Earlier Cretaceous of Jordan. *Freiberger Forschungsheft.* 474:39-57.

Sinkankas, J. 1966. *Gemstones of North America*, Vol. II. New York: Van Nostrand Reinhold.

Sinkankas, J. 1972. *Gemstone and mineral data book.* Tucson: Geoscience Press.

Sinkankas, J. 1997. *Gemstones of North America.* Vol. III. Tucson: Geoscience Press.

Sivkov, V. V., and B.V. Chubarenko. 1997. Influence of amber mining on the concentration and chemical composition of suspended sedimentary matter (Sambian Peninsula, southeast Baltic). *Marine Georesources and Geotechnology* 15(2):115–126.

Skalski, A. W. 1973. Studies on the Lepidortera from fossil resins. Part II. General remarks and descriptions of new genera and species in the families Tineidae and Oecophoridae from the Baltic amber. In *Acta Paleontologica Polonica.* Warsaw. 19(1)153–60. Part I. In *Prace Muzeum Ziemi*, 1977. Warsaw, 26:3–24.

Skalski, A. W., and A. Veggiani. 1988. Fossil resins in Sicily and northern Apennines; geology and organic content. In *Abstracts of the sixth international conference on amber and amber-bearing sediments* (20–21October 1988) Warsaw: Museum of the Earth, p. 29b. Also: *Prace Muzeum Ziemi* 1990: 41, 37-49.

Slare, Dr. 1715. Experiments and observations upon Oriental and other Bezoar Stones. In J. Evans, 1968, *Magical jewels.* New York: Dover Publications.

Slotta, R., and M. Ganzelewski, eds. 1996. *Bernstein, Tränen der Götter.* Bochum, Germany: Deutsches Bergbau-Museum.

Spahr, U. 1981. Systematischer Katalog der Bernstein- und Kopel-Käfer (Coleoptera). Stuttgarter Beiträge zer Naturkunde (Series B) 80.

Spekke, A. 1957. *The ancient amber routes and the geographical discovery of the eastern Baltic.* Stockholm: M. Goppers.

Srebrodolskiy, B. I. 1975. *Amber in sulfur deposits.* Doklady Akademii Nauk S.S.S.R. 223(1–6):204–5.

Srebrodolskiy, B. I. 1980*a. Jantar Ukrainy.* In Ukrainian. Kiev: Naukova Dumka.

Srebrodolskiy, B. I. 1980*b.* Accumulation of amber. *Doklady, Academy of Sciences of the USSR, Earth Science Section,* 253:184–6.

Srebrodol'skiy, B. I. 1984. *Amber.* Moscow: Nauka.

Srebrodol'skiy, B. I., and E. A. Glebovskaya. 1971. Amber from the Yazov sulfur deposits. (Lviv, Derzh. Univeristy, Lvov, USSR) In *Doklady Akademii Nauk Ukrain SSR.* 33(12):1081–3.

Stach, E. J. 1972. Owady bezskrzydłé (Apterygota) z bursztynu bałtyckiego (Apterygota [Insecta] from the Baltic amber). In Polish. *Przegląd Zoologiczny* 16(4)416–20.

Stevens, B. C. 1976. *The collector's book of snuff bottles.* New York: Weatherhill.

Stout, E., C. W. Beck, and B. Kosmowska-Ceranowicz. 1995. Gedanite and Gedano-Succinite. In *Amber, Resinite, and Fossil Resins*, K. B. Anderson and J. B. Crelling, eds. Washingston D.C.:American Chemical Society, 130-48.

Strong, D. E. 1966. *Catalogue of the carved amber in the department of Greek and Roman Antiquities.* British Museum, Department of Greek and Roman Antiquities. London: The Trustees of the British Museum.

Strunz, H. 1966. *Mineralogische Tabellen.* Leipzig: Akademische Verlagsgesellschaft Geest und Portig.

Strzelczyk, G. 1991. *Występowanie bursztynu w utworach trzeciorzedowych w rejonie Górki Lubartowskiej (gmina Niedźwiada, Ostrówek, województwo lubelski).* Typescript in Polish.

Stuart, M. 1923. Dating the amber-bearing beds of Burma. *Records of the Geological Survey of India* 54:1–12.

Sturms, E. 1953. Der ostbaltische Bernsteinhandel in der vorschristlichen Zeit. In *Jahrbuch des Baltischen Forschungsinstituts, Commentationes Balticae.* Bonn, Germany. (1954) 1:167–205.

Sullivan, M. 1977. The fashion outlook. *Modern Jewelry* (Fall):58.

Suprichev, V. 1978. (Amber—talisman, medicine, ornament). In Russian. *Nauka i Tecknika* (Dec):20–2.

Sviridov, N. V., and V. V. Sivkov. 1992. The use of the seismoacustic data for the near-bottom currents investigation in the south-west Baltic. *Oceanology* 32(5):941–7.

Swann, Charles. 1994. Gardens and toads or, Milton—another fly in Marianne Moore's amber? *Notes and Queries* 41(September):376–7.

Szwedo, J., and R. Kulicka. 1999. Auchenorrhyncha (Insecta, Homoptera) in Balitc amber from the collection of the Museum of the Earth. *Estudios del Museo de Ciencias Naturales de Alava* 14 (2):179-197.

Szadziewski, R. 1992. Biting midges (Diptera, Ceratopogonidae) from Miocene Saxonian amber. *Acta Zoologica Cracoviensia* 35(3):603–56.

Szejnert, M. 1977. *Traffic on the amber route.* Poland.

Tabaczyńska, E. 1999. A Thousand Years of amber-Craft in Gdańsk. *Investigations into Amber, Proceedings of the International Interdisciplinary Symposium Baltic amber and Other Fossil Resins (2-6 September 1997)* Gdańsk: Polish Academy of Sciences, Muzeum Ziemi and the Archaeological Museum in Gdańsk 177-181.

Tabaczyńska, E. 1999. Amber as a subject of Archaeological research: the experiences of Polish-Italian Collaboration. *Investigations into Amber, Proceedings of the International Interdisciplinary Symposium Baltic amber and Other Fossil Resins (2-6 September 1997)* Gdańsk: Polish Academy of Sciences, Muzeum Ziemi and the Archaeological Museum in Gdańsk 191-194.

Thomas, B. R. 1970. Modern and fossil plant resins. In *Phytochemical phylogeny.* J. B. Harborne, ed. London: Academic Press, 59–79.

Thomas, G. M., and G. O. Poinar, Jr. 1988. A fossil Aspergillus from Eocene Dominican amber. *Journal of Paleontology* 62(1):141–3.

Tratman, E. K. 1950. Amber from paleolithic deposits: Goughs Cave. *Proceedings of the University of Bristol Spelaeocological Society* 6(3).

Treptow, E. 1900. *Beregbau und Huttenwesen.* Leipzig: Verläg von Otto Spamer.

Trophimov, W. S. 1974. *Amber.* Moscow: Nauka.

Trusted, M. 1985. *Catalogue of European Ambers in the Victoria and Albert Museum.* London: Trustees of the Victoria and Albert Museum.

Tyrrell, J. B. 1893. Summary report on the operations of the Geological Survey, for the year 1890. *Annual Report of the Geological Survey of Canada* (New series) 5:30A–31A.

Urban, W. 1994. The Teutonic Knights and Baltic chivalry. *The Historian* 56 (Spring):519–30.

Urbański, T., T. Glinka, and E. Wesolowska. 1976. On chemical composition of Balitc amber. *Bulletin of the Academy of Political Science.* Sec: Chemical 24(8):625–9.

Van Der Sleen, W. G. N. 1973. *A handbook on beads.* York, Pa..: Liberty Cap Books.

Vargas, G., and M. Vargas. 1969. *Faceting for amateurs.* Palm Desert, Calif.:Vargas.

Vávra, N. 1982. Bernstein und andere fossile Harze. *Zeitschrift der Deutschen Gemmologie Gesellshaft* 31(4):213–54.

Vávra, N. 1984. "Reich an armen Fundstellen": Übersicht über die fossilen Harze Osterreichs. In *Bernstein-Neuigkeiten. Stuttgarter Beiträge zer Naturkunde*, D. Schlee, ed., 18:9–14.

Vávra, N. 1990. Gas liquid chromatography: An effective tool for the chemical characterization of fossil resins. *Prace Muzeum Ziemi* 41:3–14.

Vávra, N. 1993. Chemical characterizations of fossil resins ("amber") -a critical review of methods, problems and possibilities: Determination of mineral species, botanical sources and geographical attraction. *Proceedings of a Symposium held in Neukirchen am Grossvenediger* (Salzburg, Austria) September 1990. V. Hock and F. Koller, eds. Vienna: Epidot, 147–57.

Vávra, N., V. Bouska, and Z. Dvorak. 1997. Duxite and its geochemical biomarkers ("chemofossils") from Bilina open-cast mine in the North Bohemian Basin (Miocene, Czech Republic.) *Neues Jahrbuch für Geologie Paleantologie Mb.* Stuttgart: E. Schweizerbart sche Verlägsbuch handlung 4:223–43.

Velikiy, N. M. 1975. Amber finds on the northwest shore of the Aral Sea. *Dodklady Akademii Nauk S.S.S.R.,* Earth Science Section, Washington, D.C. (1976) 221:164–6.

Vickery, V. R., and G. O. Poinar. 1994. Crickets (Grylloptera: Grylloidea) in Dominican amber. *Canadian Entomologist* 126(1):13–22.

Villiers, E. 1973. *The book of charms.* New York: Simon and Schuster.

Vogel, G. 1996. Viewing velvet worms in amber—paleontology. *Science* 273:1340.

Waggoner, B. M. 1993. Naegleria-like cysts in Cretaceous amber from central Kansas. *Journal of Eukaryotic Microbiology* 40(1):97–100.

Waggoner, B. M. 1994a. An aquatic microfossil assemblage from Cenomanian amber of France. *Lethaia* 27(1):77–84.

Waggoner, B. M. 1994*b*. Fossil microorganisms from Upper Cretaceous amber of Mississippi. *Review of Palaeobotany & Palynology* 80(1–2):75–84.

Waggoner, B. M. 1996. Bacteria and protists from Middle Cretaceous amber of Ellsworth County, Kansas. *PaleoBios* 17(1):20-26.

Waggoner, B. M., and G. O. Poinar, Jr. 1992. A fossil myxomycete plasmodium from Eocene-Oligocene amber of the Dominican Republic. *Journal of Protozoology* 39(5):639–42.

Walden, K. and H. M. Robertson. 1997. Amcient DNA from amber fossil bees? Molecular Biology & Evlution 14: 1075-1077.

Walker, T. L. 1934. Chemawinite or Canadian amber. *University of Toronto Studies, Geological Series* 36:5–10.

Wang, X. S., H. N. Poinar, G. O. Poinar, Jr., and J. Bada. 1995. Amino acids in the amber matrix and in entombed insects. In *Amber, resinite and fossil resins*, K. B. Anderson and J. C. Crelling, eds. Washington, D.C.: American Chemical Society.

Wapińska, A. 1967. Materiały do wczesnośredniowiecznego bursztyniarstwa gdańskiego. In *Gdańsk wczesnośredniowieczny. Gdański* 6:83–100.

Webster, R. 1975. *Gems, their sources, descriptions and identification.* Hamden, Conn.: Archon Books.

Webster, R. 1976. *Practical gemology.* London: N.A.G. Press.

Weigel, Ch. 1698. Kunstler und Handwerker. Regensburg. In J. Grabowska, 1983. *Polish amber.* Warsaw: Interpress Publications.

Weinstein M. 1946. *Precious and semi-precious stones.* New York: Pitman.

Weinstein, M. 1958. *The world in jewel stones.* New York: Sheridan House.

Weitschat, W., W. Wichard, and F Pfeil. 2002. Atlas of Plants and Animals in Baltic Amber. English version Verlag F. Pfeil. Munchen, Germany. 1998. *Atlas der Pflanzen und Tiere im Baltischen Betnstein.* Munchen.

Wert, C. A., and M. Weller. 1988. The polymeric nature of amber. *Bulletin of the American Physics Society* 33:497.

Wheeler, R E. M. 1955. *Rome, beyond the imperial frontier.* London: G. Bell & Sons.

Wielowiejski, P. 1991. Pracownie obróbki bursztynu z okresu wpływów rzymskich na obszarze kultury przeworskiej. (The amber workshops from the Roman influence period in the Przeworsk culture area). In Polish. *Kwartalnik Historuii Kultury Materialnej.* Warsaw. 39:317–61.

Williams, N. 1995. The trials and tribulations of cracking the prehistoric code—ancient DNA. *Science* 269:923–4.

Williams, S. W. 1901. *The middle kingdom.* New York: Scribners.

Williamson, G. 1932. *The book of amber.* London: Ernest Benn.

Wilson, A. C. 1985. The molecular basis of evolution. *Scientific American* 253:164–73.

Wilson, C. J., H. K. Mahanty, and T. A. Jackson. 1992. Adhesion of bacteria (*Serratia sp.*) to the foregut of grass grub (*Costelytra zealandica* [White]) larvae and its relationship to the development of amber disease. *Biocontrol Science & Technology* 2(1):59–64.

Wilson, E. O. 1971. *The insect societies.* Cambridge: Belknap Press.

Wilson, E. O. 1985. Invasion and extinction in the West Indian ant fauna: evidence from the Dominican amber. *Science* 229(4710):265–7.

Wilson, E. O., F. M. Carpenter, and W. L. Brown. 1967. The first Mesozoic ants, with the description of a new subfamily. *Psyche* 74:1–19. Also in *Science* 157:1038–40.

Wolff, R. J. 1990. A new species of *Thiodina* (Araneae: Salticidae) from Dominican amber. *Acta Zoologica Fennica* 190:405–8.

Yushkin, N. R., and N. Y. Sergeyeva. 1974. Textures of amber from Yugorskiy Peninsula. *Doklady Akademii Nauk S.S.S.R., Earth Science Section* 216(1–6):152–3.

Zaddach, E. G. 1860. Über die Bernstein- und Braunkohlenlager des Samlands. Schriften der Königlichen Physikalisch - Ökonomischen Gesellscgaft zu Königsberg 1:101–8.

Zaddach, E. G. 1867. *Das Tertiärgebirge Samlands.* Königsberg: Schriften der Physikalische-Ökonomische Gesellschaft 8:85–197. Also in 1868, Vol. 9.

Zaddach, E. G. 1868. Amber its origin and history, geology of Samland. *Quarterly Journal of Science* 5:167–85.

Zahl, P. A. 1977. Amber, golden window of the past. *National Geographic* 152(3):423–35.

Zalewska, Z. 1974. *Amber in Poland, a guide to the exhibition.* Warsaw: Wydawnictwa Geologiczne. Also 1964. Burszstyn w Polsce. Przewodnik po wystawie. Warszawa. In Polish.

Zamoyski, A. 1996. *The Polish way. A thousand-year history of the Poles and their culture.* New York: Hippocrene Books.

Zherichin (Zherikhin), V. V., and I. D. Sukacheva. 1973. On Cretaceous insect bearing ambers (retinites) of northern Siberia. In Russian. In *Reports of the 24th Annual Readings in Memory of N. A. Kholodkovsky.* Leningrad: Nauka, 3–48.

INDEX